# PE EXAM PREPARATION

# ELECTRICAL ENGINEERING
## A REFERENCED REVIEW FOR THE PE EXAM

Fourth Edition

James H. Bentley, PE & Hesham E. Shaalan, PhD, PE, Contributing Author

**KAPLAN) AEC EDUCATION**

This publication is designed to provide accurate and authoritative information in regard to the subject matter covered. It is sold with the understanding that the publisher is not engaged in rendering legal, accounting, or other professional service. If legal advice or other expert assistance is required, the services of a competent professional person should be sought.

**President:** Roy Lipner
**Vice President & General Manager:** David Dufresne
**Vice President of Product Development and Publishing:** Evan M. Butterfield
**Editorial Project Manager:** Laurie McGuire
**Director of Production:** Daniel Frey
**Production Editor:** Caitlin Ostrow
**Creative Director:** Lucy Jenkins

Copyright 2008 by Dearborn Financial Publishing, Inc.®

Published by Kaplan AEC Education
30 South Wacker Drive
Chicago, IL 60606-7481
(312) 836-4400
www.kaplanaecengineering.com

All rights reserved. The text of this publication, or any part thereof, may not be reproduced in any manner whatsoever without written permission in writing from the publisher.

Printed in the United States of America.

08  09  10  10  9  8  7  6  5  4  3  2  1

# CONTENTS

Introduction vii

**CHAPTER 1**

## Fundamental Concepts of Electrical Engineering 1

ELECTRICAL QUANTITIES 2
OHM'S LAW 5
CIRCUIT ELEMENTS DEFINED 6
CIRCUIT ELEMENT VALUES 7
COULOMB'S LAW 9
SERIES AND PARALLEL COMBINATIONS 10
WYE-DELTA (Y-Δ) TRANSFORMATION 11
COMPLEX ALGEBRA 11
COMPLEX NOTATION 12
CIRCUIT ELEMENT EQUATIONS 12
TRANSIENTS 13
TRANSFER FUNCTIONS 16
LAPLACE TRANSFORM 16
KIRCHHOFF'S LAWS 17
THEVENIN'S THEOREM 18
MAXIMUM POWER TRANSFER THEOREM 18
MAXIMUM POWER TRANSFER THEOREM COROLLARY 19
NORTON'S EQUIVALENT CIRCUIT 19
VOLTAGE DIVISION AND SUPERPOSITION 20
MAGNETIC CIRCUIT TERMS 21
DETERMINANTS 22
RESONANCE 24
IDEAL TRANSFORMER 27
FOURIER ANALYSIS 28
WAVEFORMS 28

**CHAPTER 2**

## Basic Circuits 33

RESISTANCE 34
WORK, ENERGY, AND POWER 35
COULOMB'S LAW 35
NETWORKS 36
CAPACITORS 40

AC CIRCUITS 43
COMPENSATING CIRCUITS 46
VOLTMETERS 48
IMPEDANCE TRANSFORMATION 49
AMMETERS 50
RESONANCE 51
MAXIMUM POWER 60
TRANSIENTS 65
INSULATION 69
WAVEFORMS 69
BLACK BOX ANALYSIS 70

### CHAPTER 3

# Power 73
SINGLE-PHASE POWER 73
POLYPHASE POWER 76
POWER FACTOR CORRECTION 83
TRANSMISSION LINE CALCULATIONS 87
WATTMETER MEASUREMENTS 92
SHORT CIRCUIT CALCULATIONS 96

### CHAPTER 4

# Machinery 99
dc MACHINES 99
ac MACHINES 108
MAGNETIC DEVICES 128

### CHAPTER 5

# Control Theory 131
BASIC FEEDBACK SYSTEMS TERMS 132
SINGULARITY FUNCTIONS 133
SECOND-ORDER SYSTEMS 134
POLES AND ZEROS 136
PARTIAL FRACTIONS 136
STABILITY 139
TRANSFER FUNCTION 144
COMPENSATION 145
BODE ANALYSIS 150
ROOT LOCUS 154

### CHAPTER 6

# Electronics 159
DEPENDENT SOURCES 159
DIODES 159

TRANSISTORS 162
BIASING AND STABILITY 180
FIELD EFFECT TRANSISTORS 185
OPERATIONAL AMPLIFIERS 189
AMPLIFIER CLASS 193
POWER SUPPLY CIRCUITS 195

## CHAPTER 7

# Communications 199

LOW-FREQUENCY TRANSMISSION 199
RF TRANSMISSION 203
ATTENUATION 220
ANTENNAS 224

## CHAPTER 8

# Logic 229

DEFINITION OF TERMS, POSTULATES, PROPERTIES 230

## APPENDIX A

# Recommended References for Further Review 247

## APPENDIX B

# Problems by Topic 249

# Index 253

# Introduction

**OUTLINE**

**HOW TO USE THIS BOOK   VII**

**BECOMING A PROFESSIONAL ENGINEER   VIII**
Education ■ Fundamentals of Engineering (FE/EIT) Exam ■ Experience ■ Professional Engineer Exam

**ELECTRICAL ENGINEERING PROFESSIONAL ENGINEER EXAM   IX**
Examination Development ■ Examination Structure ■ Exam Dates ■ Exam Procedure ■ Exam-Taking Suggestions ■ Exam Day Preparations ■ What to Take to the Exam ■ Examination Scoring and Results

## HOW TO USE THIS BOOK

*Electrical Engineering: A Referenced Review for the PE Exam* is designed as a brief, problem-oriented preparation for the Principles and Practice of Engineering (PE) exam. If you have limited time for review or if you prefer to review concepts through problems and solutions, rather than narrative presentation, this book will help you optimize your time and effort.

Each chapter provides brief textual review of major equations, terms, concepts, and analytical methods that are fundamental to the PE exam. Chapters 2 through 8 move you quickly into review via problem solving. In general, the following approach is recommended for using this book:

1. Cover the solution portion and solve the problem independently on scratch paper.

2. Compare your answer to the solution presented in the text.

3. If your solution is correct, go to the next problem. Solve all the problems sequentially, in the order presented in the book.

4. If your answer is incorrect, review the material indicated at the end of the solution. The review material includes content from this book and one or more textbooks referenced in Appendix A.

5. If your first answer was incorrect, rework the problem after reviewing the references.

This method should help you minimize time spent on analysis techniques you already understand well. At the same time, it should help you identify areas where additional review time and effort can yield improvement in your problem solving.

A subset of the problems in this book is multiple choice to give you a feel for the format of the actual exam. Incorrect answer choices are designed to be plausible based on wrong assumptions or math errors an examinee might make. For each such problem, an answer rationale explains why the wrong answers are wrong—that is, what calculation mistakes lead to them.

The book does not include national electric code problems because they apply to a limited number of engineers. Examinees working in this field will have the necessary experience and handbooks to solve any of these problems, if they should appear in the exam.

## BECOMING A PROFESSIONAL ENGINEER

Achieving registration as a professional engineer entails four distinct steps: (1) education, (2) the Fundamentals of Engineering/Engineer-In-Training (FE/EIT) exam, (3) professional experience, and (4) the professional engineer (PE) exam, more formally known as the Principles and Practice of Engineering exam. These steps are described in the following sections.

### Education

The obvious appropriate education is a BS degree in electrical engineering from an accredited college or university. This is not an absolute requirement. Alternative, but less acceptable, education is a BS degree in something other than electrical engineering, a degree from a non-accredited institution, or four years of education but no degree.

### Fundamentals of Engineering (FE/EIT) Exam

Most people are required to take and pass this eight-hour multiple-choice examination. Different states call it by different names (Fundamentals of Engineering, E.I.T., or Intern Engineer), but the exam is the same in all states. It is prepared and graded by the National Council of Examiners for Engineering and Surveying (NCEES). Review materials for this exam are found in other Kaplan AEC books such as *Fundamentals of Engineering: FE/EIT Exam Preparation*.

### Experience

Typically one must have four years of acceptable experience before being permitted to take the PE exam (California requires only two years). Both the length and character of the experience will be examined. It may, of course, take more than four years to acquire four years of acceptable experience.

### Professional Engineer Exam

The second national exam is called Principles and Practice of Engineering by NCEES, but just about everyone else calls it the Professional Engineer or PE exam. All states, plus Guam, the District of Columbia, and Puerto Rico, use the same NCEES exam.

# ELECTRICAL ENGINEERING PROFESSIONAL ENGINEER EXAM

The reason for passing laws regulating the practice of engineering is to protect the public from incompetent practitioners. Most states require engineers working on projects involving public safety to be registered, or to work under the supervision of a registered professional engineer. In addition, many private companies encourage or require engineers in their employ to pursue registration as a matter of professional development. Engineers in private practice who wish to consult or serve as expert witnesses typically also must be registered. There is no national registration law; registration is based on individual state laws and is administered by boards of registration in each of the states. You can find a list of contact information for and links to the various state boards of registration at the Kaplan AEC Web site: *www.kaplanaecengineering.com*. This list also shows the exam registration deadline for each state.

## Examination Development

Initially the states wrote their own examinations, but beginning in 1966 the NCEES took over the task for some of the states. Now the NCEES exams are used by all states. This greatly eases the ability of an electrical engineer to move from one state to another and achieve registration in the new state.

The development of the electrical engineering exam is the responsibility of the NCEES Committee on Examinations for Professional Engineers. The committee is composed of people from industry, consulting, and education, plus consultants and subject matter experts. The starting point for the exam is a task analysis survey, which NCEES does at roughly 5- to 10-year intervals. People in industry, consulting, and education are surveyed to determine what electrical engineers do and what knowledge is needed. From this NCEES develops what it calls a "matrix of knowledge" that forms the basis for the exam structure described in the next section.

The actual exam questions are prepared by the NCEES committee members, subject matter experts, and other volunteers. All people participating must hold professional registration. Using workshop meetings and correspondence by mail, the questions are written and circulated for review. Although based on an understanding of engineering fundamentals, the problems require the application of practical professional judgment and insight.

## Examination Structure

The PE exam tests an individual's experience and knowledge in the field; it is intended to show the applicant's ability to apply sound engineering principles and judgment to the solution of problems encountered in practice.

The exam is organized into breadth and depth sections.

The morning breadth exam consists of 40 multiple-choice questions covering basic electrical engineering; electronics, electronic circuits, and components; controls and communication systems; and power. You will have four hours to complete the breadth exam.

There are three afternoon depth exams focusing on the following topics respectively:

- Computers
- Electronics, Controls, and Communications
- Power

You can choose the depth exam you wish to take; the obvious choice is whichever one best matches your training and professional practice. You will have four hours to answer the 40 multiple-choice questions that make up the depth exam.

Both the breadth and depth questions include four possible answers (A, B, C, D) and are objectively scored by computer.

For more information on the topics and subtopics and their relative weights on the breadth and depth portions, visit the NCEES Web site at *www.ncees.org*.

## Exam Dates

The National Council of Examiners for Engineering and Surveying (NCEES) prepares Electrical Engineering Professional Engineer exams for use on a Friday in April and October of each year. Some state boards administer the exam twice a year in their state, whereas others offer the exam once a year. The scheduled exam dates for the next ten years can be found on the NCEES Web site (*www.ncees.org/exams/schedules/*).

People seeking to take a particular exam must apply to their state board several months in advance.

## Exam Procedure

Before the morning four-hour session begins, proctors will pass out an exam booklet, answer sheet, and mechanical pencil to each examinee. The provided pencil is the only writing instrument you are permitted to use during the exam. If you need an additional pencil during the exam, a proctor will supply one.

Fill in the answer bubbles neatly and completely. Questions with two or more bubbles filled in will be marked as incorrect, so if you decide to change an answer, be sure to erase your original answer completely.

The afternoon session will begin following a one-hour lunch break.

In both the morning and afternoon sessions, if you finish more than 15 minutes early you may turn in your booklet and answer sheet and leave. In the last 15 minutes, however, you must remain to the end of the exam in order to ensure a quiet environment for those still working and an orderly collection of materials.

## Exam-Taking Suggestions

People familiar with the psychology of exam taking have several suggestions for people as they prepare to take an exam.

1. Exam taking involves, really, two skills. One is the skill of illustrating knowledge that you know. The other is the skill of exam taking. The first may be enhanced by a systematic review of the technical material. Exam-taking skills, on the other hand, may be improved by practice with similar problems presented in the exam format.

2. Since there is no deduction for guessing on the multiple-choice problems, answers should be given for all of them. Even when one is going to guess, a

logical approach is to attempt to first eliminate one or two of the four alternatives. If this can be done, the chance of selecting a correct answer obviously improves from 1 in 4 to 1 in 3 or 1 in 2.

3. Plan ahead with a strategy. Which is your strongest area? Can you expect to see several problems in this area? What about your second strongest area? What is your weakest area?

4. Plan ahead with a time allocation. Compute how much time you will allow for each of the subject areas in the breadth exam and the relevant topics in the depth exam. You might allocate a little less time per problem for those areas in which you are most proficient, leaving a little more time in subjects that are difficult for you. Your time plan should include a reserve block for especially difficult problems, for checking your scoring sheet, and to make last-minute guesses on problems you did not work. Your strategy might also include time allotments for two passes through the exam—the first to work all problems for which answers are obvious to you, and the second to return to the more complex, time-consuming problems and the ones at which you might need to guess. A time plan gives you the confidence of being in control and keeps you from making the serious mistake of misallocation of time in the exam.

5. Read all four multiple-choice answers before making a selection. An answer in a multiple-choice question is sometimes a plausible decoy—not the best answer.

6. Do not change an answer unless you are absolutely certain you have made a mistake. Your first reaction is likely to be correct.

7. Do not sit next to a friend, a window, or other potential distractions.

## Exam Day Preparations

The exam day will be a stressful and tiring one. This will be no day to have unpleasant surprises. For this reason we suggest that an advance visit be made to the examination site. Try to determine such items as

1. How much time should I allow for travel to the exam on that day? Plan to arrive about 15 minutes early. That way you will have ample time, but not too much time. Arriving too early, and mingling with others who also are anxious, will increase your anxiety and nervousness.

2. Where will I park?

3. How does the exam site look? Will I have ample workspace? Where will I stack my reference materials? Will it be overly bright (sunglasses), cold (sweater), or noisy (earplugs)? Would a cushion make the chair more comfortable?

4. Where are the drinking fountains and lavatory facilities?

5. What about food? Should I take something along for energy in the exam? A bag lunch during the break probably makes sense.

## What to Take to the Exam

The NCEES guidelines say you may bring only the following reference materials and aids into the examination room for your personal use:

1. Handbooks and textbooks, including the applicable design standards. Specific titles that you probably will find useful include *CRC Standard Mathematical Tables* and *Reference Data for Radio Engineers*. In addition, a general electrical engineering handbook with which you are familiar is a good idea.

2. Bound reference materials, provided the materials remain bound during the entire examination. The NCEES defines "bound" as books or materials fastened securely in their covers by fasteners that penetrate all papers. Examples are ring binders, spiral binders and notebooks, plastic snap binders, brads, screw posts, and so on.

3. A battery-operated, silent, nonprinting, noncommunicating calculator from the NCEES list of appoved calculators. For the most current list, see the NCEES Web site (*www.ncees.org*). You also need to determine whether or not your state permits preprogrammed calculators. Bring extra batteries for your calculator just in case; many people feel that bringing a second calculator is also a very good idea.

At one time NCEES had a rule barring "review publications directed principally toward sample questions and their solutions" in the exam room. This set the stage for restricting some kinds of publications from the exam. *State boards may adopt the NCEES guidelines, or adopt either more or less restrictive rules.* Thus an important step in preparing for the exam is to know what will—and will not—be permitted. If possible you should obtain a written copy of your state's policy for the specific exam you will be taking. Occasionally there has been confusion at individual examination sites, so a copy of the exact applicable policy will not only allow you to carefully and correctly prepare your materials, but will also ensure that the exam proctors will allow all proper materials that you bring to the exam.

As a general rule you should plan well in advance what books and materials you want to take to the exam. Then they should be obtained promptly so you use the same materials in your review that you will have in the exam.

### *License Review Books*

The review books you use to prepare for the exam are good choices to bring to the exam itself. After weeks or months of studying, you will be very familiar with their organization and content, so you'll be able to quickly locate the material you want to reference during the exam. Keep in mind the caveat just discussed—some state boards will not permit you to bring in review books that consist largely of sample questions and answers.

### *Textbooks*

If you still have your university textbooks, they are the ones you should use in the exam, unless they are too out-of-date. To a great extent the books will be like old friends with familiar notation.

## Bound Reference Materials

The NCEES guidelines suggest that you can take any reference materials you wish, so long as you prepare them properly. You could, for example, prepare several volumes of bound reference materials, with each volume intended to cover a particular category of problem. One way to use this book would be to cut it up and insert portions of it in your individually prepared bound materials. Use tabs so specific material can be located quickly. If you do a careful and systematic review of electrical engineering and prepare a lot of well-organized materials, you just may find that you are so well prepared that you will not have left anything of value at home.

## Other Items

In addition to the reference materials just mentioned, you should consider bringing the following to the exam:

- *Clock*—You must have a time plan and a clock or wristwatch.

- *Exam assignment paperwork*—Take along the letter assigning you to the exam at the specified location. To prove you are the correct person, also bring something with your name and picture.

- *Items suggested by advance visit*—If you visit the exam site, you probably will discover an item or two that you need to add to your list.

- *Clothes*—Plan to wear comfortable clothes. You probably will do better if you are slightly cool.

- *Box for everything*—You need to be able to carry all your materials to the exam and have them conveniently organized at your side. Probably a cardboard box is the answer.

## Examination Scoring and Results

The questions are machine-scored by scanning. The answers sheets are checked for errors by computer. Marking two answers to a question, for example, will be detected and no credit will be given.

Your state board will notify you whether you have passed or failed roughly three months after the exam. Candidates who do not pass the exam the first time may take it again. If you do not pass you will receive a report listing the percentages of questions you answered correctly for each topic area. This information can help focus the review efforts of candidates who need to retake the exam.

The PE exam is challenging, but analysis of previous pass rates shows that the majority of candidates do pass it the first time. By reviewing appropriate concepts and practicing with exam-style problems, you can be in that majority. Good luck!

# CHAPTER 1

# Fundamental Concepts of Electrical Engineering

**OUTLINE**

ELECTRICAL QUANTITIES  2
Energy (Work) ■ Power ■ Charge ■ Current ■ Voltage ■ Electromotive Force ■ Resistance

OHM'S LAW  5

CIRCUIT ELEMENTS DEFINED  6
Resistor ■ Inductor ■ Capacitor

CIRCUIT ELEMENT VALUES  7
Resistance ■ Inductance ■ Capacitance

COULOMB'S LAW  9

SERIES AND PARALLEL COMBINATIONS  10

WYE-DELTA (Y-$\Delta$) TRANSFORMATION  11

COMPLEX ALGEBRA  11

COMPLEX NOTATION  12

CIRCUIT ELEMENT EQUATIONS  12

TRANSIENTS  13
Single Energy Transients ■ Double Energy Transients ■ Circuit Examples

TRANSFER FUNCTIONS  16

LAPLACE TRANSFORM  16

KIRCHHOFF'S LAWS  17

THEVENIN'S THEOREM  18

MAXIMUM POWER TRANSFER THEOREM  18

# Chapter 1 Fundamental Concepts of Electrical Engineering

MAXIMUM POWER TRANSFER THEOREM COROLLARY 19

NORTON'S EQUIVALENT CIRCUIT 19

VOLTAGE DIVISION AND SUPERPOSITION 20
Voltage Division ■ Superposition

MAGNETIC CIRCUIT TERMS 21

DETERMINANTS 22
Second Order ■ Third Order

RESONANCE 24
Figure of Merit, $Q$ ■ Series Resonance ■ Parallel Resonance (Antiresonance) ■ Impedance Transformation

IDEAL TRANSFORMER 27

FOURIER ANALYSIS 28

WAVEFORMS 28
RMS Value ■ Average Value

This chapter summarizes the fundamental concepts of electrical engineering and serves as the foundation upon which subsequent chapters are based. The several branches of electrical engineering reviewed in the following chapters make use of these fundamental concepts in unique, applied situations. Each chapter contains additional formulas and concepts particular to the subject under discussion; however, there will always be an analog between that specialized material and the fundamentals presented in this first chapter. Therefore, the reader will want to refer to this chapter for definitions of terms, conversion factors, and other information useful in solving the problems presented in subsequent chapters.

No problems are presented for solution in this chapter. It is intended as a basic review of material the reader must understand in order to solve problems in the various branches of electrical engineering contained in this text as well as on the PE Examination. Careful, thorough review of these fundamental concepts will help to ensure your success in solving problems rapidly and accurately.

# ELECTRICAL QUANTITIES

The basic electrical quantities are energy, power, charge, current, voltage, resistance, and electromotive force.

The basic electrical quantities are summarized in Table 1.1 and described in more detail in the following paragraphs.

In addition to the basic units, each quantity may also be accompanied by any of the prefixes in Table 1.2 as a scaling modifier.

## Energy (Work)

Energy ($W$) is defined by the following formula:

$$W = \int_{t_1}^{t_2} P\,dt$$

Table 1.1  Basic Electrical Quantities

| Quantity | Units | Symbols | Formula | Equivalent Units |
|---|---|---|---|---|
| Energy (work) | joule | W | $W = \int P\,dt$ | watt-second |
| Power | watt | P | $P = \dfrac{dW}{dt}$ | volt-ampere or joule/second |
| Charge | coulomb | Q | $Q = CV$ | ampere-second |
| Current | ampere | I | $I = \dfrac{V}{R}$ | coulomb/second |
| Electrostatic pressure | volt | V | $V = IR$ | joule/coulomb |
| Resistance | ohm | R | $R = \dfrac{V}{I}$ | volt/ampere |

Table 1.2  Relations between Various Units of Energy

| | | |
|---|---|---|
| watt-second | = | 1 joule |
| watt-second | = | 0.239 calorie |
| watt-second | = | 0.738 foot-pound |
| kilowatt-hour | = | 3413 Btu |
| kilowatt-hour | = | 1.34 horsepower-hours |
| kilowatt-hour | = | $3.6 \times 10^6$ joules |
| electron-volt | = | $1.6 \times 10^{-9}$ joule |
| erg | = | $10^{-7}$ joule |

where
$W$ = energy (in joules) converted between times $t_1$ and $t_2$
$P$ = power in watts
$t$ = time in seconds

If power is steady, $W = Pt$.

Three units of energy most commonly used in electrical engineering are the following:

- *Joule* or *watt-second*, defined as kinetic energy; the formula for the mechanical analog is $\frac{1}{2}mv^2$, where $m$ is mass and $v$ is velocity.
- *Kilowatt-hour*, a unit used in electrical power system calculations.
- *Electron-volt*, a unit used in calculations of electron behavior, the amount of kinetic energy acquired by an electron accelerated by a potential difference of one volt.

## Power

Power is the time rate of doing work. For a constant current $I$ maintained through any load having a voltage $V$ across it, the power is $P = IV$. With time-varying current $i$ and voltage $v$, the average power is

$$P_{av} = \frac{1}{t}\int_0^t iv\,dt$$

Table 1.3  Scaling Modifiers

| Symbol | Prefix | Multiplier |
|---|---|---|
| m | milli- | $10^{-3}$ |
| $\mu$ | micro- | $10^{-6}$ |
| n | nano- | $10^{-9}$ |
| p | pico- | $10^{-12}$ |
| k | kilo- | $10^{3}$ |
| M | mega- | $10^{6}$ |
| G | giga- | $10^{9}$ |

The units for power are obtained from energy units when the unit of time is specified. Typical units and conversion factors include the following:

watt = 0.239 calorie/second = 0.000948 Btu/second

joule per second = 1 watt

foot-pound per second = 1.356 watts

horsepower = 746 watts = 550 foot-pound/second

## Charge

Electric charge or quantity of electricity $Q$ is that amount of electricity passed through a circuit during a specified time interval by an electric current. The basic unit of charge is the *coulomb* and is equal to the quantity of electricity transported in one second across any cross section of a circuit by a current of one ampere. A coulomb is also equal to the charge possessed by $6.24 \times 10^{18}$ electrons.

## Current

Electric current is the rate of flow of electrons through a conductor. Current may be classed according to the manner in which it changes with time:

- *Direct current:* a unidirectional current that may vary in amount with time but never reverses direction.

- *Pulsating current:* a direct current that pulsates in magnitude periodically.

- *Continuous current:* an essentially nonpulsating direct current. Unless otherwise specified, this is the type of current referred to when the term *direct current* is used.

- *Alternating current:* a current that changes direction periodically. Normally, the net flow of current is zero. Electrons move slightly back and forth past any fixed cross section of conductor without any progression along the conductor.

The basic unit of current is the ampere:

1 ampere = 1 coulomb per second

## Voltage

Voltage refers to the quantity of energy that is gained or lost when a charge is moved from one point to another in an electric circuit. This quantity is called *potential difference* and is measured in *volts*. The potential difference between points $a$ and $b$ is expressed by the formula:

$$V_{ab} = \frac{W}{Q} = \frac{\text{joule}}{\text{coulomb}}$$

where
$Q$ = coulombs transferred
$W$ = joules lost or gained by $Q$ during transfer

Another way of expressing potential difference is by considering $dw$, the gain or loss in energy, and $dq$, the charge of a very small particle. Thus,

$$V_{ab} = \frac{dw}{dq}$$

If $n$ charged particles are transferred from $a$ to $b$ in an interval of time $dt$, the total charge is $ndq$ and the total energy is $ndw$. Thus,

$$V = \frac{ndw}{ndq} = \frac{ndw/dt}{ndq/dt} = \frac{\text{watts}}{\text{amperes}} = \frac{P}{I}$$

## Electromotive Force

Electromotive force (emf) relates to the physical process by which energy is changed from nonelectrical form into electrical form, in the presence of a force that tends to separate electric charges. This action occurs in electric generators, in thermocouples, and in chemical cells. Emf is defined the same way as potential difference:

$$E = \frac{dw}{dq} = \frac{P}{I} = V$$

The difference between the two is generally defined as follows:

$E$ relates to energy being given up or generated.

$V$ relates to energy being consumed, or a potential difference between two terminals.

## Resistance

The ratio of potential difference to current is called *resistance* ($R$). Thus,

$$R = \frac{V}{I}$$

This is an expression of *Ohm's law*; the basic unit of resistance is the *ohm*. The resistor is a common circuit element having resistance. There are also other circuit elements, defined later, that react to impede current flow; these too are expressed quantitatively in ohms.

# OHM'S LAW

Ohm's law defines the relationship between the voltage across and the current through a resistance (Figure 1.1). Thus,

$$V = IR$$
$$I = V/R$$
$$R = V/I$$

The resistance must be constant under all conditions. In ac circuits other reactive elements, such as capacitance and inductance, may also be present. Ohm's law applies to these elements as well, as long as they are constant.

# Chapter 1 Fundamental Concepts of Electrical Engineering

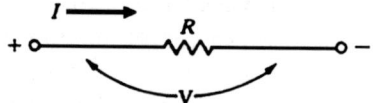

**Figure 1.1** Ohm's law relates voltage, current, and resistance

Ohm's law does not hold if the circuit elements are nonlinear or the current is not steady-state, but it holds for non–steady-state conditions if linear. Non–steady-state current conditions are discussed in this chapter in the section Transients.

## CIRCUIT ELEMENTS DEFINED

The three basic passive circuit elements are the resistor, the inductor, and the capacitor.

Characteristics of the three basic passive circuit elements are summarized in Table 1.4 and defined in the following paragraphs. Discussion of circuit elements is based on Ohm's law.

### Resistor

A resistor is a device in which the flow of electric current always produces heat and nothing else. Since in a resistor all the electric power is consumed in generating heat, the current through a resistor must always flow from plus to minus. If the current is reversed, the potential difference must reverse; if the current is zero, the potential difference must be zero.

### Inductor

An inductor is a device capable of storing and giving up magnetic energy. A pure inductor dissipates no heat or energy. When the current $i$ in an electric circuit varies with time, a self-induced voltage is produced according to the formula

$$V = -L\frac{di}{dt} \{\text{polarity is such as to oppose } \textit{change} \text{ in current}\}$$

**Table 1.4** Circuit Element Characteristics

| Element | Schematic Symbol | Current Through | Voltage Drop | Power Dissipation | Stored Energy | Units |
|---|---|---|---|---|---|---|
| resistor | | $I = \dfrac{V}{R}$ | $V = IR$ | $V^2/R$ or $I^2R$ | zero | ohm $\Omega, k\Omega, M\Omega$ |
| inductor | | $I = \dfrac{\phi}{L} = \dfrac{1}{L}\int V\,dt$ | $V = -L\dfrac{dI}{dt}$ | zero | $W = \dfrac{1}{2}LI^2$ | henry H, mH, $\mu$H |
| capacitor | | $I = C\dfrac{dV}{dt}$ | $V = \dfrac{1}{C}\int I\,dt$ | zero | $W = \dfrac{1}{2}CV^2$ | farad F, $\mu$F, pF |

in which the proportionality constant $L$ is the **inductance** or the **coefficient of self-induction**, expressed in henrys. In ac circuits inductance is termed reactive or imaginary, as opposed to resistance, which is real. This subject is discussed further in this chapter in the section Complex Algebra.

### Capacitor

A capacitor is a device capable of storing or giving up electric energy. The charge on a capacitor depends upon the voltage across it and its capacitance, as described by the formula

$$Q = CV$$

where
$Q$ = charge in coulombs
$C$ = capacitance in farads
$V$ = voltage in volts

In ac circuits, capacitance, like inductance, is termed reactive or imaginary. In writing circuit equations, capacitance is given an imaginary sign 180° out of phase with any circuit inductance.

## CIRCUIT ELEMENT VALUES

Values of the three basic circuit elements may be calculated from the information given in this section.

### Resistance

The resistance of a section of conductor of uniform cross section (Figure 1.2) is

$$R = \rho l / A$$

where
$A$ = cross-sectional area (square meters or circular mils*)
$l$ = length (meters or feet)
$R$ = resistance (ohms)
$\rho$ = resistivity of material (ohm-meters) = $\dfrac{1}{\text{conductivity}}$

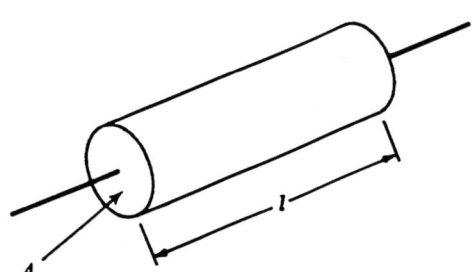

**Figure 1.2** Resistance of a conductor

---

*The area of a circle one mil (0.001 inch) in diameter is one circular mil; the area of any circle in circular mils equals the square of its diameter in mils.

Table 1.5  Values of Resistivity for Typical Conductive Materials

| Material | Resistivity, ohm-meters | Resistivity, ohm-cir mils / ft |
|---|---|---|
| aluminum | $2.6 \times 10^{-8}$ | 17 |
| copper | $1.7 \times 10^{-8}$ | 10 |
| cast iron | $9.7 \times 10^{-8}$ | 58 |
| lead | $22 \times 10^{-8}$ | 132 |
| silver | $1.6 \times 10^{-8}$ | 9.9 |
| steel | $(11 - 90) \times 10^{-8}$ | 66–540 |
| tin | $11 \times 10^{-8}$ | 69 |
| nichrome | $100 \times 10^{-8}$ | 602 |

## Inductance

The inductance of a coil (Figure 1.3) is

$$L = KN^2$$

where
 $N$ = number of turns
 $L$ = inductance in henries
 $K$ = a constant dependent upon geometry and material

Formulas for specific shapes are as follows:

### Solenoid with Nonmagnetic Core

$$L = \frac{4\pi^2 \times 10^{-7} N^2 r^2}{S} \text{ henrys}$$

where
 $S$ = length of solenoid
 $r$ = radius of solenoid
 $N$ = number of turns

### Toroid of Rectangular Section with Nonmagnetic Core

$$L = 2N^2 b \left( \ln \frac{g+a}{g-a} \right) 10^{-9} \text{ henrys}$$

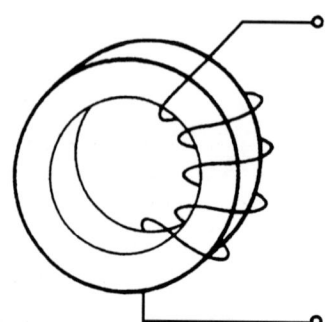

Figure 1.3  Coil inductance

where
    $b$ = axial length of core, in cm
    $g$ = mean diameter of ring, in cm
    $a$ = radial depth of core, in cm
    $N$ = number of turns

### Toroid of Circular Section with Nonmagnetic Core

$$L = 2\pi N^2 (g - \sqrt{g^2 - d^2})10^{-9} \text{ henrys}$$

where
    $d$ = diameter of cross section, in cm
    $g$ = mean diameter of ring, in cm
    $N$ = number of turns

When every flux line is linked by every stream line of electric current, as at the center of long solenoids and in toroids, and the permeability of the entire circuit is independent of current, the constant $K$ is replaced by $R$, which is the reluctance of the complete magnetic circuit.

If two coils exist in a series circuit and there is mutual coupling inductance $M$ between them, the total inductance is either

$$L = L_1 + L_2 + 2M \quad \text{for additive mutual inductance}$$

or

$$L = L_1 + L_2 - 2M \quad \text{for bucking (subtractive) mutual inductance}$$

## Capacitance

The capacitance of two parallel plates (Figure 1.4) is

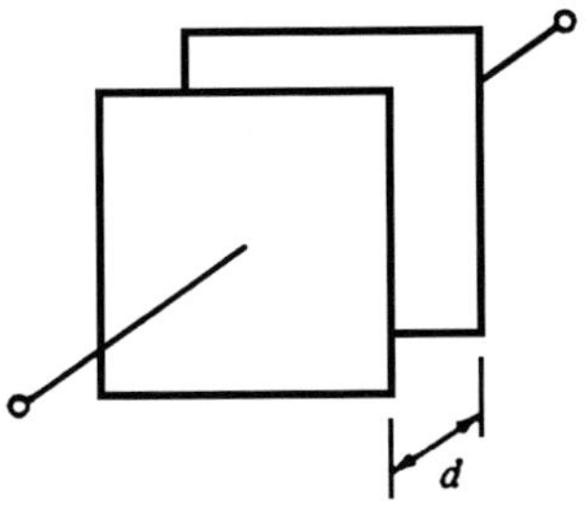

$$C = \frac{\varepsilon_r \varepsilon_v A}{d} = \frac{DA}{Ed}$$

where
    $\varepsilon_v = \dfrac{10^{-9}}{36\pi}$ farads/meter = 8.84 pf/m
    $\varepsilon_r$ = relative permittivity of dielectric (dielectric constant)
    $A$ = common area, in square meters
    $d$ = plate spacing, in meters
    $C$ = capacitance, in farads
    $D$ = electric flux density, in coulombs per square meter
    $E$ = electric field intensity, in volts per meter

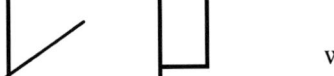

**Figure 1.4** Capacitance of two parallel plates

# COULOMB'S LAW

Coulomb's law is often used as the basis for developing the theory of electrostatic fields. It defines the force on a charge in an electric field or the force between two charges.

In the field of a charge $+Q$ concentrated at a point, the total flux, $\phi$, through a sphere of radius $r$ is

$$\phi = 4\pi r^2 D$$

But flux is also equal to the charge, $Q$, so

$$Q = 4\pi r^2 D$$

and $D = Q/4\pi r^2$, the flux density at distance $r$ caused by a point charge, $Q$. Electric field intensity, $E$, at this radius is

$$E = \frac{Q}{4\pi \varepsilon r^2}$$

If a charge, $Q_2$, is placed in the field of a charge, $Q_1$, the force, $F$, on $Q_2$ is

$$F = EQ_2 = \frac{Q_1 Q_2}{4\pi \varepsilon r^2}$$

where
  $F$ = force in newtons
  $Q$ = charge in coulombs
  $\varepsilon$ = permittivity of the medium = $8.85 \times 10^{-12}$ for air
  $r$ = distance between charged particles in meters

One electron has a charge of $1.6 \times 10^{-19}$ coulomb; $6.28 \times 10^{18}$ electrons have the charge of one coulomb.

## SERIES AND PARALLEL COMBINATIONS

The effect of connecting resistors, inductors, or capacitors in series or in parallel with their own kind is shown in Table 1.6.

Table 1.6  Series-parallel Combinations

| Circuit Element | Series | Parallel |
|---|---|---|
| Resistor $R_1, R_2$ | $R = R_1 + R_2$ | $R = \dfrac{R_1 R_2}{R_1 + R_2}$ or $R = \left(\dfrac{1}{R_1} + \dfrac{1}{R_2}\right)^{-1}$ |
| Inductor $L_1, L_2$ | $L = L_1 + L_2$ | $L = \dfrac{L_1 L_2}{L_1 + L_2}$ or $L = \left(\dfrac{1}{L_1} + \dfrac{1}{L_2}\right)^{-1}$ |
| Capacitor $C_1, C_2$ | $C = \dfrac{C_1 C_2}{C_1 + C_2}$ or $C = \left(\dfrac{1}{C_1} + \dfrac{1}{C_2}\right)^{-1}$ | $C = C_1 + C_2$ |

# Complex Algebra

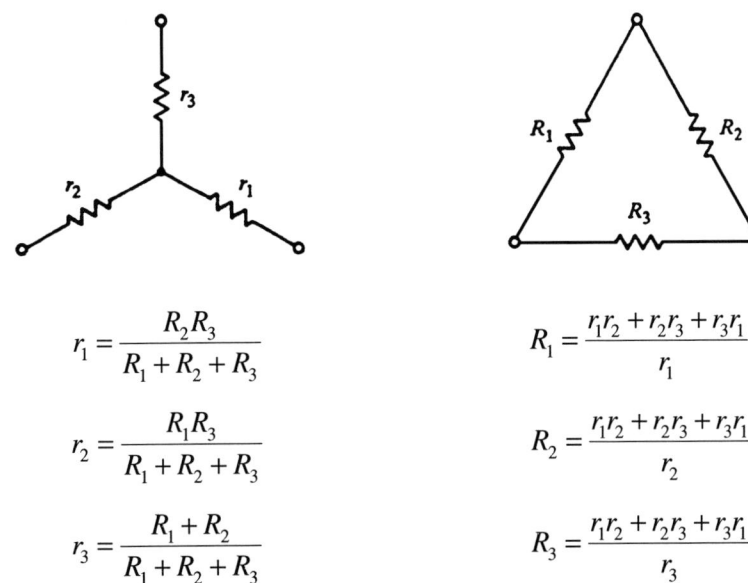

$$r_1 = \frac{R_2 R_3}{R_1 + R_2 + R_3} \qquad R_1 = \frac{r_1 r_2 + r_2 r_3 + r_3 r_1}{r_1}$$

$$r_2 = \frac{R_1 R_3}{R_1 + R_2 + R_3} \qquad R_2 = \frac{r_1 r_2 + r_2 r_3 + r_3 r_1}{r_2}$$

$$r_3 = \frac{R_1 + R_2}{R_1 + R_2 + R_3} \qquad R_3 = \frac{r_1 r_2 + r_2 r_3 + r_3 r_1}{r_3}$$

**Figure 1.5** Wye-delta transformation for simplifying networks

In a series circuit, the same current flows through each and every element; the total voltage drop is the algebraic sum of all of the individual voltage drops.

In a parallel circuit, the same voltage appears across each and every element; the total current is the algebraic sum of all of the individual currents.

## WYE-DELTA (Y-Δ) TRANSFORMATION

In solving circuit networks, it is often desirable to reduce them to a simpler form. The procedure is sometimes complicated, however, because the connections of circuit elements are such that the network cannot be reduced by series and parallel combinations. The Y-Δ transformation sometimes is a useful technique; it is summarized in Figure 1.5.

## COMPLEX ALGEBRA

The following is a brief review of complex algebraic notation as applied to electrical engineering. See Figure 1.6.

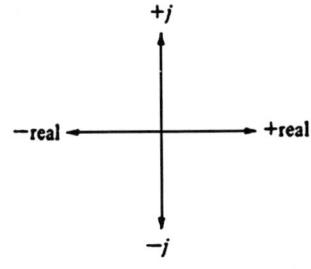

**Figure 1.6** Graph for complex numbers

$j = \sqrt{-1}$ = imaginary component, referred to as the $j$-operator

$$(a + jb) + (c + jd) = (a + c) + j(b + d)$$

complex conjugate of $a + jb = \overline{a + jb} = a - jb$

$$(a + jb)(c + jd) = (ac - bd) + j(ad + bc)$$

$$\frac{a + jb}{c + jd} = \frac{(a + jb)(c - jd)}{c^2 + d^2} = \frac{(ac + bd) + j(bc - ad)}{c^2 + d^2}$$

$$A\angle\theta = A\cos\theta + jA\sin\theta = Ae^{j\theta}$$

$$(A\angle\theta)(B\angle\alpha) = AB\angle\theta + \alpha$$

$$\frac{A\angle\theta}{B\angle\alpha} = \frac{A}{B}\angle\theta - \alpha$$

$$A\angle\theta + B\angle\alpha = (A\cos\theta + B\cos\alpha) + j(A\sin\theta + B\sin\alpha) = C\angle\delta$$

where
$$C = \sqrt{(\Sigma \text{real})^2 + (\Sigma \text{imaginary})^2}$$
$$\delta = \arctan\left[\frac{\Sigma \text{imaginary}}{\Sigma \text{real}}\right]$$

$$(A\angle\theta)^m = A^m \angle m\theta$$
$$(A\angle\theta)^{1/m} = A^{1/m} \angle \frac{\theta}{m} + \frac{2\pi n}{m}$$

where $n = 0, 1, 2, \ldots, (m-1)$.

# COMPLEX NOTATION

The several forms of complex notation may be summarized as follows:

rectangular form:    $Z = R \pm jX$

polar form:    $Z = |Z| \angle \pm\theta$, where $\theta = \arctan X/R$ in degrees

exponential form:    $Z = |Z| e^{\pm j\theta}$, where $\theta$ is expressed in radians

trigonometric form: $Z = |Z|(\cos\theta \pm j\sin\theta)$

# CIRCUIT ELEMENT EQUATIONS

In performing circuit analysis, a convenient notational format is needed to manipulate the three circuit elements.

In a purely inductive ac circuit the current lags the voltage by 90°. In a purely capacitive circuit, the current leads the voltage by 90°. The sum of a resistance and an inductance is $R + jX_L$ and is written as an impedance:

$$Z = R + jX_L \text{ ohms}$$

The sum of a resistance and a capacitance is $R - jX_C$ and is written as an impedance:

$$Z = R - jX_C \text{ ohms}$$

These two equations can be manipulated as required using complex algebraic relationships. The actual values of $X_L$ and $X_C$ may be calculated from the following formulas:

$$X_L = 2\pi f L = \omega L$$
$$X_C = \frac{1}{2\pi f C} = \frac{1}{\omega C}$$

In complex notation, these are written

$$jX_L = j\omega L$$
$$jX_C = j\frac{1}{\omega C} = -\frac{1}{j\omega C}$$

In certain cases it is more convenient to work with the reciprocal of impedance and its real and imaginary components. Such cases include parallel circuits and current (rather than voltage) sources. The following relationships hold:

$$\frac{1}{Z} = Y = \text{admittance} = \frac{1}{\text{impedance}}$$

$$\frac{1}{R} = G = \text{conductance} = \frac{1}{\text{resistance}}$$

$$-\frac{1}{X} = B = \text{susceptance} = \frac{-1}{\text{reactance}}$$

Thus,

$$Y = G + jB \text{ siemens}$$
$$Z = R + jX \text{ ohms}$$

# TRANSIENTS

In a preceding section, only circuits where Ohm's law applies were considered. These circuits are completely independent of time. In every practical circuit, however, there are two reactive elements present, no matter how small, which manifest themselves when current is changing. The effect that these two elements, inductance and capacitance, have in a transient situation is discussed next.

## Single Energy Transients

Circuits containing only one type of reactive element are readily analyzed using a generalized transient response formula of the form

$$f(t) = f_{ss} + (f_0 - f_{ss})e^{-t/\tau}$$

where
$f_{ss}$ = steady state final value of $f(t), f(t \to \infty)$
$f_0$ = initial value of $f(t), f(t = 0^+)$
$\tau$ = circuit time constant

This formula may be applied to any voltage or current in the circuit, even with initial nonzero values of inductor current or capacitor voltage, as long as only a single time constant is present (i.e., the circuit is first order).

Formulas specific to series $RL$ and $RC$ circuits are summarized as follows:

## RL Circuit

1. Inductance is associated with the magnetic field of the current.

2. Inductance has the mathematical property $v = L\,di/dt$.

3. $i = I_{ss} + (I_0 - I_{ss})e^{-t/\tau}$

where

$I_{ss} = V/R$
$\tau = L/R$ seconds

4. Stored energy is $W = \frac{1}{2}Li^2$

### RC Circuit

1. Capacitance is associated with the electric field.
2. Capacitance has the mathematical property $i = C\,dv/dt$.
3. $v = V_{ss} + (V - V_{ss})e^{-t/\tau}$

where $\tau = RC$ seconds

4. $i = dq/dt$
5. Charge is $Q = CV$.
6. Stored energy is $W = \tfrac{1}{2}CV^2$.

### Double Energy Transients

Double energy, or second order, systems are those in which energy can be stored in two separate forms. These circuits contain $R$, $L$, and $C$. Their analysis can become quite complex because of damped oscillations and complex equations. They are usually most easily solved using Laplace transforms.

For an $RLC$ series circuit the loop equation for voltage is

$$V = L\frac{di}{dt} + Ri + \frac{1}{C}\int_0^t i\,dt$$

### Circuit Examples

Transient analysis of circuits containing reactive elements may be performed using any of the following methods:

1. Transient response formula for single energy circuits
2. Differential equations
3. Laplace transforms
4. Computer solutions

Examples of some specific circuits follow.

### VL Circuit

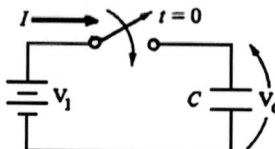

$V_L = L\dfrac{dI}{dt} = V_1, \qquad \dfrac{dI}{dt} = \dfrac{V_1}{L}$

### VC Circuit

$V_C = V_1$ after switch closed
$I(t) \to \infty$ upon closing switch
$I(t) \to 0$ for $t > 0^+$

## VRL Circuit

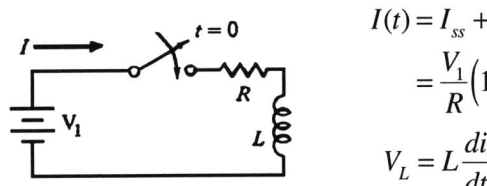

$$I(t) = I_{ss} + (I_0 - I_{ss})e^{-t/\tau}$$
$$= \frac{V_1}{R}\left(1 - e^{-(R/L)t}\right)$$
$$\begin{cases} I_0 = 0 \\ I_{ss} = V_1/R \\ \tau = L/R \end{cases}$$

$$V_L = L\frac{di}{dt} = L\frac{V_1}{R}\frac{R}{L}e^{-(R/L)t} = V_1 e^{-(R/L)t}$$

## VRC Circuit

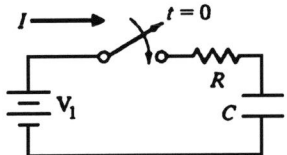

$$I(t) = I_{ss} + (I_0 - I_{ss})e^{-t/\tau}$$
$$= \frac{V_1}{R}e^{-t/RC}$$
$$\begin{cases} I_0 = V_1/R \\ I_{ss} = 0 \\ \tau = RC \end{cases}$$

$$V_C(t) = \frac{1}{C}\int_0^t I\, dt = \frac{V_1}{RC}\int_0^t e^{-t/RC}\, dt = -V_1 e^{-t/RC}\Big|_0^t = V_1(1 - e^{-t/RC})$$

Note that $V_C(t)$ may be found directly by using the transient formula

$$V_C(t) = V_{ss} + (V_0 - V_{ss})e^{-t/\tau}$$
$$= V_1(1 - e^{-t/RC})$$

## VC/CR Circuit

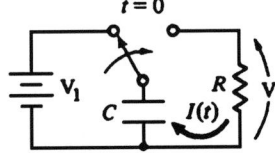

$$V_C(t) = V_{ss} + (V_0 - V_{ss})e^{-t/\tau}$$
$$= V_1 e^{-t/RC}$$
$$\begin{cases} V_0 = V_1 \\ V_{ss} = 0 \\ \tau = RC \end{cases}$$

$$I(t) = I_{ss} + (I_0 - I_{ss})e^{-t/\tau}$$
$$= \frac{V_1}{R}e^{-t/RC}$$
$$\begin{cases} I_0 = V_1/R \\ I_{ss} = 0 \\ \tau = RC \end{cases}$$

## RCC Circuit

$$W = \tfrac{1}{2}CV^2; \quad \text{for } t < 0;\ V_1 = V_0, V_2 = 0$$

$$I(t) = I_{ss} + (I_0 - I_{ss})e^{-t/\tau}$$
$$= \frac{V_1}{R}e^{-t/\tau}$$
$$\begin{cases} I_0 = V_1/R \\ I_{ss} = 0 \\ \tau = R\dfrac{C_1 C_2}{C_1 + C_2} \end{cases}$$

$$V_2(t) = \frac{1}{C_2}\int_0^t I\, dt = \frac{V_1}{RC_2}\int_0^t e^{-t/\tau}\, dt = \frac{-\tau V_1}{RC_2}e^{-t/\tau}\Big|_0^t = \frac{C_1 V_1}{C_1 + C_2}(1 - e^{-t/\tau})$$

Initial energy $= \tfrac{1}{2}C_1 V_1^2$

$$\text{final energy} = \frac{1}{2}(C_1+C_2)V_{ss}^2 = \frac{1}{2}(C_1+C_2)\left(\frac{C_1V_1}{C_1+C_2}\right) = \frac{1}{2}\frac{C_1V_1^2}{\left(1+\frac{C_2}{C_1}\right)}$$

$$\text{lost energy (dissipated in R)} = \frac{1}{2}C_1V_1^2 - \frac{1}{2}\frac{C_1V_1^2}{\left(1+\frac{C_2}{C_1}\right)} = \frac{1}{2}\frac{C_1V_1^2}{\left(1+\frac{C_1}{C_2}\right)}$$

If $C_1 = C_2$, then one-half of the energy is lost.

$$\frac{\text{final energy}}{\text{initial energy}} = \frac{C_1}{C_1+C_2}$$

$$\frac{\text{lost energy}}{\text{initial energy}} = \frac{C_2}{C_1+C_2}$$

## TRANSFER FUNCTIONS

One of the most important characteristics of a component or a circuit is the relationship between the input and output signals. This relationship is expressed by the transfer function, which is defined as the ratio of the output signal divided by the input signal. In general, a transfer function is the Laplace transform of a ratio of variables. The variables can be impedance, current ratio, or voltage ratio. The Laplace transformation converts the equations from the time domain to the frequency domain. Transfer functions are used for transient and steady-state analysis. For example, if a component has an input voltage $V_{in}$ and output voltage $V_{out}$, then the frequency domain transfer function of the component is defined as follows:

$$G(s) = \frac{V_{out}(s)}{V_{in}(s)}$$

## LAPLACE TRANSFORM

The Laplace transform is a transformation technique relating time functions to frequency dependent functions of a complex variable. The Laplace transform is defined as follows:

$$\mathcal{L}[f(t)] = F(s) = \int_0^\infty f(t)e^{-st}dt$$

The inverse Laplace transform is written in the following notation:

$$\mathcal{L}^{-1}[F(s)] = f(t)$$

The following is an example of a Laplace transform:

$$\mathcal{L}[e^{-t}] = \int_0^\infty e^{-t}e^{-st}dt = \left.\frac{1e^{-(s+1)t}}{-(s+1)}\right|_0^\infty = \frac{1}{s+1}$$

**Table 1.7**  Laplace Transforms

| Description | Laplace Transform $F(s) = \mathcal{L}f(t)$ | Time Function $f(t)$ |
|---|---|---|
| unit pulse | 1 | $\delta(t)$ |
| unit step | $1/s$ | $U(t)$ |
| unit ramp | $1/s^2$ | $t$ |
| polynomial | $n!/s^{n+1}$ | $t^n$ |
| exponential | $1/(s-a)$ | $e^{at}$ |
| sine wave | $1/(s^2 + \omega^2)$ | $\dfrac{1}{\omega}\sin\omega t$ |
| cosine wave | $s/(s^2 + \omega^2)$ | $\cos\omega t$ |
| damped sine wave | $1/[(s+a)^2 + \omega^2]$ | $\dfrac{1}{\omega}e^{at}\sin\omega t$ |
| damped cosine wave | $(s+a)/[(s+a)^2 + \omega^2]$ | $e^{at}\cos\omega t$ |

**Table 1.8**  Circuit Element Symbology

| Circuit Element | $Z(s)$ | $Z(j\omega)$ | $X(j\omega)$ |
|---|---|---|---|
| resistor | $R$ | $R$ | $R$ |
| inductor | $sL$ | $j\omega L$ | $\omega L$ |
| capacitor | $\dfrac{1}{sC}$ | $\dfrac{1}{j\omega C}$ | $\dfrac{1}{\omega C}$ |

Table 1.7 lists several of the more common Laplace transforms.

Table 1.8 shows how the three basic circuit elements are symbolized in the *s*-domain and time domain.

# KIRCHHOFF'S LAWS

Two fundamental simple laws of the electric circuit have received the name of Kirchhoff's laws. These are stated as follows:

*First Law.* The amount of current flowing away from a point in a circuit is equal to the amount flowing to that point. In short,

$$\sum \text{currents entering a node} = 0$$

*Second Law.* The difference of electric potential between any two points is the same regardless of the path along which it is measured. In short,

$$\sum \text{voltage drops (or rises) around a closed path} = 0$$

The number of independent equations that can be written using Kirchhoff's first law is *one* less than the number of *nodes* or junction points at which two currents join to form a third. The number of independent equations that can be written using Kirchhoff's second law is equal to the number of branches minus the number of independent node equations; a *branch* is any section of a circuit that directly joins two nodes without passing through a third node.

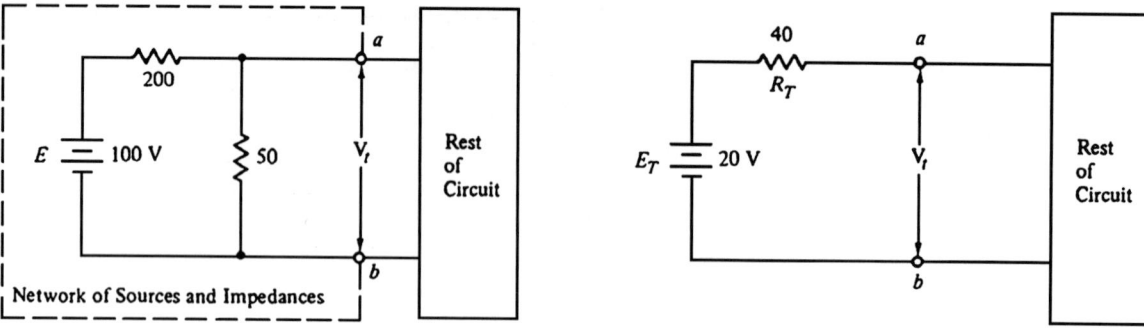

**Figure 1.7** Converting a circuit with Thevenin's theorem

# THEVENIN'S THEOREM

Thevenin's theorem is a handy tool in the solution of both dc and ac circuits. It applies only to the terminal voltage and current conditions of a linear two-terminal network not magnetically coupled to an external network. For dc circuits, it states that one can replace a two-terminal network by a voltage source $E_T$ and a resistance $R_T$ connected in series.

For example, the circuit on the left of Figure 1.7 may be converted to that shown on the right.

The two circuits are equivalent if both of the following are true:

1. $E_T$ is made equal to $V_t$ on open circuit; thus, both circuits have equal open-circuit voltages.

2. $R_T$ is given a value such that when the terminals $a - b$ of each are short-circuited, the currents through the short circuits of both are equal.

Thus, for the above example,

$$E_T = 100 \frac{50}{200 + 50} = 20 \text{ volts}$$

$$R_T = \frac{200 \times 50}{200 + 50} = 40 \text{ ohms}$$

# MAXIMUM POWER TRANSFER THEOREM

Using the Thevenin equivalent of a voltage source circuit, it is easy to study the characteristics of the rest of the circuit. With a "Thevenized" circuit,

$$V_t = E_T - I_t R_T$$
$$P_t = V_t I_t = E_T I_t - I_t^2 R_T$$

Maximum power occurs at a value of $I_t = \frac{1}{2} E_T / R_T$ and at a value of $V_t = E_T/2$, so that maximum power is $P_{max} = (E_T)^2 / 4R_T$. This is obtained by differentiating the above equation for $P_t$ with respect to $I_t$ and setting the result equal to zero. The resistance of the load, for maximum power transfer, is

$$R_{L \max p} = \frac{V_t}{I_t} + \left(\frac{E_T}{2}\right)\left(\frac{2R_T}{E_T}\right) = R_T$$

This result demonstrates the **maximum power transfer theorem**, which states that the maximum power is delivered to a load by a two-terminal linear network when that load is so adjusted that the terminal voltage is half its open-circuit value.

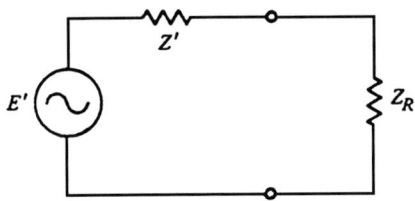

**Figure 1.8** Corollary to the maximum transfer theorem

## MAXIMUM POWER TRANSFER THEOREM COROLLARY

The maximum power transfer theorem, restated, says that maximum power will be delivered by a network to an impedance $Z_R$ if $Z_R$ is the complex conjugate of the $Z$'s of the network, measured looking back into the terminals of the network.

A corollary states that if only the absolute magnitude and not the angle of $Z_R$ may be varied, then the greatest power output will be delivered from the network if the absolute magnitude of $Z_R$ is made equal to the absolute magnitude of $Z'$, as shown in Figure 1.8:

$$|Z_R| = |Z'|$$

The amount of power delivered by matching magnitudes will be somewhat less than the amount possible if both magnitude and angle are adjusted to the conjugate condition.

## NORTON'S EQUIVALENT CIRCUIT

In the solution of many electric circuits, it is often more advantageous to consider constant-current sources rather than the constant-voltage sources thus far considered. The comparison between a Norton equivalent circuit and a Thevenin equivalent circuit for a two-port network is illustrated in Figure 1.9.

$$E_T = \frac{A_N}{G_N}, \quad R_T = \frac{1}{G_N}, \quad A_N = E_T G_N$$

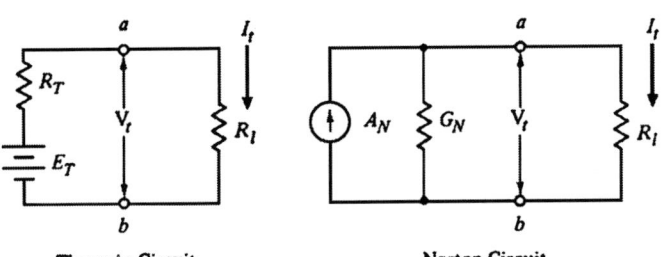

**Thevenin Circuit**     **Norton Circuit**

**Figure 1.9** Thevenin and Norton circuits compared

where $A_N$ is a constant-current source through short-circuited terminals.

$$I_t = A_N - V_t G_N$$

$$V_t = \frac{A_N}{G_N} - \frac{I_t}{G_N}$$

Note that these two sources are equivalent only in the current, voltage, and power they deliver to the terminals $a - b$. They are not equivalent in the amount of internal power they consume within themselves.

# VOLTAGE DIVISION AND SUPERPOSITION

Following are two time-saving methods of analyzing circuits.

### Voltage Division

The voltage $V_l$ across the load terminals of a circuit may be determined in the manner shown in Figure 1.10.

### Superposition

The response of a linear network to a number of excitations applied simultaneously is equal to the sum of the responses of the network when each excitation is applied individually, as shown in Figure 1.11.

$$V_X = \frac{Z_2 \| Z_3}{Z_1 + Z_2 \| Z_3} E_1 + \frac{Z_1 \| Z_3}{Z_2 + Z_1 \| Z_3} E_2$$

where

$$Z_2 \| Z_3 = \frac{Z_2 Z_3}{Z_2 + Z_3}$$

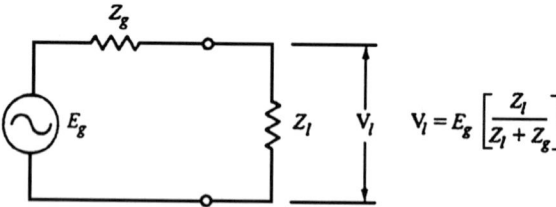

Figure 1.10  Voltage division

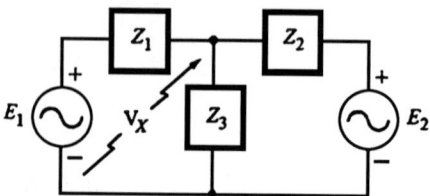

Figure 1.11  Superposition

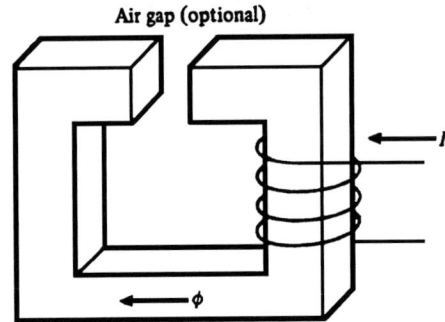

**Figure 1.12** Magnetic circuit

## MAGNETIC CIRCUIT TERMS

Figure 1.12 shows a typical magnetic circuit diagram. In SI units, the following terms apply:

$F$ = magnetomotive force (amp turns)

$\phi$ = flux (webers)

$H$ = magnetic field intensity (amp turns/meter)

$B$ = flux (magnetic field) density (weber/square meter)

$\mu_r$ = permeability (dimensionless)

$\Re$ = reluctance

$I$ = current (amps)

$N$ = number of complete turns about core

$A$ = cross-sectional area (square meters)

$l$ = length of magnetic path (meters)

$\mu_v$ = permeability of free space = $4\pi \times 10^{-7}$ webers/amp turn-meter

Some relationships between terms include the following:

$\phi = F/R$ (similar to $I = V/R$)

$F = NI$

$H = NI/l$

$V = N d\phi/dt$

$\Re = l/\mu \mu_v A$

$\mu = B/H$

$\Re_{tot} = \Re_1 + \Re_2$ (reluctances combine like resistances)

There is actually more than one metric system of units, and multiple metric systems are sometimes used in magnetic field measurements. One is called the CGS, which stands for centimeter-gram-second. The other was originally known as the MKS system, which stood for meter-kilogram-second. This system was eventually

**Table 1.9** Magnetic Circuit Symbols, Units, and Interrelationships

| Symbol | SI Units | CGS Units | CGS Equation | CGS Units/ SI Units | SI Equation |
|---|---|---|---|---|---|
| $F$ | amp turn | Gilbert | $F = 0.4\,NI$ | 1.257 | $F = NI$ |
| $H$ | amp turn/meter | Oersted | $H = F/l$ | 0.01257 | $H = F/l$ |
| $\phi$ | weber | Maxwell or line | $\phi = F/R$ | $10^8$ | $\phi = F/R$ |
| $B$ | weber/meter$^2$ or tesla | gauss | $B = \phi/A$ | $10^4$ | $B = \phi/A$ |
| $\mathcal{R}$ | amp turn per weber | Gilbert per maxwell | $\mathcal{R} = \dfrac{l}{A\mu_r} = \dfrac{Hl}{BA}$ | $1.257 \times 10^{-8}$ | $\mathcal{R} = \dfrac{l}{A\mu_r\mu_v}$ |
| $A$ | meter$^2$ | centimeter$^2$ | | $10^4$ | |
| $L$ | meter | centimeter | | $10^2$ | |

adopted as the international standard named SI (Système International). Table 1.9 expresses these terms in different units and illustrates their interrelationships.

# DETERMINANTS

In solving circuit networks, node and loop equations are often written using Kirchhoff's laws. This results in $n$ equations with $n$ unknowns. An easy way of solving for the unknowns is to use determinants when the equations for voltage, current, and impedance are properly ordered. The following are the determinant solutions to second-order ($n = 2$) and third-order ($n = 3$) sets.

## Second Order

Circuit equations:

$$V_1 = Z_A I_1 + Z_B I_2$$
$$V_2 = Z_C I_1 + Z_D I_2$$

Determinant form:

$$\begin{bmatrix} V_1 \\ V_2 \end{bmatrix} = \begin{bmatrix} Z_A & Z_B \\ Z_C & Z_D \end{bmatrix} \begin{bmatrix} I_1 \\ I_2 \end{bmatrix}$$

Determinant solution:

$$I_1 = \frac{\begin{bmatrix} V_1 \\ V_2 \end{bmatrix}\begin{bmatrix} Z_B \\ Z_D \end{bmatrix}}{\begin{bmatrix} Z_A \\ Z_C \end{bmatrix}\begin{bmatrix} Z_B \\ Z_D \end{bmatrix}} = \frac{V_1 Z_D - V_2 Z_B}{Z_A Z_D - Z_B Z_C} = \frac{V_1 Z_D - V_2 Z_B}{D}$$

$$I_2 = \frac{\begin{bmatrix} Z_A \\ Z_C \end{bmatrix}\begin{bmatrix} V_1 \\ V_2 \end{bmatrix}}{D} = \frac{V_2 Z_A - V_1 Z_C}{D}$$

## Third Order

Circuit equations:

$$V_1 = Z_A I_1 + Z_B I_2 + Z_C I_3$$
$$V_2 = Z_D I_1 + Z_E I_2 + Z_F I_3$$
$$V_3 = Z_G I_1 + Z_H I_2 + Z_I I_3$$

Determinant form:

$$\begin{bmatrix} V_1 \\ V_2 \\ V_3 \end{bmatrix} = \begin{bmatrix} Z_A & Z_B & Z_C \\ Z_D & Z_E & Z_F \\ Z_G & Z_H & Z_I \end{bmatrix} \begin{bmatrix} I_1 \\ I_2 \\ I_3 \end{bmatrix}$$

Determinant solution:

$$I_1 = \frac{\begin{bmatrix} V_1 & Z_B & Z_C \\ V_2 & Z_E & Z_F \\ V_3 & Z_H & Z_I \end{bmatrix}}{D}$$

where

$$D = \begin{bmatrix} Z_A & Z_B & Z_C \\ Z_D & Z_E & Z_F \\ Z_G & Z_H & Z_I \end{bmatrix}$$

$$= Z_A Z_E Z_I + Z_B Z_F Z_G + Z_C Z_D Z_H - Z_G Z_E Z_C - Z_H Z_F Z_A - Z_I Z_D Z_B$$

Therefore

$$I_1 = \frac{V_1 Z_E Z_I + V_3 Z_B Z_F + V_2 Z_H Z_C - V_1 Z_H Z_F - V_2 Z_B Z_I - V_3 Z_E Z_C}{D}$$

Similarly,

$$I_2 = \frac{\begin{bmatrix} Z_A & V_1 & Z_C \\ Z_D & V_2 & Z_F \\ Z_G & V_3 & Z_I \end{bmatrix}}{D}$$

$$= \frac{V_1 Z_F Z_G + V_2 Z_A Z_I + V_3 Z_D Z_C - V_1 Z_D Z_I - V_2 Z_C Z_G - V_3 Z_F Z_A}{D}$$

$$I_3 = \frac{\begin{bmatrix} Z_A & Z_B & V_1 \\ Z_D & Z_E & V_2 \\ Z_G & Z_H & V_3 \end{bmatrix}}{D}$$

$$= \frac{V_1 Z_D Z_H + V_2 Z_B Z_G + V_3 Z_A Z_E - V_1 Z_G Z_E - V_2 Z_H Z_A - V_3 Z_B Z_D}{D}$$

# RESONANCE

Resonance is defined as the property of cancellation of reactance when inductive and capacitive reactances are in series, or cancellation of susceptance when in parallel. Under resonant conditions, reactive circuits operate at **unity power factor** with current and voltage in phase ($\theta = 0°$ and $\cos\theta = 1$). Basic electrical engineering texts deal at length with resonance. Some of the basic formulas are delineated here.

## Figure of Merit, Q

$$Q = 2\pi \times \frac{\text{maximum energy stored per cycle}}{\text{energy dissipated per cycle}}$$

For an inductor,

$$Q = \frac{\omega L}{R_s}$$

$$Q = \frac{R_p}{\omega L}$$

For a capacitor,

$$Q = \omega C R_p$$

$$Q = \frac{1}{\omega C R_s}$$

For series RLC:

$$Q = \frac{\text{series reactance}}{\text{series resistance}}$$

$$Q = \frac{\omega_r L}{R} = \frac{1}{\omega_r RC}$$

where $\omega_r$ is at the resonant frequency.

For parallel-to-series conversion, the following derivation is applied at a specific frequency:

$$Z = \frac{R(jX)}{R + jX} = \frac{jRX(R - jX)}{R^2 + X^2} = \frac{RX^2 + jXR^2}{R^2 + X^2}$$

$$\frac{RX^2}{R^2 + X^2} \qquad j\frac{XR^2}{R^2 + X^2}$$

Therefore,

$$Q = \left[\frac{XR^2}{R^2+X^2}\right]\left[\frac{R^2+X^2}{RX^2}\right] = \frac{R}{X}$$

## Series Resonance

At resonant frequency,

$$\omega_r L = \frac{1}{\omega_r C}, \quad \omega_r^2 LC = 1, \quad \omega_r = \frac{1}{\sqrt{LC}}$$

$$f_r = \frac{\omega_r}{2\pi} = \frac{1}{2\pi\sqrt{LC}}$$

$$Q = \frac{\omega_r L}{R} = \frac{1}{R}\sqrt{\frac{L}{C}}$$

The **bandwidth,** *B*, of a resonant circuit is defined as the width of the resonance curve between the two frequencies at which power in the circuit is one-half maximum power.

For a complete circuit including *R, L, C*, generator, and load,

$$B = \Delta f = \frac{f_r}{Q}$$

For matched conditions in which generator resistance equals the remainder of the circuit resistance, as shown in Figure 1.13,

$$B = \frac{2}{Q}f_r$$

At half-power point, $R = X = |X_L - X_C|$.

Series resonance yields high current and low impedance (pure resistance).

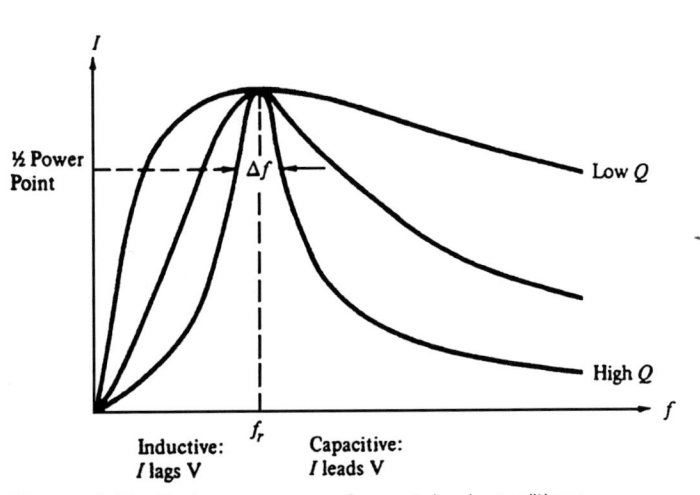

**Figure 1.13** Series resonance for matched conditions

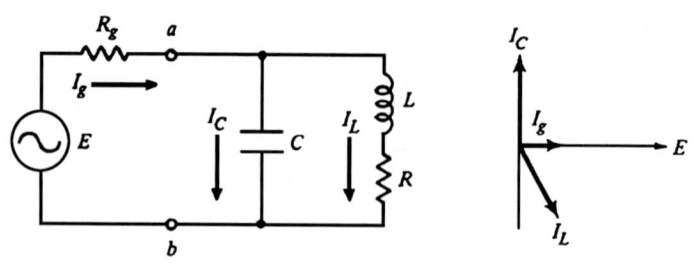

**Figure 1.14** Antiresonance

## Parallel Resonance (Antiresonance)

For antiresonance (Figure 1.14), the circuit must have unity power factor. Generator current is a minimum, and impedance is a maximum. The capacitor is assumed to have no associated shunt resistance.

$$R_{ar} = \frac{L}{CR}$$

$$f_{ar} = \frac{1}{2\pi}\sqrt{\frac{1}{LC} - \frac{R^2}{L^2}} \approx \frac{1}{2\pi}\sqrt{\frac{1}{LC}}$$

This is the same expression as for a series resonant circuit except for the small second term $(R^2/L^2)$ under the radical. Resonance is not possible for values of $R$ that make $R^2/L^2 > 1/LC$. This contrasts with the series circuit, which can be resonant for all values of $R$.

For matched conditions, where $R_g = R_{ar}$:

$$B = \frac{2}{Q} f_{ar} \quad \text{(same as for series circuit)}$$

Rearranging the formula for $f_{ar}$,

$$f_{ar} = \frac{1}{2\pi}\sqrt{\frac{1}{LC}}\sqrt{1 - \frac{R^2 C}{L}} = \frac{1}{2\pi}\sqrt{\frac{1}{LC}}\sqrt{1 - \frac{1}{Q^2}}$$

where

$$Q = \frac{1}{R}\sqrt{\frac{L}{C}}$$

for the circuit to the right of terminals $a$, $b$. If $Q > 10$, the error introduced by neglecting the radical $\sqrt{1 - (1/Q^2)}$ is less than 1%. This radical shows that resonance is not possible if $Q < 1$.

## Impedance Transformation

Using the principles of resonance, two reactances of opposite sign may be arranged as an $L$-section to transform, at a single frequency, a load resistance $R$ to provide a matched load $R_{in}$ for the generator, as shown in the following two cases.

For $R < R_g$,

**Low Pass**

$$C = \frac{1}{\omega R_{in}} \sqrt{\frac{R_{in}}{R} - 1}$$

$$L = \frac{R}{\omega} \sqrt{\frac{R_{in}}{R} - 1}$$

or

**High Pass**

$$C = \frac{1}{\omega R \sqrt{\frac{R_{in}}{R} - 1}}$$

$$L = \frac{R_{in}}{\omega \sqrt{\frac{R_{in}}{R} - 1}}$$

For $R > R_g$,

$$C = \frac{1}{\omega R} \sqrt{\frac{R}{R_{in}} - 1}$$

$$L = \frac{R_{in}}{\omega} \sqrt{\frac{R}{R_{in}} - 1}$$

or

$$C = \frac{1}{\omega R_{in} \sqrt{\frac{R}{R_{in}} - 1}}$$

$$L = \frac{R}{\omega \sqrt{\frac{R}{R_{in}} - 1}}$$

## IDEAL TRANSFORMER

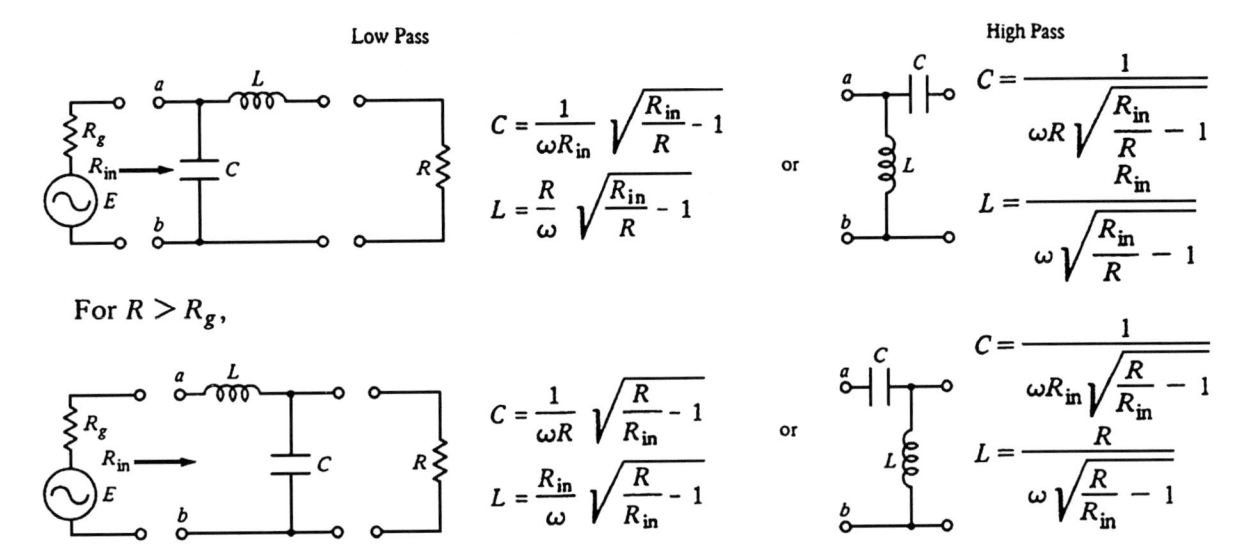

An ideal transformer is one in which there are no internal losses. In analyzing circuits, it is sometimes justifiable to ignore these losses. The following are some relationships for an ideal two-winding transformer (Figure 1.15):

$$\text{turns ratio, } a = \frac{N_p}{N_s} = \frac{V_p}{V_s}$$

$$V_p I_p = V_s I_s$$

$$\frac{I_s}{I_p} = \frac{N_p}{N_s}$$

$$I_s = \frac{V_s}{Z_l} = \frac{V_g}{Z_l} \frac{N_s}{N_p}$$

$$I_p = \frac{N_s}{N_p} I_s = \frac{V_g}{Z_l} \left(\frac{N_s}{N_p}\right)^2 = \frac{V_g}{Z_l \left(\frac{N_p}{N_s}\right)^2}$$

Secondary load reflects into primary multiplied by $(N_p/N_s)^2$.

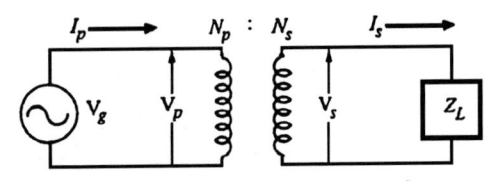

**Figure 1.15** Ideal two-winding transformer

# FOURIER ANALYSIS

Periodic functions can be conveniently analyzed by expressing them as the sum of an infinite number of sinusoidal terms. Such a sum of terms is called a Fourier series, and the process of evaluating the terms is known as Fourier analysis. Since most series tend to converge rapidly, a good approximation can usually be obtained with a limited number of sinusoidal terms. A periodic function with a period $T$ can be represented by the following Fourier series equation:

$$f(t) = \frac{a_0}{2} + \sum_{n=1}^{\infty} a_n \cos\frac{2\pi n}{T}t + \sum_{n=1}^{\infty} b_n \sin\frac{2\pi n}{T}t$$

where

$$a_0 = \frac{2}{T}\int_{t_1}^{t_1+T} f(t)dt$$

$$a_n = \frac{2}{T}\int_{t_1}^{t_1+T} f(t)\cos\frac{2\pi n}{T}t\,dt$$

$$b_n = \frac{2}{T}\int_{t_1}^{t_1+T} f(t)\sin\frac{2\pi n}{T}t\,dt$$

# WAVEFORMS

It is often desired to determine the RMS value or average value of a periodic waveform.

## RMS Value

The following formulas are used to calculate RMS value.

### General Periodic Waveform

The root mean square (also called virtual or effective) value of a *periodic* waveform is given by

$$V_{RMS} = \sqrt{\frac{1}{T}\int_0^T v^2(t)dt}$$

and is a direct measure of heating value upon a resistive load.

For example, we can write the equation for the waveform shown in Figure 1.16 as

$$V_{RMS}^2 = \frac{1}{T}\left[\int_0^{t_1}\left(V_1\frac{t}{t_1}\right)^2 dt + \int_{t_1}^{t_2} V_2^2 dt + \int_{t_2}^{t_3} V_3^2 dt + \int_{t_3}^{T} 0\, dt\right]$$

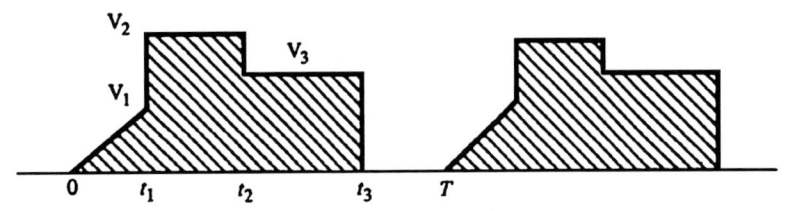

**Figure 1.16** General periodic waveform

### Sine Wave
For sine wave of $v(t) = V_{max} \sin \omega t$,

$$V_{RMS} = \frac{V_{max}}{\sqrt{2}} = 0.707 V_{max}, \quad \text{for one complete cycle}$$

### Composite Waveform

$$V_{RMS} = \sqrt{V_{1\,RMS}^2 + V_{2\,RMS}^2 + V_{3\,RMS}^2 + \cdots + V_{n\,RMS}^2}$$

where every term is at a different frequency (including dc).

## Average Value

The average value of any ac wave that is symmetrical about the zero axis is zero. However, when average value is applied to alternating quantities, it usually means the average of either the positive or negative loop of the wave. Since the average ordinate multiplied by the base is equal to the area under the curve,

$$\text{average value} = \frac{2}{T} \int_0^{T/2} i \, dt$$

where $T$ is one cycle. This equation is applicable only when the wave passes through zero at the time $t = 0$. For any other condition, the time $t_1$ at which the instantaneous value of the wave is zero must be determined and the average value found from

$$\frac{2}{T} \int_{t_1}^{t_1 + (T/2)} i \, dt$$

The average value of a sinusoid over one half-cycle is

$$I_{av} = \frac{2}{\pi} I_{max} = 0.637 I_{max}$$

More generally, the average value of one period of any periodic waveform is

$$\text{Average value} = \frac{\text{net positive area}}{\text{base}} = \text{dc value of } f(t)$$

$$= \frac{1}{T} \int_0^t f(t) \, dt,$$

where $T$ is one period or cycle.

The following brief sections apply the average value equation to several example waveforms.

### Unsymmetrical Square Wave

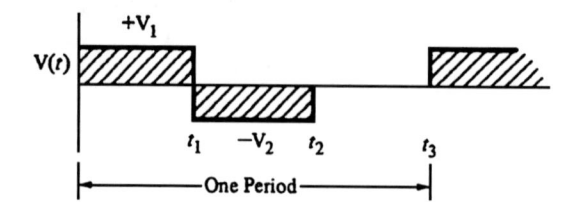

$$V_{dc} = \frac{1}{t_3}\left[\int_0^{t_1} V_1\, dt + \int_{t_1}^{t_2} -V_2\, dt + \int_{t_2}^{t_3} 0\, dt\right]$$

$$= \frac{V_1 t_1 - V_2(t_2 - t_1)}{t_3}$$

### Symmetrical Sine Wave

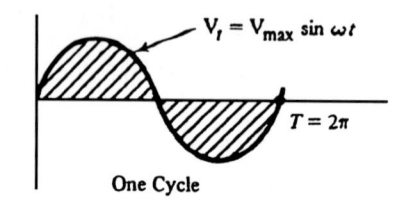

$$\omega = \frac{2\pi}{T}, \int \sin \omega t\, dt = \frac{-1}{\omega}\cos \omega t$$

$$V_{dc} = \frac{1}{T}\int_0^T V_{max} \sin \omega t\, dt = \frac{V_{max}}{T}\left[-\frac{T}{2\pi}\cos\left(\frac{2\pi}{T}\right)t\right]_0^T$$

$$= -\frac{V_{max}}{2\pi}(\cos 2\pi - \cos 0) = 0$$

### Rectified Half Wave

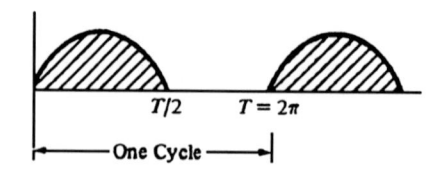

$$\omega = \frac{2\pi}{T}$$

$$V_{dc} = \frac{1}{T}\left[\int_0^{T/2} V_{max} \sin \omega t\, dt + \int_{T/2}^T 0\, dt\right]$$

$$= \frac{V_{max}}{2\pi}\left[-\cos\frac{2\pi t}{T}\right]_0^{T/2} = -\frac{V_{max}}{2\pi}(\cos \pi - \cos 0) = \frac{V_{max}}{\pi}$$

## Rectified Full Wave

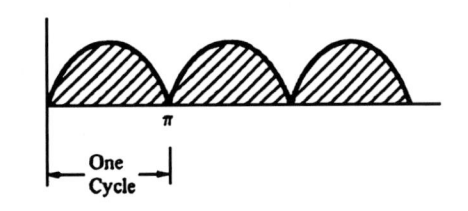

$$\omega = \frac{\pi}{T}$$

$$V_{dc} = \frac{1}{T}\int_0^T V_{max} \sin \omega t\, dt = -\frac{V_{max}}{\pi}\left[\cos\frac{\pi}{T}t\right] = \frac{2V_{max}}{\pi}$$

# CHAPTER 2

# Basic Circuits

**OUTLINE**

RESISTANCE  34

WORK, ENERGY, AND POWER  35

COULOMB'S LAW  35

NETWORKS  36

CAPACITORS  40

AC CIRCUITS  43

COMPENSATING CIRCUITS  46

VOLTMETERS  48

IMPEDANCE TRANSFORMATION  49

AMMETERS  50

RESONANCE  51

MAXIMUM POWER  60

TRANSIENTS  65

INSULATION  69

WAVEFORMS  69

BLACK BOX ANALYSIS  70
General Two-Port (Four-Terminal) Network ■ Open-Circuit Impedance Parameters ■ Short-Circuit Admittance Parameters ■ Hybrid Parameters

This chapter contains basic circuit problems with solutions that use one or more of the fundamental concepts presented in chapter 1.

Follow these steps throughout the remainder of the text:

1. As you begin each problem, place a sheet of paper over the solution section of the problem.

2. Read the problem carefully.

3. Work the problem on scratch paper. You may refer to concepts and formulas contained in chapter 1 if necessary. (Use the index to locate needed information rapidly.)

4. Evaluate your solution to the problem by comparing it to that given in the text.

5. If your solution is correct, go to the next problem; if it is incorrect, review the material indicated at the end of the solution. The review material includes content from this book and one or more textbooks referenced in Appendix A.

6. If your first answer was incorrect, rework the problem after reviewing the references.

## RESISTANCE

The problem in this section illustrates the basics of combining series and parallel resistances into a single equivalent resistance.

### PROBLEM 2.1  Series-Parallel Resistance

The equivalent resistance of the resistance network shown in Exhibit 2.1 is:

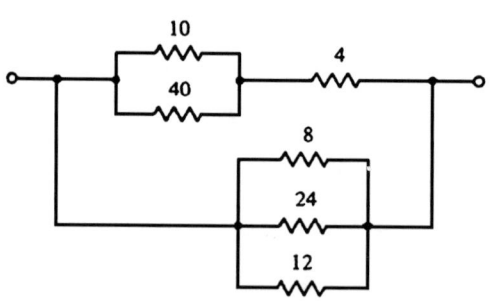

    a. $3\ \Omega$          b. $2.18\ \Omega$
    c. $0.24\ \Omega$     d. $4\ \Omega$

Exhibit 2.1

*Solution*

By combining parallel resistors, the network reduces to Exhibit 2.1a.

$$R_1 = \frac{10 \times 40}{10 + 40} = \frac{400}{50} = 8$$

$$R_2 = \left[\frac{1}{8} + \frac{1}{24} + \frac{1}{12}\right]^{-1} = 4$$

Exhibit 2.1a

This circuit finally reduces to

$$R = \frac{(8+4)(4)}{8+4+4} = \frac{48}{16} = 3\ \text{ohms}$$

The correct answer is a.

*Answer Rationale*

Incorrect solution b is the result of dividing 40 by 50 instead of 400 by 50 in calculating $R_1$. Incorrect solution c is the result of not taking the reciprocal of the values added in calculating $R_2$. Incorrect solution d is the result of not adding the second 4-ohm value in the denominator while calculating $R$ in the last step.

If your answer is correct, go on to the next section.

If your answer is not correct, review the subject of series and parallel combinations in chapter 1. For more details, see chapter 1 in reference 1 and chapter 7 in reference 2.

## WORK, ENERGY, AND POWER

The problem in this section illustrates the application of three fundamental engineering parameters.

### PROBLEM 2.2  Work, Energy, and Power

A hoist driven by a three-phase electric motor has a 4000-pound load capacity.
  a. How much energy in joules and power in kW is required to raise a full-capacity load 100 feet in one minute?
  b. What motor capacity in HP is required to drive the hoist if the hoisting machinery is only 75% efficient?

*Solution*

a.
$$W = \int_0^t p\,dt = pt = \left[\frac{\text{lb-ft}}{\text{s}}\right]\text{s} = \text{lb-ft}$$

1 watt-second = 1 joule = 0.738 lb-ft

$$W = 4000 \text{ lb} \times 100 \text{ ft} \times \frac{\text{joule}}{0.738 \text{ lb-ft}} = 5.42 \times 10^5 \text{ joule}$$

$$P = \frac{W}{t} = \frac{5.42 \times 10^5 \text{ joule}}{1 \text{ min} \times 60 \text{ s/min}} \times \frac{\text{watt-s}}{\text{joule}}$$
$$= 9033 \text{ W or } 9.033 \text{ kW}$$

b.
$$P = 9.033 \text{ kW} \times \frac{\text{HP}}{0.746 \text{ kW}} \times \frac{1}{0.75} = 16.17 \text{ HP}$$

If your answers are correct, go on to Problem 2.3.

If your answers are not correct, review the subjects of energy and power in chapter 1. For more details, see chapter 2 in reference 1 and chapter 3 in reference 3.

## COULOMB'S LAW

The problem in this section illustrates the application of Coulomb's law, as defined in chapter 1.

### PROBLEM 2.3  Coulomb's Law

The force between two electrons in air separated by a distance of 200 cm is:
  a. $1.15 \times 10^{-28}$ newton
  b. $5.75 \times 10^{-29}$ newton
  c. $5.75 \times 10^{-53}$ newton
  d. $1.81 \times 10^{-28}$ newton

## Solution

$$F = \frac{Q_1 Q_2}{4\pi\varepsilon r^2} = \frac{(1.6 \times 10^{-19})^2}{4\pi \times 8.85 \times 10^{-12} \times 2^2}$$

$$= 5.75 \times 10^{-29} \text{ newton}$$

The correct answer is b.

### Answer Rationale

Incorrect solution a is the result of not taking the square of $r$ in the denominator. Incorrect solution c is the result of using $10^{12}$ instead of $10^{-12}$ in the denominator. Incorrect solution d is the result of not multiplying by $\pi$ in the denominator.
    If your answer is correct, go on to Problem 2.4.
    If your answer is not correct, review the subject of Coulomb's law in chapter 1. For more details, see chapter 1 in reference 1 and chapter 2 in reference 2.

# NETWORKS

The problems in this section illustrate a few aspects of network analysis and amplification. The study of lattice and other types of networks is recommended.

## PROBLEM 2.4  Wye-Delta Transformation

The circuit in Exhibit 2.4 is a Wheatstone bridge with a resistance in place of the galvanometer. Find the current supplied by the battery.

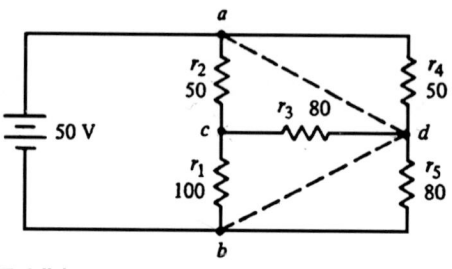

Exhibit 2.4

### Solution

Using the wye-delta transformation, the network may be reduced as shown in Exhibit 2.4a.
    Solving for current,

$$I = \frac{V}{R} = \frac{50}{69.55} = 0.72 \text{ amp}$$

If your answer is correct, go on to Problem 2.5.
    If your answer is not correct, review the subject of wye-delta transformation in chapter 1. For more details, see chapter 3 in reference 1 and chapter 17 in reference 2.

Networks

[Exhibit 2.4a circuit diagrams]

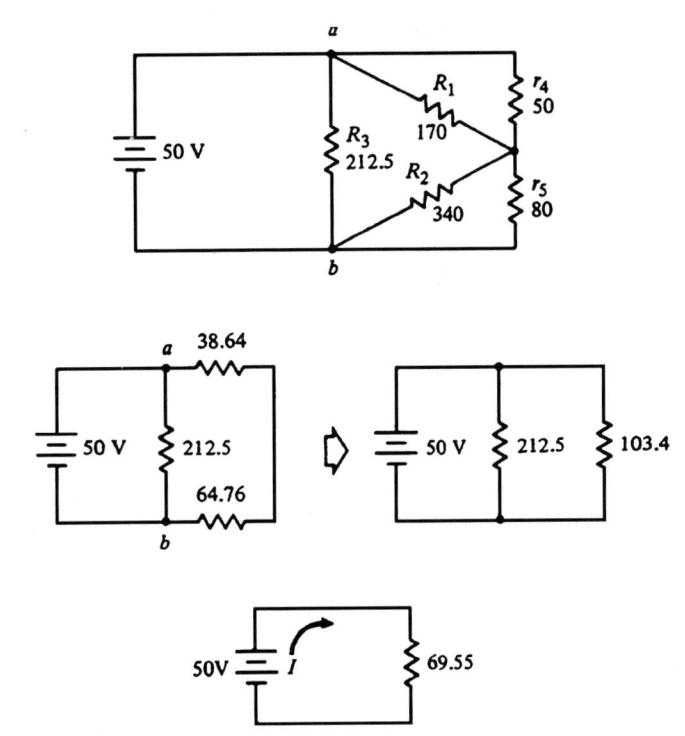

Exhibit 2.4a

## PROBLEM 2.5 Schering Bridge

The value of an unknown capacitor is to be measured using a Schering bridge. If the values of the bridge capacitors and resistors are as shown in Exhibit 2.5, what is the value of the unknown capacitor, $C_x$, and its series resistance, $R_x$?

   a. $C_x = 1$ mF and $R_x = 10$ Ω
   b. $C_x = \mu$F and $R_x = 10{,}000$ Ω
   c. $C_x = \mu$F and $R_x = 10$ Ω
   d. $C_x = \mu$F and $R_x = 0.01$ Ω

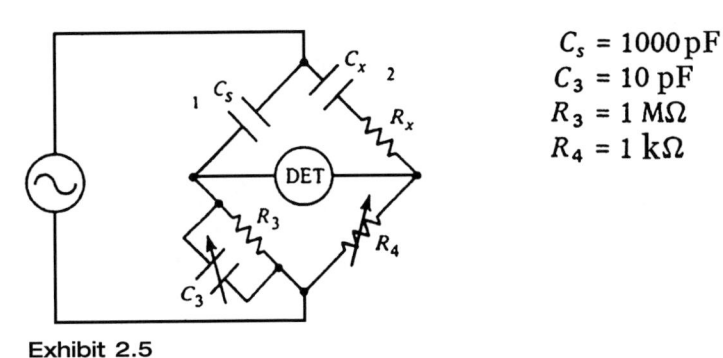

$C_s = 1000\,\text{pF}$
$C_3 = 10\,\text{pF}$
$R_3 = 1\,\text{M}\Omega$
$R_4 = 1\,\text{k}\Omega$

Exhibit 2.5

### Solution

For balanced conditions,

$$Z_1 = -j\frac{1}{\omega C_s}, \quad Z_4 = R_4, \quad Y_3 = \frac{1}{R_3} + j\omega C_3$$

$$Z_x = Z_1 Z_4 Y_3$$

$$Z_x = \left[ Rx - j\frac{1}{\omega C_x} \right] = \left[ -j\frac{1}{\omega C_s} \right] [R_4] \left[ \frac{1}{R_3} + j\omega C_3 \right]$$

Evaluating real and imaginary components,

$$Z_x = \frac{C_3 R_4}{C_s} - j\frac{R_4}{\omega C_s R_3} = R_x - jXc_x$$

$$C_x = C_s \frac{R_3}{R_4} = [10^{-9}]\frac{10^6}{10^3} = 10^{-6} \text{ or } 1\,\mu F$$

$$R_x = R_4 \frac{C_3}{C_4} = 10^3 \left[\frac{10^{-11}}{10^{-9}}\right] = 10\,\Omega$$

The correct solution is c.

*Answer Rationale*

Incorrect solution a is the result of using the incorrect value $C_s = 10^{-6}$ in calculating $C_x$. Incorrect solution b is the result of incorrectly assuming that pico means $10^{-9}$, which leads to using $C_3 = 10^{-8}$ instead of $10^{-11}$ in calculating $R_x$. Incorrect solution d is the result of not multiplying by $10^3$ in calculating $R_x$.

If your answers are correct, go on to Problem 2.6.

If your answers are not correct, review the subject of circuit element equations in chapter 1. For more details, see chapter 1 in reference 1 and chapter 8 in reference 2.

## PROBLEM 2.6  Circuit Network—Loop Current Analysis

Find the current in the capacitor branch of the network shown in Exhibit 2.6.

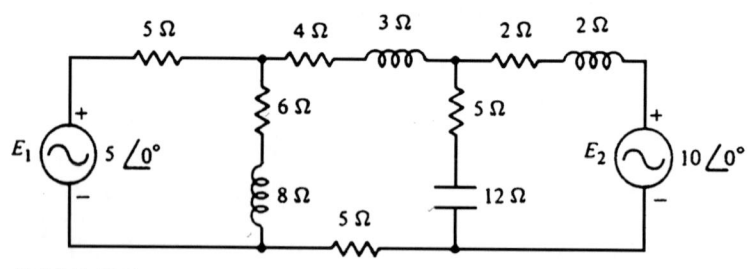

Exhibit 2.6

*Solution*

Since there are three loops, three voltage equations may be written. The problem is simplified if the capacitor branch is included in only one loop (Exhibit 2.6a), since the capacitor current is obtained directly. Therefore, write the loop equations as follows:

$$E_1 = i_1(Z_1 + Z_4) + i_2(Z_4) + i_3(0)$$
$$E_2 = i_1(Z_4) + i_2(Z_2 + Z_3 + Z_4 + Z_6) + i_3(Z_3)$$
$$E_2 = i_1(0) + i_2(Z_3) + i_3(Z_3 + Z_s)$$

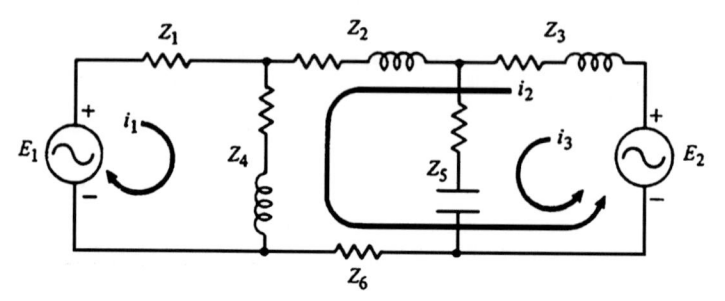

Exhibit 2.6a

Solving for $i_3$ using determinants yields

$$\begin{bmatrix} E_1 \\ E_2 \\ E_2 \end{bmatrix} = \begin{bmatrix} (Z_1+Z_4) & Z_4 & 0 \\ Z_4 & (Z_2+Z_3+Z_4+Z_6) & Z_3 \\ 0 & Z_3 & Z_3+Z_5 \end{bmatrix} \begin{bmatrix} i_1 \\ i_2 \\ i_3 \end{bmatrix}$$

$$i_3 = \frac{\begin{bmatrix} (Z_1+Z_4) & Z_4 & E_1 \\ Z_4 & (Z_2+Z_3+Z_4+Z_6) & E_2 \\ 0 & Z_3 & E_2 \end{bmatrix}}{\begin{bmatrix} (Z_1+Z_4) & Z_4 & 0 \\ Z_4 & (Z_2+Z_3+Z_4+Z_6) & Z_3 \\ 0 & Z_3 & Z_3+Z_5 \end{bmatrix}}$$

$$= \frac{\begin{bmatrix} (11+j8) & (6+j8) & 5 \\ (6+j8) & (17+j13) & 10 \\ 0 & (2+j2) & 10 \end{bmatrix}}{\begin{bmatrix} (11+j8) & (6+j8) & 0 \\ (6+j8) & (17+j13) & (2+j2) \\ 0 & (2+j2) & (7-j10) \end{bmatrix}}$$

$$i_3 = \frac{10(11+j8)(17+j13) + 5(2+j2)(6+j8) - 10(6+j8)(6+j8) - 10(2+j2)(11+j8)}{(11+j8)(17+j13)(7-j10) - (6+j8)(6+j8)(7-j10) - (11+j8)(2+j2)(2+j2)}$$

$$= \frac{\begin{array}{c} 10(13.6\angle 36°)(21.4\angle 37.4°) + 5(2.83\angle 45°)(10\angle 53.1°) - 10(10\angle 53.1°)(10\angle 53.1°) \\ -10(2.83\angle 45°)(13.6\angle 36°) \end{array}}{\begin{array}{c}(13.6\angle 36°)(21.4\angle 37.4°)(12.21\angle -55°) - (10\angle 53.1°)(10\angle 53.1°)(12.21\angle -55°) \\ -(13.6\angle 36°)(2.83\angle 45°)(2.83\angle 45°)\end{array}}$$

$$= \frac{2910.4\angle 73.4° + 141.5\angle 98.5° - 1000\angle 106.2° - 384.88\angle 81°}{3553.6\angle 18.4° - 1221\angle 51.2° - 108.9\angle 126°}$$

$$= \frac{831.47 + j2789.1 - 19.94 + j140.09 + 278.99 - j960.29 - 60.21 - j380.14}{3371.93 + j1121.69 - 765.08 - j951.57 + 64.01 - j88.1}$$

$$= \frac{1030.31 + j1588.76}{2670.85 + j82.02} = \frac{1893.59\angle 57.04°}{2672.11\angle 1.76°} = 0.71\angle 55.28° \text{ amp}$$

*Alternate Solution*

Select three different loop equations as shown in Exhibit 2.6b.

$$5 = i_1(11+j8) - i_2(6+j8) \qquad \text{(i)}$$
$$0 = -i_1(6+j8) + i_2(20-j1) - i_3(5-j12) \qquad \text{(ii)}$$
$$-10 = 0 - i_2(5-j12) + i_3(7-j10) \qquad \text{(iii)}$$

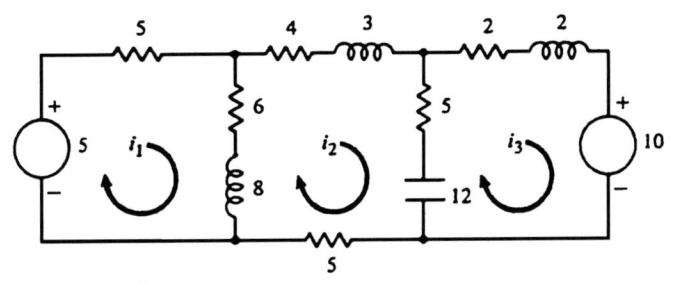

Exhibit 2.6b

Using the first two equations to eliminate $i_1$ yields

$$5(6+j8) = i_1(11+j8)(6+j8) - i_2(6+j8)(6+j8)$$
$$\underline{0 = i_1(11+j8)(6+j8) = i_2(20-j1)(11+j8) - i_3(5-j12)(11+j8)}$$
$$5(6+j8) = i_2[(20-j1)(11+j8) - (6+j8)(6+j8)] - i_3(5-j12)(11+j8)$$
$$30 + j40 = i_2[(20.02\angle -2.86°)(13.6\angle 36.03°) - (10\angle 53.1°)(10\angle 53.1°)]$$
$$- i_3(13\angle -67.38°)(13.6\angle 36.03°)$$
$$50\angle 53.13° = i_2[272.27\angle +33.17° - 100\angle 106.2°] - i_3(176.8\angle -31.35°)$$
$$50\angle 53.13° = i_2(227.9 + j148.97 + 27.9 - j96.03) - i_3(176.8\angle -31.35°)$$
$$50\angle 53.13° = i_2(255.8 + j52.94) - i_3(176.8\angle -31.35°)$$
$$50\angle 53.13° = i_2(261.22\angle 11.69°) - i_3(176.8\angle -31.35°) \quad \text{(iv)}$$

Using equations (iii) and (iv) to eliminate $i_2$ yields

$$(5-j12)50\angle 53.13° = i_2(261.22\angle 11.69°)(5-j12) - i_3(176.8\angle -31.35°)(5-j12)$$
$$\underline{-10(261.22\angle 11.69°) = -i_2(261.22\angle 11.69°)(5-j12) + i_3(7-j10)(261.22/11.69°)}$$
$$(13\angle -67.38°)(50\angle 53.13°) - 2612.2\angle 11.69°$$
$$= i_3[(261.22\angle 11.69°)(12.21\angle -55.01°) - (13\angle -67.38°)(176.8\angle -31.35°)]$$

$$650\angle -14.25 - 2612.2\angle 11.69° = i_3(3189.5\angle -43.32° - 2298.4\angle -98.73°)$$
$$630 - j160 - 2558.02 - j529.27 = i_3(2320.47 - j2188.2 + 348.85 + j2271.77)$$
$$-1928.02 - j689.27 = i_3(2669.31 - j83.54)$$
$$2047.52\angle -160.33° = i_3(2670.62\angle 1.79°)$$
$$i_3 = \frac{2047.52\angle -160.33°}{2670.62\angle 1.79} = 0.77\angle -162.12°$$

Solving for $i_2$ using $i_3$ and equation (iii) yields

$$-10 = -i_2(13\angle -67.38°) + (0.77\angle -162.12°)(12.21\angle -55.01°)$$
$$i_2 = \frac{10 + 9.4\angle -217.13°}{13\angle -67.38°} = \frac{10 - 7.49 + j5.67}{13\angle -67.38°} = \frac{2.51 + j5.67}{13\angle -67.38°} = \frac{6.2\angle 66.12°}{13\angle -67.38°}$$
$$= 0.48\angle 133.5°$$

Solving for capacitor current provides the required answer:

$$i_C = i_2 - i_3 = 0.48\angle 133.5° - 0.77\angle -162.12°$$
$$= -0.33 + j0.35 + 0.73 + j0.24 = 0.4 + j0.58 = 0.71\angle 55.62° \text{ amp}$$

If your answer is correct, go on to Problem 2.7.

If your answer is not correct, review the subjects of Kirchhoff's laws and determinants in chapter 1. For more details, see chapter 1 in reference 1 and chapter 17 in reference 2.

# CAPACITORS

The problems in this section illustrate the fundamental properties of capacitor energy and charge.

### PROBLEM 2.7 Capacitor Energy and Charge

The current waveform shown in Exhibit 2.7 approximates a current impulse. If this pulse is applied to a capacitor, determine:

Exhibit 2.7

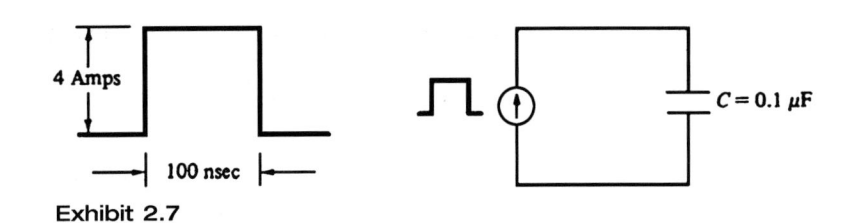

a. How much energy is applied to the capacitor.
b. How much charge it accepts.

*Solution*

Equations required for solution are

$$W = \frac{1}{2}CV^2$$

$$Q = CV$$

$$V = \frac{1}{C}\int_0^T i\, dt \quad \text{(the integral represents the area under the current impulse curve)}$$

Solving for capacitor voltage,

$$V = \frac{1}{10^{-7}}\int_0^{10^{-7}} 4\, dt = \frac{4\times 10^{-7}}{10^{-7}} = 4 \text{ volts}$$

Therefore,

a. $$W = \frac{10^{-7}}{2}(4)^2 = 8\times 10^{-7} \text{ joule}$$

b. $$Q = 4\times 10^{-7} \text{ coulomb}$$

If your solution is correct, go on to Problem 2.8.

If your solution is not correct, review the subjects of energy and circuit elements defined in chapter 1. For more details, see chapter 1 in reference 1 and chapter 10 in reference 2.

## PROBLEM 2.8 Two-Capacitor Charge Transfer

In the circuit shown in Exhibit 2.8, the two capacitors are charged to the given initial conditions prior to switch closure. The value of the resistor is $0 < R < \infty$.

Initial conditions:

$$V_{C_1} = 100 \text{ volts} \quad V_{C_2} = 25 \text{ volts}$$

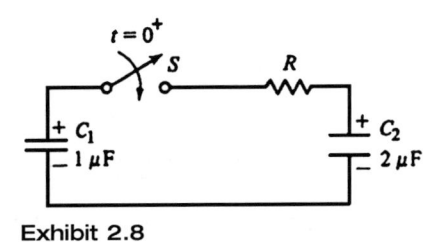

Exhibit 2.8

Determine for the two capacitors the final values for:
a. Voltage
b. Charge
c. Energy
d. The energy dissipated as heat in the resistor after the switch has been closed for a long time

*Solution*

After the switch is closed, current will flow in the circuit until the two capacitors reach the same voltage (somewhere between 25 and 100). The length of time current flows is dependent, in part, upon the value of $R$. No matter what value $R$ is, it will eventually dissipate a fixed amount of energy as heat. Since no energy is being added to the circuit, by the principle of **conservation of charge,** the charge in the circuit will always remain unchanged; it will merely be redistributed after the switch is closed.

Initial conditions:

$$Q_1 = C_1 V_1 = 10^{-6} \times 100 = 100 \times 10^{-6} \text{ coulomb}$$

$$Q_2 = C_2 V_2 = 2 \times 10^{-6} \times 25 = 50 \times 10^{-6} \text{ coulomb}$$

$$Q_{tot} = Q_1 + Q_2 = 150 \times 10^{-6} \text{ coulomb}$$

$$W_1 = \frac{1}{2} C_1 V_1^2 = \frac{1}{2} \times 10^{-6} \times 10^4 = 5 \times 10^{-3} \text{ joule}$$

$$W_2 = \frac{1}{2} C_2 V_2 = \frac{1}{2} \times 2 \times 10^{-6} \times 625 = 0.625 \times 10^{-3} \text{ joule}$$

$$W_{tot} = W_1 + W_2 = 5.625 \times 10^{-3} \text{ joule}$$

Final conditions:

$$Q_{tot} = 150 \times 10^{-6} \text{ coulomb} = Q_1 + Q_2 = C_1 V_1 + C_2 V_2 = (C_1 + C_2) V$$

due to conservation of charge and

$$V_1 = V_2$$

a. $$V = \frac{Q_{tot}}{C_1 + C_2} = \frac{150 \times 10^{-6}}{3 \times 10^{-6}} = 50 \text{ volts}$$

b. $Q_1 = 50 C_1 = 50 \times 10^{-6}$ coulomb
$Q_2 = 50 C_2 = 100 \times 10^{-6}$ coulomb

c. $$W_1 = \frac{1}{2} C_1 V^2 = \frac{1}{2} \times 10^{-6} \times 2500 = 1.25 \times 10^{-3} \text{ joule}$$

$$W_2 = \frac{1}{2} C_2 V^2 = \frac{1}{2} \times 2 \times 10^{-6} \times 2500 = 2.5 \times 10^{-3} \text{ joule}$$

$$W_{tot} = W_1 + W_2 = 3.75 \times 10^{-3} \text{ joule}$$

d. Energy lost in the resistor is the difference between the initial total capacitor energy and the final total capacitor energy. Therefore,

$$W_R = (5.625 - 3.75) 10^{-3} = 1.875 \times 10^{-3} \text{ joule}$$

If your solutions are correct, go on to Problem 2.9.

If your solutions are not correct, review the subjects of energy, circuit elements defined, and Coulomb's law in chapter 1. For more details, see chapter 1 in reference 1 and chapter 10 in reference 2.

### PROBLEM 2.9 Power, Energy, and Charge

In the steady-state circuit shown in Exhibit 2.9, calculate:
- a. Voltage across the capacitor
- b. Stored charge
- c. Energy
- d. Current delivered by the battery
- e. Power delivered by the battery

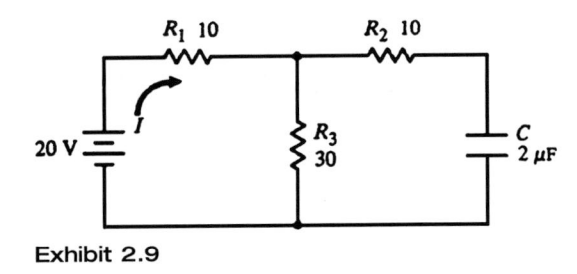

Exhibit 2.9

*Solution*

Since conditions are steady-state, there is no current flow in the capacitor loop. Therefore, resistors $R_1$ and $R_3$ act as a voltage divider and

a. $V = 20 \dfrac{R_3}{R_1 + R_3} = 20 \dfrac{30}{40} = 15$ volts

b. $Q = CV = 30 \times 10^{-6}$ coulomb

c. $W = \dfrac{1}{2} CV^2 = 225 \times 10^{-6}$ joule

d. $I = \dfrac{V}{R} = \dfrac{20}{40} = 0.5$ amp

e. $P = I^2 R = (0.5)^2 40 = 10$ watts

If your solutions are correct, go on to Problem 2.10.

If your solutions are not correct, review the subjects of energy, power, and circuit elements defined in chapter 1. For more details, see chapter 1 in reference 1 and chapter 10 in reference 2.

## AC CIRCUITS

The problems in this section illustrate that Ohm's law works for ac circuits as well as dc circuits.

### PROBLEM 2.10 Unknown Device

You have been given a potted device and asked to determine the type and value of elements it contains. When the device is connected to 120 volts at 60 Hz,

## Chapter 2 Basic Circuits

Exhibit 2.10

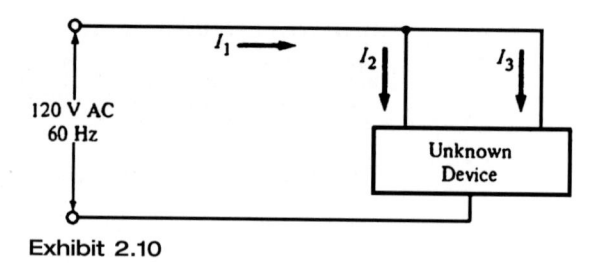

2 amps ac flow in each of the three paths, as shown in Exhibit 2.10. Assuming all elements are linear, passive, and bidirectional, their type and value are:
   a. $R_2 = 30\ \Omega$ and $L_2 = 0.14$ H; $R_3 = 30\ \Omega$ and $L_3 = 0.14$ H
   b. $R_2 = 30\ \Omega$ and $C_2 = 321\ \mu F$; $R_3 = 30\ \Omega$ and $L_3 = 0.14$ H
   c. $R_2 = 30\ \Omega$ and $C_2 = 51.1\ \mu F$; $R_3 = 30\ \Omega$ and $C_3 = 51.1\ \mu F$
   d. $R_2 = 30\ \Omega$ and $C_2 = 51.1\ \mu F$; $R_3 = 30\ \Omega$ and $L_3 = 0.14$ H

*Solution*

In order for all three currents to be the same magnitude, the current phasor diagrams must form an equilateral triangle as shown in Exhibit 2.10a (assume that $I_1$ is at 0° with respect to the ac source and that branch 2 is capacitive).

$$I_1 = I_2 \angle +60° + I_3 \angle -60° = 2\angle 0°\text{ amps}$$

$$Z_2 = \frac{120\angle 0°}{2\angle +60°} = 60\angle -60° = 30 - j51.96$$

$$R_2 = 30\ \Omega$$

$$X_{C_2} = 51.96\ \Omega$$

$$C_2 = \frac{1}{2\pi f X_{C_2}} = \frac{1}{377 \times 51.96} = 51.1\ \mu F$$

$$Z_3 = \frac{120\angle 0°}{2\angle -60°} = 60\angle 60° = 30 + j51.96$$

$$R_3 = 30\ \Omega$$

$$X_{L_s} = 51.96\ \Omega$$

$$L_3 = \frac{X_{L_s}}{2\pi f} = \frac{51.96}{377} = 0.14\ \text{H}$$

The correct answer is d.

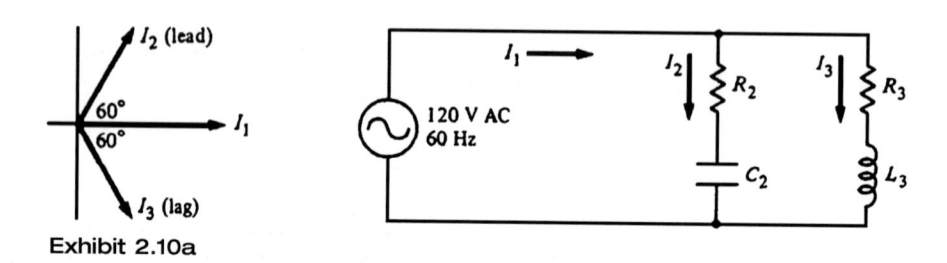

Exhibit 2.10a

*Answer Rationale*

Incorrect solution a is the result of using the incorrect positive sign instead of negative for the imaginary component of $Z_2$, which leads to assuming that it represents an inductance. Incorrect solution b is the result of not multiplying $2\pi$ in the denominator while calculating $C_2$. Incorrect solution c is the result of using the incorrect negative sign instead of positive for the imaginary component of $Z_3$, which leads to assuming that it represents a capacitance.

If your answers are correct, go on to Problem 2.11.

If your answers are not correct, review the subjects of complex algebra and circuit element equations in chapter 1. For more details, see chapter 1 in reference 1 and chapter 16 in reference 2.

## PROBLEM 2.11  Parallel Branches

In the circuit shown in Exhibit 2.11, the power dissipated by the load is:
  a.  $P = 2448$ watts
  b.  $P = 3171$ watts
  c.  $P = 2446$ watts
  d.  $P = 1584$ watts

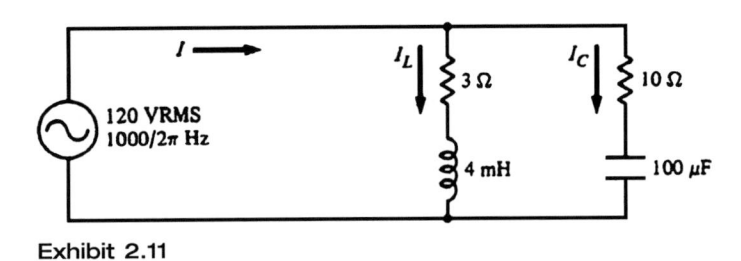

Exhibit 2.11

*Solution*

$$X_L = 2\pi f L = 2\pi \times \frac{1000}{2\pi} \times 4 \times 10^{-3} = 4\,\Omega$$

$$X_C = \frac{1}{2\pi f C} = \frac{1}{2\pi \times \frac{1000}{2\pi} \times 100 \times 10^{-6}} = 10\,\Omega$$

$$Z_L = 3 + j4 = 5\angle 53.13°\,\Omega$$
$$Z_C = 10 - j10 = 14.14\angle -45°\,\Omega$$
$$I_C = \frac{E}{Z_C} = \frac{120\angle 0°}{14.14\angle -45°} = 8.49\angle 45° = 6 + j6 \text{ amps}$$
$$I_L = \frac{E}{Z_L} = \frac{120\angle 0°}{5\angle 53.13°} = 24\angle -53.13° = 14.4 - j19.2 \text{ amps}$$
$$I = I_L + I_C = 20.4 - j13.2 = 24.3\angle -32.91°$$
$$P = VI\cos\theta = 120 \times 24.3\cos 32.91° = 2448 \text{ watts}$$

The correct answer is a.

*Answer Rationale*

Incorrect solution b is the result of not taking the reciprocal in calculating $X_C$, which leads to the wrong value of 0.1 ohm. Incorrect solution c is the result of incorrectly using a negative sign for the 45-degree angle of $I_c$. Incorrect solution d is the result of using the sine of the angle 32.91 instead of the cosine in the final step of calculating $P$.

If your answer is correct, go on to Problem 2.12.

If your answer is not correct, review the subjects of complex algebra and circuit element equations in chapter 1. For more details, see chapter 1 in reference 1 and chapter 16 in reference 2.

## COMPENSATING CIRCUITS

The problems in this section apply to a variety of electrical engineering specialties including control theory and transient analysis. Use is made of partial fractions (see chapter 5) and the Laplace transform.

### PROBLEM 2.12  Simple Lag Circuit

Derive the equation for output voltage across the capacitor of the simple lag circuit shown in Exhibit 2.12 after the switch is closed. Assume the capacitor is initially uncharged.

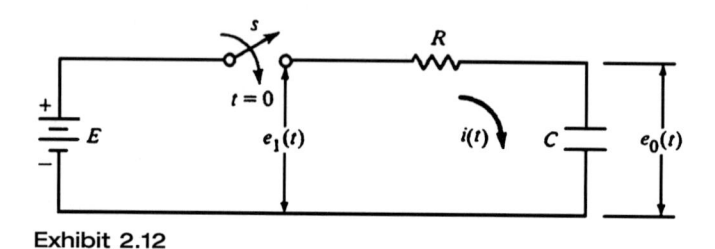

Exhibit 2.12

*Solution*

$$e_1(t) = Ri(t) + \frac{1}{C}\int_0^t i\,dt \xrightarrow{\mathcal{L}} E_1(s) = I(s) = \left[R + \frac{1}{Cs}\right]$$

$$e_0(t) = \frac{1}{C}\int_0^t i\,dt \xrightarrow{\mathcal{L}} E_0(s) = \frac{I(s)}{Cs}$$

The transfer function of a circuit is

$$\frac{\text{output}}{\text{input}} = \text{T.F.}$$

$$\frac{E_0(s)}{E_1(s)} = \left[\frac{I(s)}{Cs}\right]\left[\frac{1}{I(s)(R+1/Cs)}\right] = \frac{1}{RCs+1} = \frac{1/RC}{s+1/RC}$$

When $S$ is closed at $t = 0$, a transient unit step is input.

$$\mathcal{L}e_1(t) = e_1(s) = \frac{E}{s}$$

$$E_0(s) = E_1(s) \times \text{T.F.} = \left[\frac{E}{s}\right]\left[\frac{1/RC}{s+1/RC}\right]$$

Using partial fractions,

$$\left[\frac{E}{s}\right]\left[\frac{1/RC}{s+1/RC}\right] = \frac{AE}{s} + \frac{B/RC}{s+1/RC} = \frac{AE(s+1/RC) + Bs/RC}{s(s+1/RC)}$$

Canceling common denominators,

$$\frac{E}{RC} = AE(s+1/RC) + \frac{Bs}{RC}$$

Solving for $A$ and $B$ yields

Let $s = 0$, $\quad \dfrac{E}{RC} = \dfrac{AE}{RC}, \quad A = 1$

Let $s = -\dfrac{1}{RC}$, $\quad \dfrac{E}{RC} = -\dfrac{B/RC}{RC}, \quad B = -ERC$

Substituting,

$$E_0(s) = \frac{AE}{s} + \frac{B/RC}{s+1/RC} = \frac{E}{s} - \frac{ERC/RC}{s+1/RC} = E\left[\frac{1}{s} - \frac{1}{s+1/RC}\right]$$

Taking the inverse $\mathcal{L}$ (see Exhibit 2.12a),

$$e_0(t) = E(1 - e^{-t/RC})$$

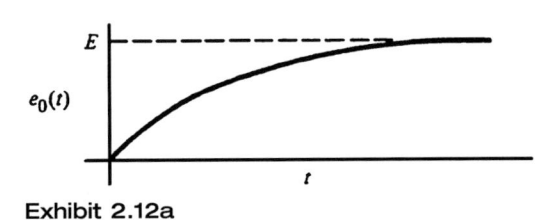

**Exhibit 2.12a**

If your answer is correct, go on to Problem 2.13.

If your answer is not correct, review the subjects of Laplace transform and transfer functions in chapter 1. For more details, see chapter 1 in reference 1 and chapter 10 in reference 2.

## PROBLEM 2.13  Simple Lead Circuit

What is the equation for output voltage across the resistor of the simple lead circuit in Exhibit 2.13 after the switch is closed? Assume the capacitor is initially uncharged.

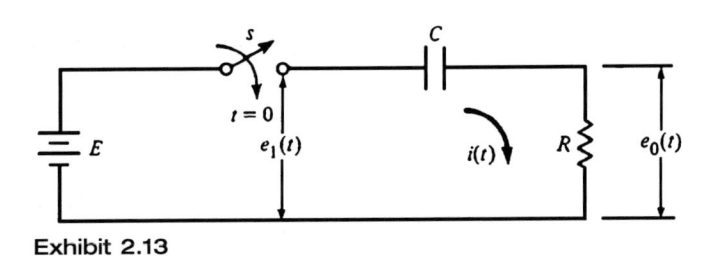

**Exhibit 2.13**

## Solution

The derivation of the voltage response across the resistor is similar to that of the preceding problem. Therefore, some shortcuts are taken here.

Using Kirchhoff's voltage law,

$$e_1(t) = \frac{1}{C}\int_0^t i\,dt + Ri(t)$$

$$e_0(t) = Ri(t)$$

$$\text{Transfer function} = \frac{E_0(s)}{E_1(s)} = \frac{s}{s + 1/RC} = \text{T.F.}$$

$$\text{at } t = 0, \quad E_1(s) = \frac{E}{s}$$

$$E_0(s) = \frac{E_s}{s(s + 1/RC)}$$

Taking the inverse $\mathcal{L}$ (see Exhibit 2.13a),

$$e_0(t) = Ee^{-t/RC}$$

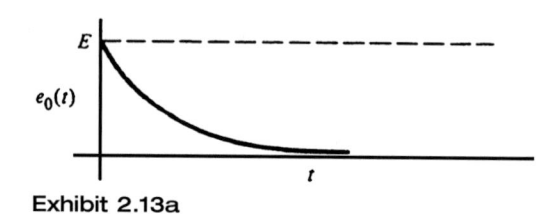

**Exhibit 2.13a**

If your answer is correct, go on to Problem 2.14.

If your answer is not correct, review the solution to Problem 2.12. For more details, see chapter 1 in reference 1 and chapter 10 in reference 2.

## VOLTMETERS

This problem illustrates the practical application of using a microammeter and some precision resistors to create a voltmeter.

### PROBLEM 2.14 Voltmeter Design

Given a microammeter having a 50 microamp movement and an internal resistance of 10 $\Omega$, design a voltmeter with ranges of 10 VDC, 50 VDC, 100 VDC, and 500 VDC.

### Solution

The voltmeter circuit is shown in Exhibit 2.14.

A 50 microamp movement yields a sensitivity of

$$\frac{1}{50 \times 10^{-6}} = 20{,}000 \text{ ohms/volt}$$

Next calculate the precision resistors.

Exhibit 2.14

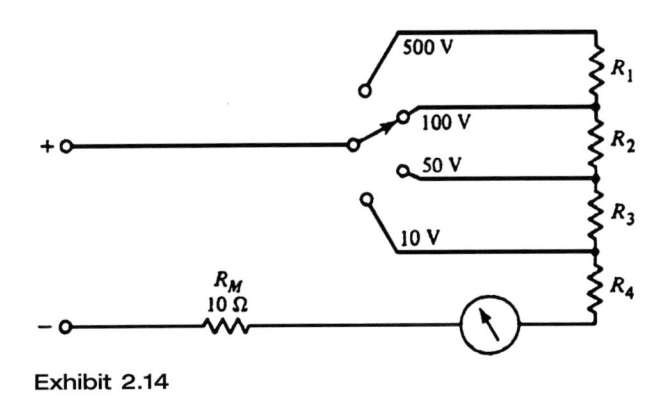

Resistance of the 10-volt range, $R_{10}$, is

$$R_{10} = R_4 + R_M = (10 \text{ volts})(20,000 \, \Omega/\text{volt}) = 200,000 \, \Omega$$
$$R_4 = 200,000 - 10 = 199,990 \, \Omega$$

Resistance of the 50-volt range, $R_{50}$, is

$$R_{50} = R_3 + R_{10} = (50 \text{ volts})(20,000 \, \Omega/\text{volts}) = 1 \, \text{M}\Omega$$
$$R_3 = 10^6 - 200,000 = 800,000 \, \Omega$$

Resistance of the 100-volt range, $R_{100}$, is

$$R_{100} = R_2 + R_{50} = (100 \text{ volts})(20,000 \, \Omega/\text{volt}) = 2 \, \text{M}\Omega$$
$$R_2 = 2 \times 10^6 - 10^6 = 1 \, \text{M}\Omega$$

Resistance of the 500-volt range, $R_{500}$, is

$$R_{500} = R_1 + R_{100} = (500 \text{ volts})(20,000 \, \Omega/\text{volt}) = 10 \, \text{M}\Omega$$
$$R_1 = 10 \times 10^6 - 2 \times 10^6 = 8 \, \text{M}\Omega$$

If your answers are correct, go on to Problem 2.15.

If your answers are not correct, recheck your calculations. For more details, see chapter 1 in reference 1 and chapter 7 in reference 2.

## IMPEDANCE TRANSFORMATION

There are several methods for matching circuits by means of impedance transformation. Chapter 6 illustrates the use of a stub in parallel with a high-frequency transmission line. The problem in this section illustrates an in-line method using an L-section.

### PROBLEM 2.15   Impedance Transformation

For the circuit shown in Exhibit 2.15, the load-section parameters needed in order to match the load to the generator for maximum power transfer are:

    a.  $L = 0.0179$ H and $C = 2.48$ μF
    b.  $L = 0.0338$ H and $C = 2.48$ μF
    c.  $L = 0.0179$ H and $C = 15.59$ μF
    d.  $L = 0.0111$ H and $C = 2.48$ μF

*Exhibit 2.15*

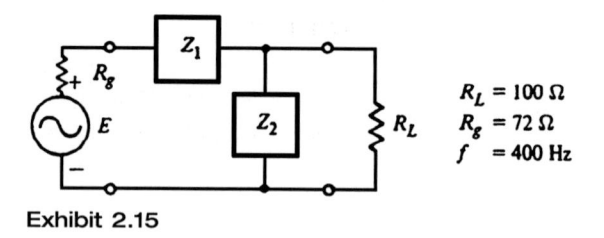

$R_L = 100\ \Omega$
$R_g = 72\ \Omega$
$f = 400$ Hz

*Solution*

Refer to the section on resonance in chapter 1. This problem fits the case in which $R_L > R_g$. Therefore, $Z_1$ is an inductor $L$ and $Z_2$ is a capacitor $C$, the values of which are calculated as follows:

$$L = \frac{R_g}{\omega}\sqrt{\frac{R_L}{R_g}-1} = \frac{72}{2\pi \times 400}\sqrt{\frac{100}{72}-1} = 0.0179\text{ H}$$

$$C = \frac{1}{\omega R_L}\sqrt{\frac{R_L}{R_g}-1} = \frac{1}{2\pi \times 400 \times 100}\sqrt{\frac{100}{72}-1} = 2.48\ \mu\text{F}$$

The correct answer is a.

*Answer Rationale*

Incorrect solution b is the result of not subtracting 1 from $\frac{R_L}{R_g}$ in calculating $L$. Incorrect solution c is the result of not multiplying by $2\pi$ in the denominator while calculating $C$. Incorrect solution d is the result of not taking the square root of $\frac{R_L}{R_g}-1$ in calculating $L$.

If your answers are correct, go on to Problem 2.16.

If your answers are not correct, review the subject of impedance transformation in chapter 1. For more details, see chapter 1 in reference 1 and chapter 15 in reference 2.

# AMMETERS

The problem in this section illustrates the difference in ac and dc current measurement.

## PROBLEM 2.16   ac and dc Ammeters

In the circuit shown in Exhibit 2.16, an ac and a dc source are connected in series with a 100-ohm load. Ammeters $A_1$ and $A_2$ are also in series with the circuit to measure ac and dc current. $A_1$ is a true rms ammeter, and $A_2$ is a D'Arsonval-type ammeter.

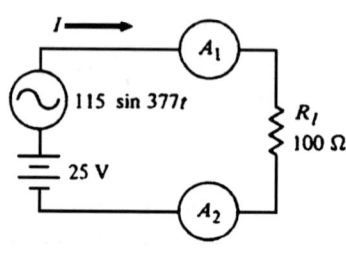

*Exhibit 2.16*

Determine:
a. The reading of $A_1$
b. The reading of $A_2$
c. The power dissipated in the load

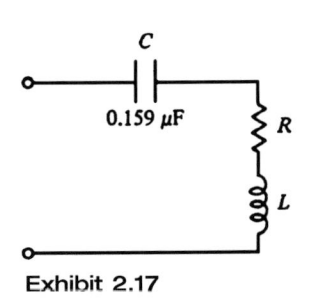

*Solution*

ac voltmeters and ammeters for power frequencies are designed to indicate effective value, even for nonsinusoidal waveforms. Instruments for higher frequencies also give the effective value for sine waves, but for other waveforms, the reading is not the effective reading. The composite current $I$ has an ac component and a dc component. These two components are calculated as follows:

$$I_{ac} = \frac{115}{100\sqrt{2}} = 0.813 \text{ amp}$$

$$I_{dc} = \frac{25}{100} = 0.25 \text{ amp}$$

a. Meter $A_1$ reads $I_{RMS}$, which is equal to the square root of the sum of the squares of the current components:

$$A_1 = I_{RMS} = \sqrt{I_{ac}^2 + I_{dc}^2} = \sqrt{0.813^2 + 0.25^2}$$
$$= \sqrt{0.66 + 0.0625} = \sqrt{0.724} = 0.85 \text{ amp}$$

b. Meter $A_2$ reads $I_{dc} = 0.25$ amp.

c. Power dissipated in the load is

$$P_L = I_{RMS}^2 R_L = 0.85^2 \times 100 = 72.4 \text{ watts}$$

If your answers are correct, go to Problem 2.17.

If your answers are not correct, review the subject of waveforms in chapter 1. For more details, see chapter 1 in reference 1 and chapters 13 and 19 in reference 2.

# RESONANCE

The subject of resonance is quite broad. This section contains a variety of problems illustrating this subject.

### PROBLEM 2.17  Series Resonance

A series resonant circuit as shown in Exhibit 2.17 has a resonant frequency of 3000 Hz. The values of the coil inductance $L$ and the maximum value of coil resistance $R$ if the bandwidth is not to exceed 100 Hz are:
a. $L = 0.017$ H and $R = 11.12$ Ω
b. $L = 333.7$ H and $R = 11.12$ Ω
c. $L = 0.017$ H and $R = 3711$ Ω
d. $L = 0.017$ H and $R = 333.6$ Ω

Exhibit 2.17

*Solution*

a.  At resonant frequency, $\omega L = 1/\omega C$

$$L = \frac{1}{\omega^2 C} = \frac{1}{(2\pi \times 3000)^2 \times 0.159 \times 10^{-6}} = 0.017 \text{ H}$$

b.  Bandwidth is $B = f_r/Q = 100$ Hz

$$Q = \frac{3000}{100} = 30$$

$$Q = \frac{1}{R}\sqrt{\frac{L}{C}}$$

Therefore,

$$R = \frac{1}{Q}\sqrt{\frac{L}{C}} = \frac{1}{30}\sqrt{\frac{0.0177}{0.159 \times 10^{-6}}} = 11.12 \, \Omega$$

The correct answer is a.

*Answer Rationale*

Incorrect solution b is the result of not taking the square of $(2\pi \times 3000)$ in calculating $L$. Incorrect solution c is the result of not taking the square root of $\frac{L}{C}$ while calculating $R$. Incorrect solution d is the result of not multiplying by $\frac{1}{30}$ in calculating $R$.

If your answers are correct, go on to Problem 2.18.

If your answers are not correct, review the subject of series resonance in chapter 1. For more details, see chapter 1 in reference 1 and chapter 20 in reference 2.

## PROBLEM 2.18  Passive Circuit Calculation

A coil is required to produce parallel resonance at 60 Hz in a circuit with a 50-kVAR-rated capacitor at 4800 volts. The inductance of the coil is:
  a. $L = 1222$ H
  b. $L = 1.22$ H
  c. $L = 173{,}797$ H
  d. $L = 7.68$ H

*Solution*

$$\frac{50{,}000 \text{ VAR}}{4{,}800 \text{ V}} = 10.42 \text{ amps}$$

$$X_C = \frac{4{,}800}{10.42} = 461 \, \Omega$$

For parallel resonance: $|B_L| = |B_C|$

For the special case of nondissipative circuits:

$$\frac{1}{\omega L} = \omega C, \quad \text{or } |X_L| = |X_C|$$

Therefore, $|X_L| = |X_C| = 461 \, \Omega$.

$$L = \frac{X_L}{\omega} = \frac{461}{377} = 1.22 \, \text{H}$$

The correct answer is b.

*Answer Rationale*

Incorrect solution a is the result of incorrectly using 50 instead of 50,000 in the numerator while calculating the current in the first step. Incorrect solution c is the result of using the wrong equation $L = \omega X_L$ in calculating $L$. Incorrect solution d is the result of incorrectly dividing by 60 hertz instead of $2\pi \times 60$ in the denominator while calculating $L$ in the last step.

If your answer is correct, go on to Problem 2.19.

If your answer is not correct, review the subjects of circuit element equations and resonance in chapter 1. For more details, see chapter 1 in reference 1 and chapter 19 in reference 2.

### PROBLEM 2.19  Series-Parallel Resonant Filter

A filter is a device that passes currents of a certain frequency and offers high impedance to currents of another frequency. Many complex reactive networks have been developed as active filters over the years. (They are described in detail in textbooks and handbooks.)

The circuit in Exhibit 2.19 is required to pass 60 kHz with minimum impedance and must block 30 kHz current as effectively as possible. To meet these requirements, the values of $C_2$ and $L_0$ are:

a. $C_2 = 0.006976 \, \mu\text{F}$ and $L_0 = 0.362 \, \text{mH}$
b. $C_2 = 26.294 \, \mu\text{F}$ and $L_0 = 0.362 \, \text{mH}$
c. $C_2 = 2.48 \, \mu\text{F}$ and $L_0 = 0.345 \, \text{mH}$
d. $C_2 = 0.026294 \, \mu\text{F}$ and $L_0 = 0.362 \, \text{mH}$

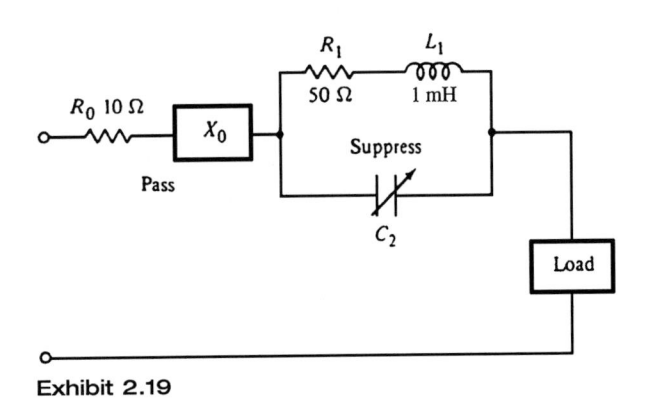

Exhibit 2.19

*Solution*

This filter is designed in two steps: first determine $C_2$ in the parallel antiresonant circuit for the suppress frequency; then solve for the reactance required for pass frequency in the series resonant branch.

1. Calculate $C_2$ to produce parallel resonance:

$$X_1 = 2\pi f L_1 = 2\pi \times 30 \times 10^3 \times 10^{-3} = 188.5 \ \Omega$$
$$Z_1 = R_1 + jX_1 = 50 + j188.5 = 195 \angle 75.14°$$
$$Y_1 = \frac{1}{Z_1} = 0.0051278 \angle -75.14° \text{ siemens} = 0.001315 - j0.0049563$$
$$= G_1 - jB_1$$

For parallel resonance, $|B_{C_2}| = |B_1|$

$$B_{C_2} = 0.0049563 \text{ siemens} = 2\pi f C_2 \text{ siemens}$$
$$X_{C_2} = \frac{1}{B_{C_2}} = 201.762 \ \Omega$$
$$C_2 = \frac{0.0049563}{2\pi \times 30 \times 10^3} = 2.6294 \times 10^{-8} = 0.026294 \ \mu F$$

The parallel branch now has an admittance $Y_P$ and impedance $Z$ at 30 kHz if

$$Y_P = 0.001315 \text{ siemens}$$
$$Z_P = 760.4 \angle 0° \ \Omega$$

2. Calculate $X_0$ and its capacitance or inductance to produce series resonance. Recalculating the parallel circuit parameters at the pass frequency of 60 kHz yields

$$X_L = 2\pi f L = 2\pi \times 60 \times 10^3 \times 10^{-3} = 377 \ \Omega$$
$$X_C = \frac{1}{2\pi f C_2} = \frac{1}{2\pi \times 60 \times 10^3 \times 2.6294 \times 10^{-8}} = 100.88 \ \Omega$$

The new parallel impedance becomes

$$Z_P = \frac{(50 + j377)(-j100.88)}{50 + j377 - j100.88} = \frac{(380.3 \angle 82.45°)(100.88 \angle -90°)}{50 + j276.12}$$
$$= \frac{38364.78 \angle -7.55°}{280.61 \angle 79.74°} = 136.72 \angle -87.29° = 6.45 - j136.57$$

Since $Z_P$ is capacitive at 60 kHz, $X_0$ must be inductive to make the total circuit purely resistive.

$$X_0 = j136.57 = 2\pi f L_0$$
$$L_0 = \frac{136.57}{2\pi \times 60 \times 10^3} = 0.362 \text{ mH}$$

The correct answer is d.

*Answer Rationale*

Incorrect solution a is the result of incorrectly using the value of $G_1$ instead of $B_1$ in the numerator when calculating $C_2$. Incorrect solution b is the result of not multiplying by $10^3$ in the denominator while calculating $C_2$. Incorrect solution c is the result of incorrectly adding the angle 79.74 to −7.55 instead of subtracting it in calculating the resulting angle of $Z_p$ for $L_0$.

If your answer is correct, go to Problem 2.20.

If your answer is not correct, review the subjects of circuit element equations and resonance in chapter 1. For more details, see chapter 1 in reference 1 and chapters 20 and 23 in reference 2.

### PROBLEM 2.20   Series Resonant Filter

A small company is experiencing an annoying 780-Hz tone on its telephone circuits. It has been determined that the tone is due to power and telephone lines running parallel for a long distance. For some reason, the thirteenth harmonic of the power line is being induced into the telephone circuits at an intolerable level. To correct the problem, a three-phase delta-connected shunt will be installed on the power line that will provide 15 ± 1% ohms to the harmonic and 10,000 ± 1% ohms to the fundamental 60-Hz frequency.

Determine the parametric values of the series-resonant shunt that will meet these specifications.

*Solution*

The schematic for one phase of the shunt is shown in Exhibit 2.20.

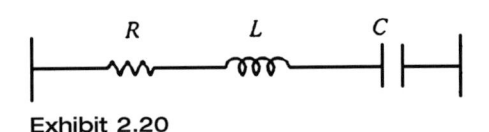

**Exhibit 2.20**

At 780 Hz, the lowest impedance is desired, which is achieved by series resonance. Therefore,

$$X_L = X_C \quad \text{and} \quad R = 15 \ \Omega$$

At 60 Hz, high impedance is desired. At low frequencies, $X_C$ offers the highest impedance. Therefore, $X_C$ should be close to the desired 10,000 ohms. We can try a value of 10,060 ohms for $X_C$ at 60 Hz ($X_L$ will subtract from this to bring it closer to 10,000 ohms) and see if the result is within the 1% tolerance.

$$\text{At} \quad 60 \text{ Hz}, \quad C = \frac{1}{2\pi f X_C} = \frac{1}{377 \times 10,060} = 0.264 \ \mu\text{F}$$

$$\text{At} \quad 780 \text{ Hz}, \quad X_C = \frac{1}{2\pi f_C} = \frac{1}{13 \times 377 \times 0.264 \times 10^{-6}} = 774 \ \Omega$$

For series resonance: $|X_L| = |X_C| = 774$

$$L = \frac{X_L}{2\pi f} = \frac{774}{13 \times 377} = 0.158 \text{ H}$$

Checking at 60 Hz:

$$X_L = 2\pi f L = 377 \times 0.158 = 59.5 \ \Omega$$
$$Z = R + jX_L - jX_C = 15 + j59.5 - j10,060$$
$$= 15 - j10,000 = 10,000 \angle -89.9° \ \Omega$$

This is well within the permitted tolerance. To summarize, the values of R, L, and C are

$$R = 15\ \Omega$$
$$L = 0.158\ \text{H}$$
$$C = 0.264\ \mu\text{F}$$

If your answers are correct, go on to Problem 2.21.

If your answers are not correct, review the solution to Problem 2.18. For more details, see chapter 1 in reference 1 and chapters 20 and 23 in reference 2.

### PROBLEM 2.21  RLC Bandpass Filter

A series RLC circuit is driven by a fixed-amplitude swept-frequency generator. It is desired that the current be maximum at 1500 Hz and down 3 dB at 1400 and 1600 Hz. The $Q$ of the circuit and the values of $R$ and $C$ if $L = 50$ mH are:

   a. $Q = 7.5$, $R = 62.85\ \Omega$, and $C = 0.198\ \mu\text{F}$
   b. $Q = 7.5$, $R = 29630\ \Omega$, and $C = 0.225\ \mu\text{F}$
   c. $Q = 7.5$, $R = 62.85\ \Omega$, and $C = 0.225\ \mu\text{F}$
   d. $Q = 7.5$, $R = 471.4\ \Omega$, and $C = 0.225\ \mu\text{F}$

*Solution*

At the series resonant frequency (where maximum current exists),

$$X_L = 2\pi fL = 2\pi \times 1500 \times 50 \times 10^{-3} = 471\ \Omega$$

$$|X_C| = |X_L| = 471 = \frac{1}{2\pi fC}$$

$$C = \frac{1}{2\pi fX_C} = \frac{1}{2\pi \times 1500 \times 471} = 0.225\ \mu\text{F}$$

$$Q = \frac{fr}{B} = \frac{1500}{200} = 7.5$$

$$Q = \frac{1}{R}\sqrt{\frac{L}{C}},\quad R = \frac{1}{Q}\sqrt{\frac{L}{C}} = \frac{1}{7.5}\sqrt{\frac{50 \times 10^{-3}}{0.225 \times 10^{-6}}} = 62.85\ \Omega$$

The correct answer is c.

*Answer Rationale*

Incorrect solution a is the result of incorrectly using the value of 1600 Hz in calculating $X_L$. Incorrect solution b is the result of not taking the square root of $L/C$ while calculating $R$. Incorrect solution d is the result of not multiplying by 1/7.5 in calculating $R$.

If your answers are correct, go on to Problem 2.22.

If your answers are not correct, review the subject of resonance in chapter 1. For more details, see chapter 1 in reference 1 and chapter 23 in reference 2.

### PROBLEM 2.22  Damped RLC Circuit

The circuit in Exhibit 2.22 represents a system in which a dc voltage is periodically applied, causing a current pulse that never goes negative. The capacitor is discharged

# Resonance

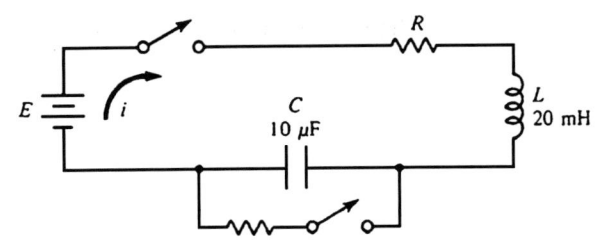

**Exhibit 2.22**

after each cycle. If the values of $L$ and $C$ are 20 mH and 10 $\mu$F, the minimum value of $R$ in order to achieve the desired result is:

- a. $R = 4000 \ \Omega$
- b. $R = 89.4 \ \Omega$
- c. $R = 2.83 \ \Omega$
- d. $R = 44.7 \ \Omega$

## Solution

The desired current waveform for one period is shown in Exhibit 2.22a.

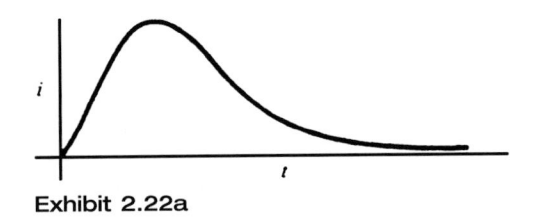

**Exhibit 2.22a**

The circuit must be critically damped if $R$ is to be a minimum while the current is always positive.

Applying Kirchhoff's voltage law yields

$$E(t) = iR + L\frac{di}{dt} + \frac{1}{C}\int i\, dt$$

In Laplace form,

$$E(s) = I(s)\left[R + sL + \frac{1}{sC}\right]$$

The characteristic equation is

$$R + sL + \frac{1}{sC} = 0$$

or

$$s^2 L + sR + \frac{1}{C} = 0$$

Applying the quadratic equation to solve for the roots of the equation yields

$$s = -\frac{R}{2L} \pm \sqrt{\frac{R^2}{4L^2} - \frac{1}{LC}}$$

This is of the form

$$-\alpha \pm \sqrt{\alpha^2 - \omega_r^2}$$

where

$$\omega_r = \frac{1}{\sqrt{LC}} = \text{resonant frequency}$$

$$\alpha = \frac{R}{2L} = \text{attenuation}$$

For critical damping, the expression under the radical must equal zero. Therefore,

$$\alpha^2 = \omega_r^2$$

or

$$\frac{R}{2L} = \frac{1}{\sqrt{LC}}$$

Solving for $R$,

$$R = \frac{2L}{\sqrt{LC}} = 2\sqrt{\frac{L}{C}} = 2\sqrt{\frac{20 \times 10^{-3}}{10 \times 10^{-6}}} = 89.4 \, \Omega$$

The correct answer is b.

### Answer Rationale

Incorrect solution a is the result of not taking the square root of $L/C$ while calculating $R$. Incorrect solution c is the result of incorrectly using $10 \times 10^{-3}$ in the denominator while calculating $R$. Incorrect solution d is the result of not multiplying by 2 in calculating $R$.

If your answer is correct, go on to Problem 2.23.

If your answer is not correct, review the subject of resonance in chapter 1. For more details, see chapter 1 in reference 1 and chapter 12 in reference 2.

## PROBLEM 2.23  Shunt Peaking

One stage of a video amplifier (Exhibit 2.23) is used to drive a 3.3-k$\Omega$ load. Since a wider bandwidth is required, it is decided to use shunt peaking compensation. Output resistance of the amplifier is 1 M$\Omega$, and the equivalent output shunt capacitance is 27 pF.

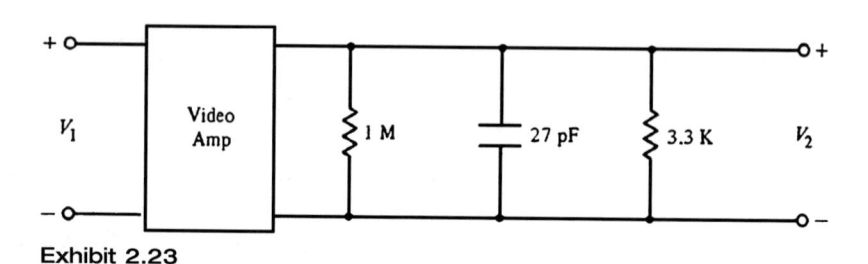

**Exhibit 2.23**

Determine the value of $L$ required to provide a normalized gain of 1 at the upper 3-dB uncompensated frequency. Then determine the overshoot and the new 3-dB bandwidth. Assume flat response to 0 Hz.

### Solution

The new circuit is shown in Exhibit 2.23a. Old and new frequency response curves are compared in Exhibit 2.23b. Martin has an excellent section on shunt peaking, and certain of his data are referenced in the solution of this problem.

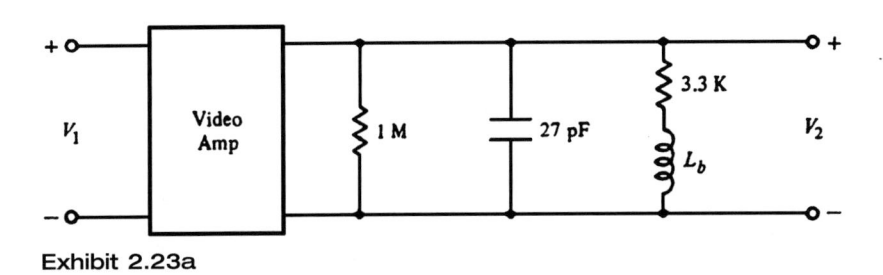

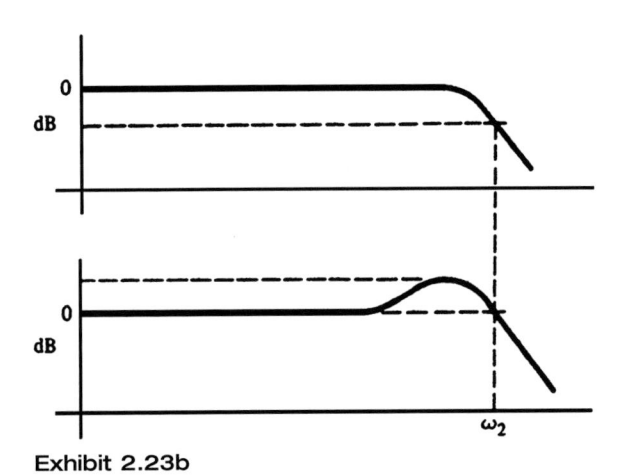

Exhibit 2.23a

Exhibit 2.23b

$$RC_T = \frac{1}{\omega_2}$$

where
  $\omega_2$ = upper cutoff frequency without $L_b$
  $C_T$ = 27 pF
  $R$ = 1 M//3.3 K = $R_2//R_L$

$$m = \frac{\omega_2 L_b}{R_L} = \text{peaking parameter} = \frac{L_b}{R_L^2 C_T}$$

$$L_b = mR_L^2 C_T = m \times (3.3 \times 10^3)^2 \times 27 \times 10^{-12}$$

A practical range for $m$ is 0.3 to 0.4.

No peaking occurs at less than 0.25, and above 0.6 overshoot becomes excessive. Therefore, a value for $m$ of 0.4 will be used here.

$$L_b = 0.4 \times (3.3 \times 10^3)^2 \times 27 \times 10^{-12} = 0.118 \text{ mH}$$

For $m = 0.4$, overshoot is 2.5%. For $m = 0.4$, the new upper cutoff frequency is $1.7\omega_2$. Therefore, the new 3-dB bandwidth is 1.7 times the original bandwidth.

If your answers are correct, go on to Problem 2.24.

If your answers are not correct, review appropriate texts on shunt peaking compensation. For more details, see chapter 1 in reference 1 and chapter 23 in reference 2.

## MAXIMUM POWER

Chapter 1 defines the maximum power transfer theorem and its corollary. The problems in this section illustrate these principles.

### PROBLEM 2.24  Thevenin Circuit–Maximum Power

The circuit in Exhibit 2.24 has properties found in electronic circuits. Convert the circuit to a Thevenin equivalent and determine the power dissipated in the load resistance when it is adjusted for maximum power transfer.

  a.  $E_T = 70$ volts, $R_T = 20$ ohms, and $P_l = 61.25$ watts
  b.  $E_T = 70$ volts, $R_T = 80$ ohms, and $P_l = 9.8$ watts
  c.  $E_T = 70$ volts, $R_T = 20$ ohms, and $P_l = 35$ watts
  d.  $E_T = 70$ volts, $R_T = 20$ ohms, and $P_l = 180$ watts

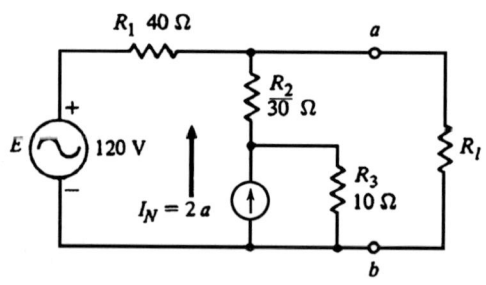

Exhibit 2.24

*Solution*

Convert the Norton circuit, consisting of $R_3$ and the current source, to a Thevenin equivalent (Exhibit 2.24a):

$$E_T = I_N R_3 = 20 \text{ volts}$$
$$R_T = R_3 = 10 \text{ } \Omega$$

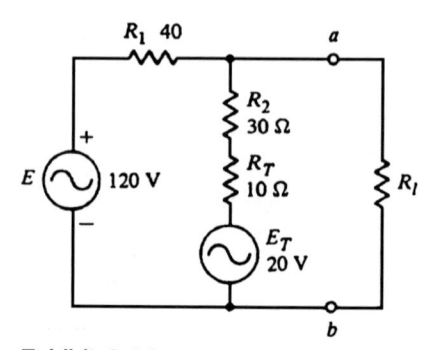

Exhibit 2.24a

Now convert the circuit to the left of terminals *ab* to a Thevenin equivalent, as shown in Exhibit 2.24b.

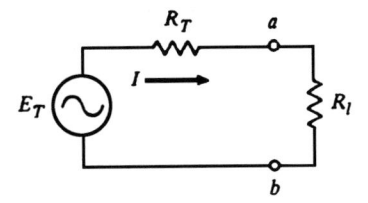

Exhibit 2.24b

$E_T$ = voltage at terminal *a* when $R_l$ is not connected. Using superposition,

$$E_T = 120\left(\frac{40}{80}\right) + 20\left(\frac{40}{80}\right) = 70 \text{ volts}$$

$$R_T = \frac{(40)(40)}{40+40} = 20 \text{ }\Omega$$

For maximum power, $R_L = R_T = 20$ $\Omega$; thus,

$$P_l = I^2 R_l = \left(\frac{E_T}{R_T + R_l}\right)^2 R_l = \left(\frac{70}{40}\right)^2 20 = 61.25 \text{ watts}$$

The correct answer is a.

*Answer Rationale*

Incorrect solution b is the result of incorrectly assuming that $R_T = 40 + 40 = 80$ ohms. Incorrect solution c is the result of not taking the square of (70/40) in calculating $P_l$. Incorrect solution d is the result of incorrectly using the value of 120 volts for $E_T$ in calculating $P_l$.

If your answers are correct, go to Problem 2.25.

If your answers are not correct, review the subject of maximum power transfer in chapter 1. For more details, see chapter 1 in reference 1 and chapter 18 in reference 2.

## PROBLEM 2.25 Ideal Transformer–Maximum Power

Power is supplied by a generator to a transformer-coupled load as shown in Exhibit 2.25. The internal impedance of the generator is $50 + j20$ ohms, and the frequency is 400 Hz. Open circuit emf of the generator is 100 VRMS. What should

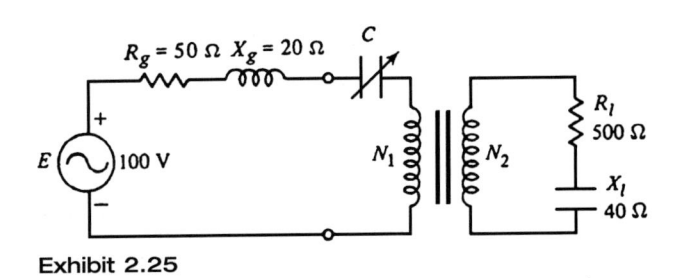

Exhibit 2.25

be the turns ratio of the transformer and the value of the tuning capacitor to achieve maximum power transfer?

a. $\dfrac{N_1}{N_2} = \dfrac{1}{10}$ and $C = 25 \ \mu F$

b. $\dfrac{N_1}{N_2} = \sqrt{\dfrac{1}{10}}$ and $C = 16.6 \ \mu F$

c. $\dfrac{N_1}{N_2} = \sqrt{\dfrac{1}{10}}$ and $C = 157 \ \mu F$

d. $\dfrac{N_1}{N_2} = \sqrt{\dfrac{1}{10}}$ and $C = 25 \ \mu F$

*Solution*

To achieve maximum power, $R_l$ must equal $R_g$ and the reactances must cancel out. Exhibit 2.25a shows the equivalent circuit referred to the transformer primary.

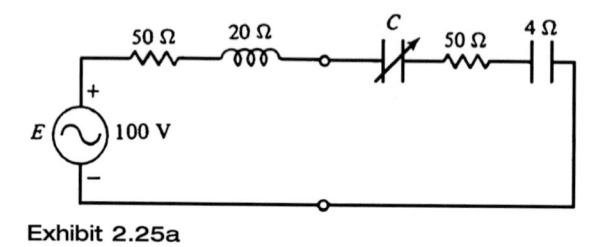

**Exhibit 2.25a**

Impedances are proportional to the square of the turns ratio. Therefore,

$$\left(\dfrac{N_1}{N_2}\right)^2 = \dfrac{R_g}{R_l} = \dfrac{50}{500} = \dfrac{1}{10} \quad \dfrac{N_1}{N_2} = \sqrt{\dfrac{1}{10}}$$

Calculating the value for the variable capacitor yields

$$C = \dfrac{1}{2\pi f X_C} = \dfrac{1}{2\pi \times 400 \times 16} = 25 \ \mu F$$

The correct answer is d.

*Answer Rationale*

Incorrect solution a is the result of not taking the square root of $(N_1/N_2)$ while calculating the turns ratio. Incorrect solution b is the result of incorrectly using $X_c = 20 + 4 = 24$ instead of $20 - 4 = 16$ in the denominator while calculating $C$. Incorrect solution c is the result of not multiplying by $2\pi$ in the denominator while calculating $C$.

If your answers are correct, go to Problem 2.26.

If your answers are not correct, review the subjects of maximum power transfer and ideal transformers in chapter 1. For more details, see chapter 1 in reference 1 and chapter 21 in reference 2.

## PROBLEM 2.26 Maximum Power Transfer

A network contains dc generators and resistors and has only two output terminals available for measurement. When the output terminals are shorted, 6 amps flow. When a 12-ohm resistor is connected between the two terminals, 3 amps flow. The maximum power that may be taken from the network at the two output terminals is:
- a. $P_L = 432$ watts
- b. $P_L = 108$ watts
- c. $P_L = 216$ watts
- d. $P_L = 36$ watts

*Solution*

The circuit shown in Exhibit 2.26 may be used to represent the network.

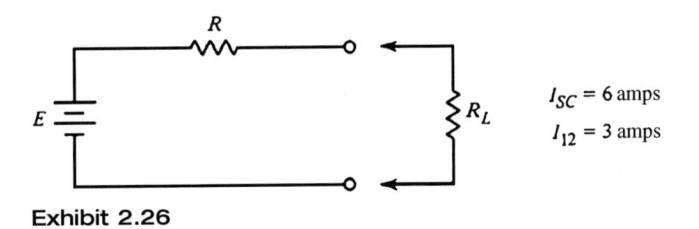

$I_{SC} = 6$ amps
$I_{12} = 3$ amps

**Exhibit 2.26**

Writing two equations,

$$(1) \quad R = \frac{E}{6}, \quad E = 6R$$

$$(2) \quad R + 12 = \frac{E}{3}, \quad E = 3R + 36$$

$$(1) - (2) \quad 0 = 3R - 36$$

$$R = 12$$
$$E = 6R = 72 \text{ volts}$$

For maximum power transfer, $R = R_L$.

$$I = \frac{E}{R + R_L} = \frac{72}{24} = 3 \text{ amps}$$

$$P_L = I^2 R_L = 9 \times 12 = 108 \text{ watts}$$

The correct answer is b.

*Answer Rationale*

Incorrect solution a is the result of using the wrong equation $I = E/R$ in calculating $I$. Incorrect solution c is the result of incorrectly using the value of $R + R_L$ instead of $R_L$ in calculating $P_l$. Incorrect solution d is the result of not taking the square of $I$ in calculating $P_l$.

If your answer is correct, go on to Problem 2.27.

If your answer is not correct, review the subject of maximum power transfer in chapter 1. For more details, see chapter 1 in reference 1 and chapter 9 in reference 2.

## PROBLEM 2.27 Maximum Power Transfer Corollary

Power at 1200 Hz is supplied to a load through a transformer by a signal generator having an internal impedance of 520 + j300. It is desired to transfer maximum power to a resistive load of 20 ohms through a step-down transformer. Open circuit voltage output of the generator is 75 volts RMS. Exhibit 2.27 shows the circuit diagram.

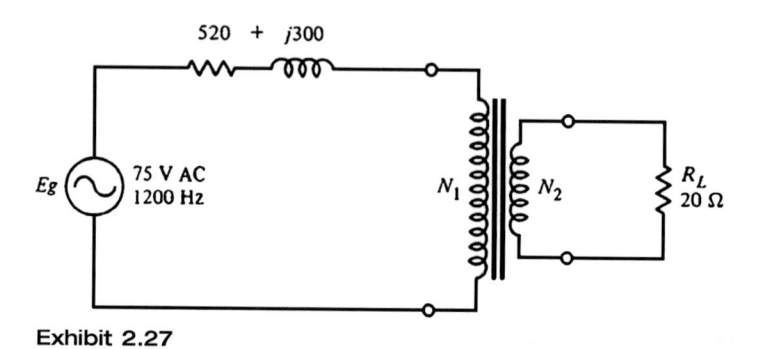

Exhibit 2.27

The transformer turns ratio and power supplied to the load are:
a. $a = \sqrt{30}$ and $P_L = 2.51$ watts
b. $a = 30$ and $P_L = 2.51$ watts
c. $a = \sqrt{30}$ and $P_L = 38.8$ watts
d. $a = \sqrt{30}$ and $P_L = 38820$ watts

### Solution

Since the resistance cannot be canceled, then maximum power transfer occurs when the equivalent load impedance equals the generator impedance.

$$Z_g = 520 + j300 = 600 \angle 30°$$
$$|Z_g| = |a^2 R_L|$$

Assuming an ideal transformer (no losses),

$$a = \sqrt{\frac{Z_g}{R_L}} = \sqrt{\frac{600}{20}} = \sqrt{30}$$

Equivalent $R_L$ referred to the primary is 600 Ω.

$$I = \frac{Eg}{Zg + R'_L} = \frac{75}{520 + 600 + j300} = \frac{75}{1159.5} = 64.7 \text{ mA}$$

$$P_L = I^2 R_L = (64.7 \times 10^{-3})^2 600 = 2.51 \text{ watts}$$

The correct answer is a.

### Answer Rationale

Incorrect solution b is the result of not taking the square root of $Z_g/R_L$ in calculating the turns ratio. Incorrect solution c is the result of not taking the square of (64.7 × $10^{-3}$) in calculating $P_L$. Incorrect solution d is the result of not multiplying by $10^{-3}$ in calculating $P_l$.

If your answers are correct, go on to the next section.

If your answers are not correct, review the subjects of Thevenin's theorem and ideal transformers in chapter 1. For more details, see chapter 1 in reference 1 and chapters 18 and 21 in reference 2.

# TRANSIENTS

The study of transients is diverse and applies to nearly all aspects of engineering. Chapter 1 contains a brief summary of the subject. Problems in this section illustrate a few aspects of this subject that are likely to apply to the PE exam.

### PROBLEM 2.28  RL Transient

In the circuit shown in Exhibit 2.28, assume that steady-state conditions exist prior to switch closure. Determine $i_2(t)$ for the period after the switch is closed.

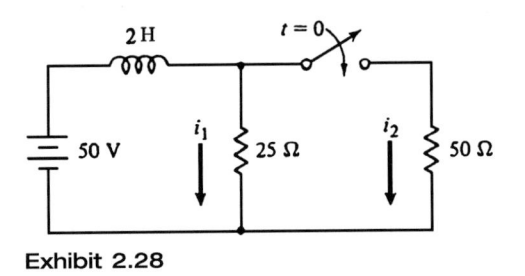

**Exhibit 2.28**

*Solution*

This problem may be solved using the general transient response formula:

$$i(t) = i_{ss} + (i_0 - i_{ss})e^{-t/\tau}$$

Immediately after the switch is closed, there is no current change through the coil. However, one-third of the current flows immediately in the second branch; the other two-thirds flow through branch 1. Initial current in branch 2 is

$$i_0 = \frac{50}{25} \times \frac{1}{3} = \frac{2}{3} \text{ amp}$$

Final current flow in branch 2 is

$$i_{ss} = \frac{50}{\frac{25 \times 50}{75}} \times \frac{1}{3} = 1 \text{ amp}$$

The time constant for the transient is

$$\tau = \frac{L}{R} = \frac{2}{16.67} = 0.12 \text{ seconds}$$

Now the transient response formula may be specified:

$$i_2(t) = 1 - \frac{1}{3}e^{-t/0.12}$$

If your answer is correct, go to Problem 2.29.

If your answer is not correct, review the subject of transients in chapter 1. For more details, see chapter 1 in reference 1 and chapter 12 in reference 2.

## PROBLEM 2.29  Double Energy Transient

In the circuit shown in Exhibit 2.29, assume steady-state conditions exist prior to switch opening. Determine $V_C(t)$ for capacitor voltage after the switch is opened.

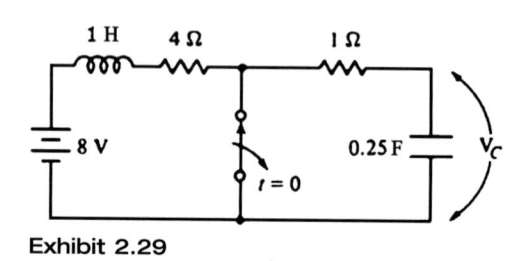

Exhibit 2.29

### Solution

This problem is best solved by means of Laplace transforms.
   Initial conditions:

$$V_C = 0$$
$$I_0 = \frac{8}{4} = 2 \text{ amps}$$

After switch is opened:

$$E = i(t)R + L\frac{di}{dt} + \frac{1}{C}\int_0^t i\,dt$$

$$\frac{E}{s} = RI(s) + L[sI(s) - I_0] + \frac{1}{Cs}I(s)$$

$$I(s)\left[R + sL + \frac{1}{Cs}\right] = \frac{E}{s} + LI_0$$

$$I(s)\left[5 + s + \frac{4}{s}\right] = \frac{8}{s} + 2$$

$$I(s) = \frac{\left(\frac{8}{s} + 2\right)s}{5s + s^2 + 4} = \frac{(2s+8)}{(s+4)(s+1)} = \frac{2}{s+1}$$

$$i(t) = 2e^{-t}$$

$$V_C = \frac{1}{C}\int_0^t i\,dt = \frac{2}{C}\int_0^t e^{-t}\,dt = 8[-e^{-t}]_0^t = 8(1 - e^{-t})$$

If your answer is correct, go to Problem 2.30.
   If your answer is not correct, review the subjects of transients and Laplace transform in chapter 1. For more details, see chapter 1 in reference 1 and chapter 12 in reference 2.

## PROBLEM 2.30 Double Energy Transient

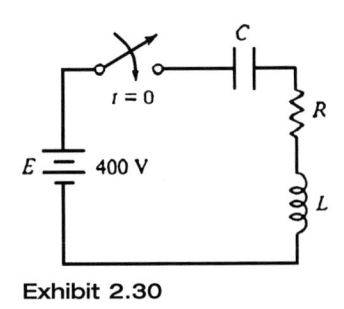

In the circuit shown in Exhibit 2.30, the capacitor is initially uncharged and the inductance has no stored energy. The resistance is finite but considered negligible for the transient state. Determine the highest instantaneous voltage that will be impressed across the capacitor.

### Solution

Initially, $I_L = 0$ and $V_C = 0$. Summing the voltages around the loop yields

$$E = iR + L\frac{di}{dt} + \frac{1}{C}\int_0^t t\, dt$$

Converting to Laplace transforms yields

$$\frac{E}{s} = \left[R + sL + \frac{1}{sC}\right]I(s)$$

$$I(s) = \frac{E}{s\left[R + sL + \frac{1}{sC}\right]} = \frac{E/L}{s^2 + s\frac{R}{L} + \frac{1}{LC}}$$

$$= \frac{E/L}{\left(s + \frac{R}{2L}\right)^2 + \left(\frac{1}{LC} - \frac{R^2}{4L^2}\right)}$$

This is the Laplace transform expression for a damped sine wave, where

$$a = -\frac{R}{2L}$$

and

$$\omega^2 = \left(\frac{1}{LC} - \frac{R^2}{4L^2}\right)$$

Converting to the time domain yields

$$i(t) = \frac{E}{\omega L}[e^{-(R/2L)t}\sin\omega t]$$

Assuming $R$ is very small, then the exponential term is negligible and the expression reduces to

$$i(t) = \frac{E}{\omega L}\sin\omega t \quad \text{and} \quad \omega^2 \approx \frac{1}{LC}$$

The voltage across the capacitor is

$$V_C = \frac{1}{C}\int_0^t i\, dt = \frac{E}{\omega LC}\int_0^t \sin\omega t\, dt$$

$$= -\frac{E}{\omega^2 LC}[\cos\omega t]_0^t = -E(\cos\omega t - 1)$$

The maximum value of $(\cos \omega t - 1) = (-1 - 1) = -2$. Therefore, the maximum value of $V_C = -E(-2) = 2E = 800$ volts.

If your answer is correct, go to Problem 2.31.

If your answer is not correct, review the subjects of transients and Laplace transform in chapter 1. For more details, see chapter 1 in reference 1 and chapter 12 in reference 2.

## PROBLEM 2.31 Transient Response

The switch in the circuit in Exhibit 2.31 has been in position $A$ for a long time. Determine the equation for $i(t)$ when the switch is moved to position $B$. Show a plot of $i(t)$ with respect to time.

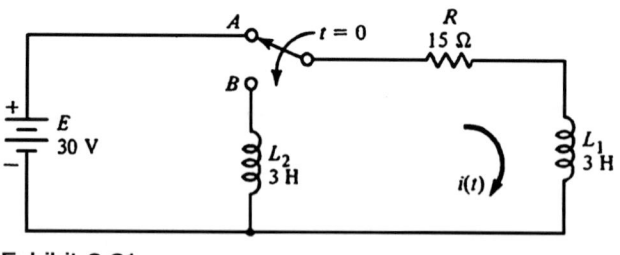

Exhibit 2.31

*Solution*

Initially,

$$i(t) = \frac{E}{R} = \frac{30}{15} = 2 \text{ amps}$$

At $t \geq 0$, excitation function $= E(s) + L_1(I_0) = 0 + 3 \times 2 = 6$

$$Z(s) = sL_1 + sL_2 + R = (L_1 + L_2)\left(s + \frac{R}{L_1 + L_2}\right) = 6(s + 2.5)$$

$$I(s) = \frac{\text{excitation function}(s)}{Z(s)} = \frac{6}{6(s+2.5)} = \frac{1}{s+2.5}$$

$$i(t) = e^{-2.5t}$$

Exhibit 2.31a shows the plot of $i(t)$.

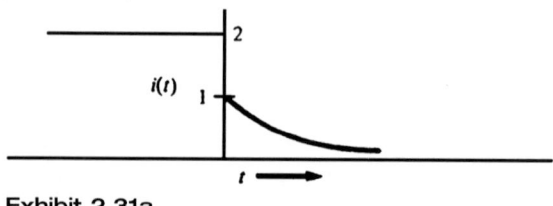

Exhibit 2.31a

If your answers are correct, go on to Problem 2.32.

If your answers are not correct, review the subject of transients in chapter 1. For more details, see chapter 1 in reference 1 and chapter 12 in reference 2.

# INSULATION

The problem in this section illustrates one aspect of electric fields, a subject of fundamental importance to electrical engineering.

### PROBLEM 2.32  Insulation

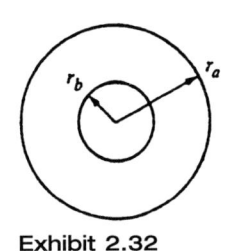

A single-conductor cable consists of a copper conductor 0.7 cm in diameter surrounded by a 0.2-cm-thick layer of polystyrene insulation and a lead sheath (Exhibit 2.32). The insulation has a dielectric constant of 2.6 and a dielectric strength of 300 kV per cm. The maximum safe potential difference that may be applied to the cable is:

- a.  $V_{ab} = 47.5$ volts
- b.  $V_{ab} = 99{,}168$ volts
- c.  $V_{ab} = -25{,}322$ volts
- d.  $V_{ab} = 47{,}560$ volts

Exhibit 2.32

*Solution*

$$V_{ab} = \frac{\Lambda}{2\pi\varepsilon} \ln \frac{r_a}{r_b} E$$

where
  $\Lambda$ = linear charge density = $2\pi r_b \varepsilon E$
  $\varepsilon$ = dielectric constant = 2.6
  $E$ = electric field intensity (= dielectric strength here) = 300 kV/cm
  $r_a$ = 0.55 cm
  $r_b$ = 0.35 cm

$$V_{ab} = \frac{2\pi r_b \varepsilon E}{2\pi\varepsilon} \ln \frac{r_a}{r_b} = r_b E \ln \frac{r_a}{r_b}$$

$$= 0.35 \text{ cm} \times \frac{300{,}000 \text{ V}}{\text{cm}} \ln \frac{0.55}{0.35} = 47{,}560 \text{ volts peak}$$

The correct answer is d.

*Answer Rationale*

Incorrect solution a is the result of incorrectly using 300 instead of 300,000 volts in calculating $V_{ab}$. Incorrect solution b is the result of incorrectly using $r_a = 0.7 + 0.2 = 0.9$ instead of 0.55 cm in calculating $V_{ab}$. Incorrect solution c is the result of incorrectly using $r_b = 0.7$ instead of 0.35 cm in calculating $V_{ab}$.

If your answer is correct, go on to Problem 2.33.

If your answer is not correct, review a text in the field of linear charge. For more details, see chapter 1 in reference 1 and chapter 25 in reference 3.

# WAVEFORMS

Determination of RMS and average values of various types of waveforms is discussed in chapter 1. The problem in this section illustrates a practical application of this type of analysis.

### PROBLEM 2.33 Waveform

A motor draws current in accordance with the periodic waveform shown in Exhibit 2.33. The cycle repeats every minute and continues for several hours. What must be the continuous current rating of the motor in order to avoid overload?

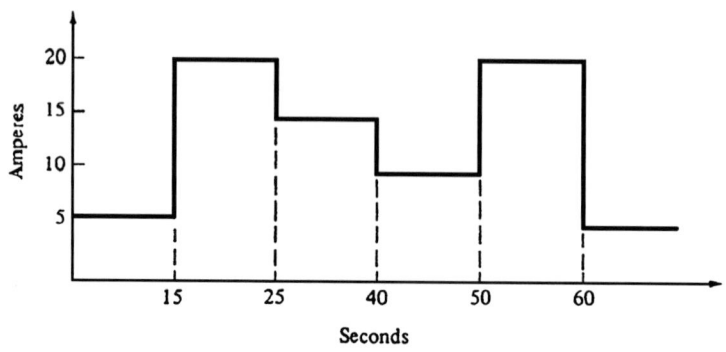

Exhibit 2.33

*Solution*

The RMS value of the curve is the maximum continuous current capacity of the motor.

$$I_{RMS} = \sqrt{\frac{1}{T}\int_0^t i^2(t)dt} =$$

$$= \sqrt{\frac{1}{60}\left[\int_0^{15} 5^2 dt + \int_{15}^{25} 20^2 dt + \int_{25}^{40} 15^2 dt + \int_{40}^{50} 10^2 dt + \int_{50}^{60} 20^2 dt\right]}$$

$$= \sqrt{\frac{1}{60}[25 \times 15 + 400 \times 10 + 225 \times 15 + 100 \times 10 + 400 \times 10]}$$

$$= \sqrt{\frac{1}{60}[12,750]} = \sqrt{212.5} = 14.58 \text{ amps}$$

If your answer is correct, go on to chapter 3.

If your answer is not correct, review the subject of waveforms in chapter 1. For more details, see chapter 1 in reference 1 and chapter 13 in reference 2.

## BLACK BOX ANALYSIS

A circuit may be represented by a black box having an input and an output with its parameters expressed in several ways, as illustrated by the following models.

### General Two-Port (Four-Terminal) Network

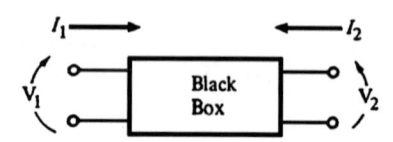

| Equations | Parameter | Units |
|---|---|---|
| $V = ZI$ | Impedance | ohms ($\Omega$) |
| $I = YV$ | Admittance | siemens (S) |
| $V_1 = f(I_1, V_2)$ $I_2 = f(I_1, V_2)$ | Hybrid | $\Omega$, S V/V, A/A |

## Open-Circuit Impedance Parameters

$$\begin{bmatrix} V_1 \\ V_2 \end{bmatrix} = \begin{bmatrix} Z_{11} & Z_{12} \\ Z_{21} & Z_{22} \end{bmatrix} \begin{bmatrix} I_1 \\ I_2 \end{bmatrix} \rightarrow \begin{matrix} V_1 = Z_{11}I_1 + Z_{12}I_2 \\ V_2 = Z_{21}I_1 + Z_{22}I_2 \end{matrix}$$

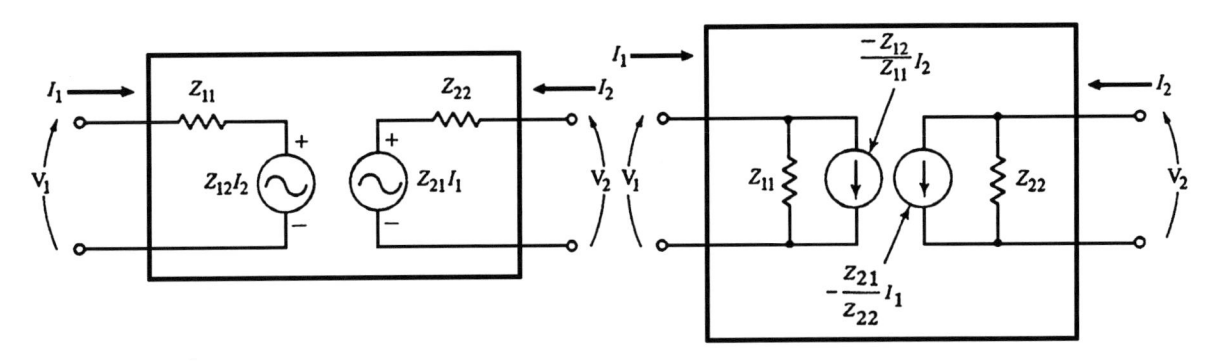

## Short-Circuit Admittance Parameters

$$\begin{bmatrix} I_1 \\ I_2 \end{bmatrix} = \begin{bmatrix} Y_{11} & Y_{12} \\ Y_{21} & Y_{22} \end{bmatrix} \begin{bmatrix} V_1 \\ V_2 \end{bmatrix} \rightarrow \begin{matrix} I_1 = Y_{11}V_1 + Y_{12}V_2 \\ I_2 = Y_{21}V_1 + Y_{22}V_2 \end{matrix}$$

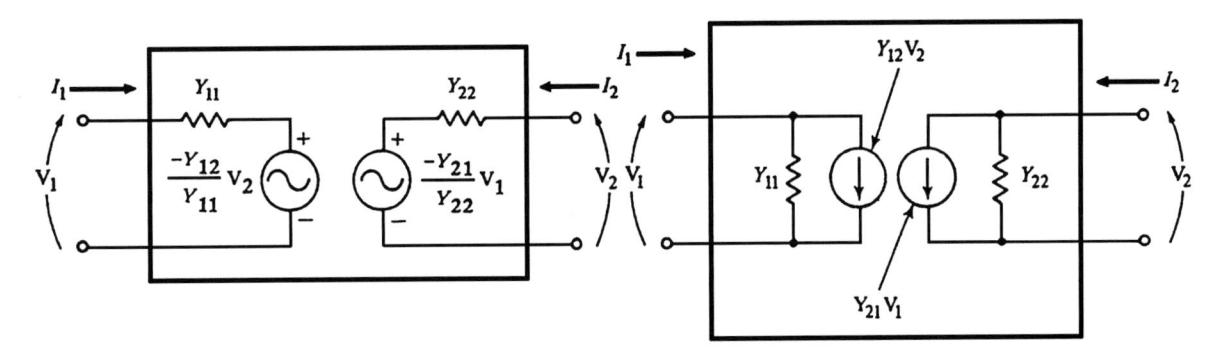

## Hybrid Parameters

$$\begin{bmatrix} V_1 \\ I_2 \end{bmatrix} = \begin{bmatrix} h_{11} & h_{12} \\ h_{21} & h_{22} \end{bmatrix} \begin{bmatrix} I_1 \\ V_2 \end{bmatrix} \rightarrow \begin{matrix} V_1 = h_{11}I_1 + h_{12}V_2 \\ I_2 = h_{21}I_1 + h_{22}V_2 \end{matrix}$$

| Parameter | Units |
|---|---|
| $h_{11}$ | $\Omega$ |
| $h_{12}$ | V/V |
| $h_{21}$ | A/A |
| $h_{22}$ | S |

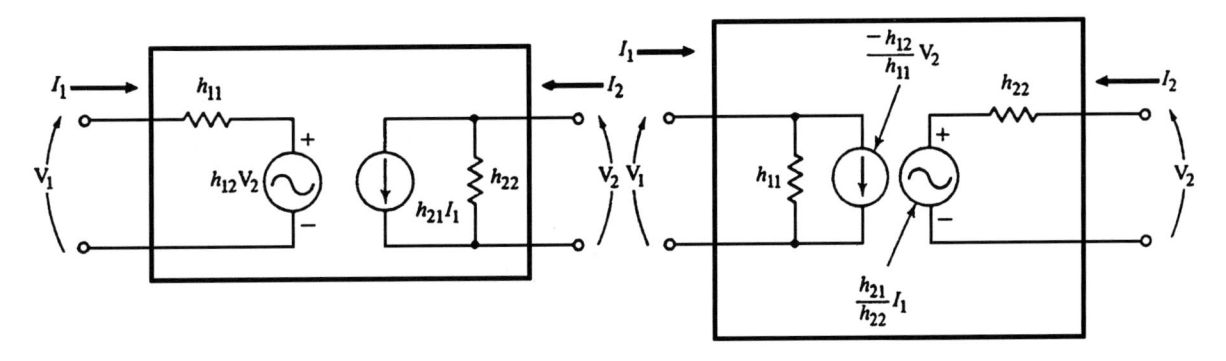

# CHAPTER 3

# Power

**OUTLINE**

SINGLE-PHASE POWER 73

POLYPHASE POWER 76

POWER FACTOR CORRECTION 83

TRANSMISSION LINE CALCULATIONS 87

WATTMETER MEASUREMENTS 92

SHORT CIRCUIT CALCULATIONS 96

This chapter covers single-phase and polyphase power distribution and loads, power factor correction, transmission line calculations at power frequencies, and wattmeter measurements. Unless otherwise stated, the material in this chapter assumes all waveforms are 60-hertz sine waves. The **vector,** or **phasor, diagram** is used to aid in solution of problems.

## SINGLE-PHASE POWER

The phase angle is a very important parameter for properly locating different alternating quantities with respect to one another. If the applied voltage is $v = V_{max} \sin \omega t$ and it is known, from the nature and magnitude of the circuit parameters, that the current comes to a corresponding point on its wave before the voltage wave by $\theta$ degrees, the current can be expressed as $i = I_{max} \sin(\omega t + \theta)$. This is an example of a positive phase angle due to an *RC* circuit which produces *leading* current and a leading power factor, where the power factor is

$$\text{PF} = \cos\theta = \frac{\text{watts}}{\text{volt-amperes}} = \frac{G}{|Y|} \quad \text{or} \quad = \frac{R}{|Z|}$$

The phasor diagram for this example is shown in Figure 3.1.

$$Z = R - jX = Z\angle-\theta°$$
$$V = IZ\angle-\theta°$$
$$I = \frac{V}{Z}\angle+\theta°$$

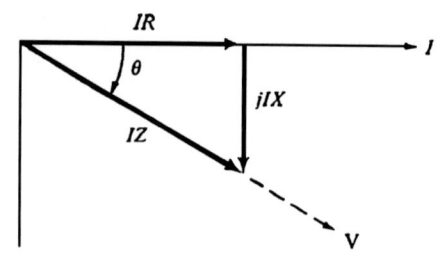

**Figure 3.1** Phasor diagram for a RC circuit

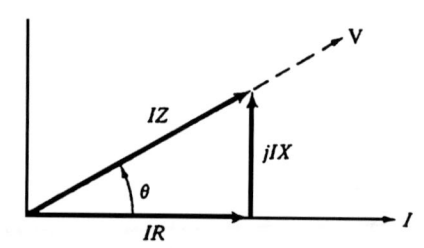

**Figure 3.2** Phasor diagram for a RL circuit

For an *RL* circuit, the phasor diagram is shown in Figure 3.2. In this case, the current *lags* the voltage by $\theta$ degrees.

$$Z = R + jX$$
$$V = IZ\angle+\theta°$$
$$I = \frac{V}{Z}\angle-\theta°$$

Power in ac circuits is either *resistive (real)* or *reactive*. Resistive (or average) power, $P$, is measured in watts, and reactive power, $Q$, is measured in vars (VAR, for volt-amperes reactive). Since vars are always equal to $VI \sin\theta$ and watts are always equal to $VI \cos\theta$, they can be thought of as being in quadrature to each other. The product $VI$ is called the *apparent power*, $S$, and is measured in volt-amperes (VA) or kilovolt-amperes (kVA), where

$$VA = \sqrt{watts^2 + vars^2}$$

This is represented by the power triangle for an *RL* circuit where current lags voltage (Figure 3.3).

This is a popular diagram for solving power factor correction problems, and will be discussed later.

For an *RC* circuit, the power phasor diagram would be drawn as shown in Figure 3.4.

These concepts are illustrated in Problem 3.1.

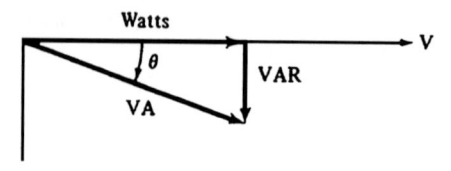

**Figure 3.3** Power triangle for an RL circuit

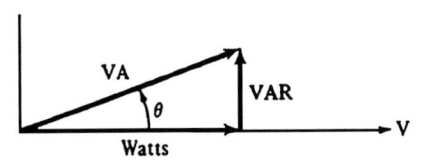

**Figure 3.4** Power triangle for an RC circuit

## PROBLEM 3.1 Single-Phase kVA and Power Factor

A machine shop has the following single-phase electrical loads:

- 50 kilowatts incandescent lighting and heating (purely resistive)
- 120 kilowatts at 0.8 power factor lagging
- 25 kilowatts at 0.9 power factor leading

The total kVA and the overall power factor are:
   a. kVA = 210∠−21.77° and PF = 0.93 lagging
   b. kVA = 220∠−27.64° and PF = 0.89 leading
   c. kVA = 210∠−21.77° and PF = 0.93 leading
   d. kVA = 210∠−21.77° and PF = 0.37 lagging

*Solution*

First, draw a power triangle of each load, observing that they are all given in kW in the problem statement (Exhibit 3.1).

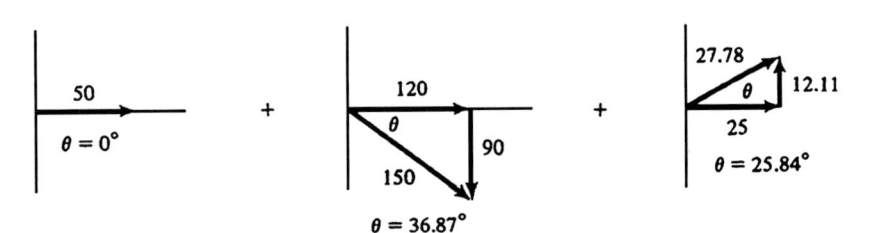

**Exhibit 3.1**

$$kVA = 50 + 120 - j90 + 25 + j12.11$$
$$= 195 - j77.89 = 210\angle-21.77$$
$$PF = \cos - 21.77° = 0.93 \text{ lagging}$$

The correct answer is a.

*Answer Rationale*

Incorrect solution b is the result of using a plus sign for the $j90$ term instead of a minus sign in calculating the kVA. Incorrect solution c is the result of not having the minus sign for the angle of 21.77 in calculating the PF. Incorrect solution d is the result of incorrectly using the sine of the angle instead of the cosine of the angle 21.77 in calculating the PF.

If your answers are correct, go on to the next section.

If your answers are not correct, review the subject of complex algebra in chapter 1 and single-phase power in this chapter. For more details, see chapter 19 in reference 2 and chapter 7 in reference 3.

## POLYPHASE POWER

Polyphase voltages are generated the same way as single-phase voltages. A polyphase system is simply several single-phase systems displaced in time phase from one another. In general, the electrical displacement between phases for a balanced, $n$-phase system is $360/n$ electrical degrees. Three-phase systems are the most common, although for certain special applications a greater number of phases are used. In general, three-phase equipment is more efficient, uses less material for a given capacity, and costs less than single-phase equipment. Also, for a fixed amount of power to be transmitted a fixed distance at a fixed line loss with a fixed voltage between conductors, three-phase makes more economical use of copper than any other number of phases.

In a balanced three-phase power system, voltages are generated 120° apart. There are two ways of connecting the three phases: one is called the delta ($\Delta$) connection; the other is called the wye (Y) connection. These are shown as in Figure 3.5.

The delta phasor diagram shows that the resultant of the voltages in any two phases is equal and opposite to the voltage in the third phase, when all voltages are equal and at 120° to one another, so that there is no net emf around the delta. Thus, $V_{ab} + V_{bc} + V_{ca} = 0$.

In the delta connection the line voltage is equal to the phase voltage. In the wye connection the line voltage is equal to $\sqrt{3}$ times the phase voltage, as shown in the wye phasor diagram. The Y-connection offers considerable advantage when

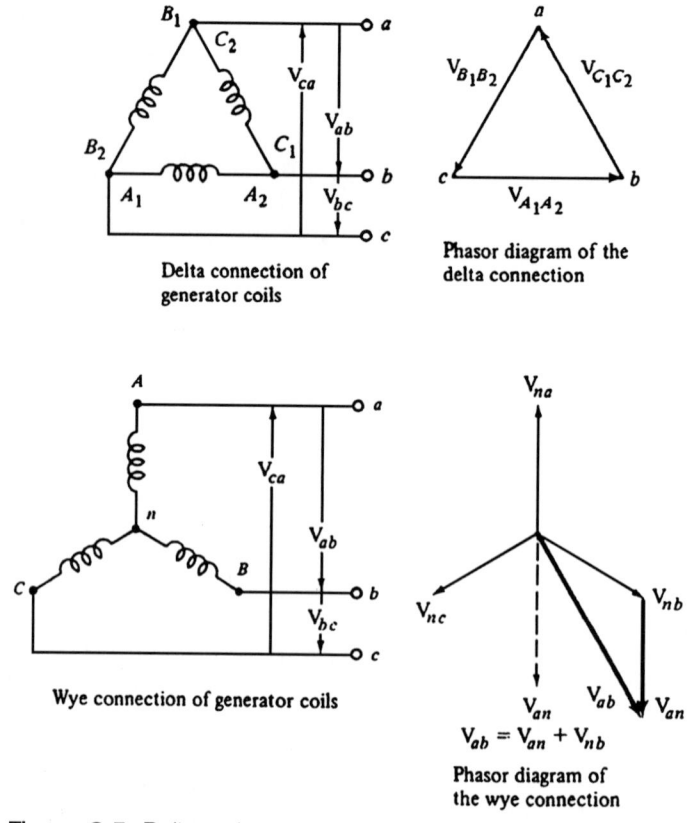

**Figure 3.5** Delta and wye connections in balanced three-phase power systems

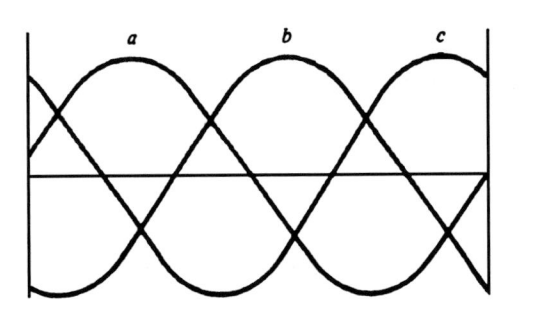

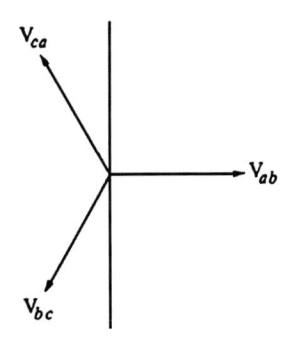

**Figure 3.6** Voltage waves from a three-phase generator

building high-voltage generators and transformers. Note that in the Y-connection each phase must carry full line current, while in the Δ-connection the phases divide the current between them, each taking $1/\sqrt{3}$ or 0.578 times the line current. Thus, for balanced three-phase,

$$VA = 3V_{line}I_{line}/\sqrt{3} = \sqrt{3}V_{line}I_{line}$$

Figures 3.6 shows the voltage waves generated by a three-phase generator. The *phase order* or *sequence* is *abc*, which is the order in which the emfs or phases *a*, *b*, and *c* come to their corresponding maximum values.

If the rotation of the generator is reversed, the phase sequence would be *acb*. In general, the phase sequence of the voltages applied to a load is fixed by the order in which the three-phase lines are connected. Interchanging any pair of the lines reverses the phase sequence. For three-phase unbalanced loads the effect is to cause a completely different set of values for line currents. Hence, when calculating such systems it is essential that phase sequence be specified.

Unless otherwise stated, the term "phase sequence" refers to *voltage* phase sequence. It should be recognized that, in unbalanced systems, the line currents and phase currents have their own phase sequence that may or may not be the same as the voltage sequence.

The following problems illustrate three-phase calculations for power factor, currents, voltages, power, and phase sequence.

### PROBLEM 3.2 Three-Phase Power Factor and Line Current

Two balanced three-phase loads in parallel with one another are described as follows:

1. 50 kVA at 0.707 PF lagging

2. $15 - j20$ ohms per phase, delta connection

If the line voltage is 220 volts, the power factor and line current of the combined loads are:
    a.  PF = 0.78 leading and $I_{line}$ = 130 amps
    b.  PF = 0.78 lagging and $I_{line}$ = 130 amps
    c.  PF = 0.62 lagging and $I_{line}$ = 130 amps
    d.  PF = 0.78 lagging and $I_{line}$ = 225 amps

*Solution*

Assuming both loads are delta connected, the load circuit is shown in Exhibit 3.2.

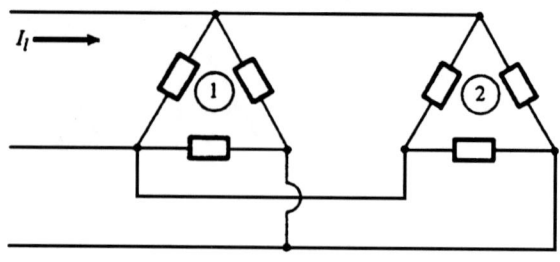

Exhibit 3.2

Load 2 is

$$\frac{V^2}{Z} = \frac{220^2}{25\angle -53.13°} = 1936 \text{ VA} @ 0.6 \text{ PF lead per phase}$$

Total three-phase load 2 is

$$3 \times 1936 = 5.808 \text{ kVA} @ 0.6 \text{ PF lead}$$
$$\text{Combined load} = 35.36 - j35.36 + 3.485 + j4.646 = 38.83 - j30.71$$
$$= 49.5 \text{ kVA} @ -38.3°$$
$$\text{PF} = \cos -38.3° = 0.78 \text{ lagging}$$

$$I_{line} = \frac{VA}{\sqrt{3}V_{line}} = \frac{49{,}517}{\sqrt{3} \times 220} = 130 \text{ amps}$$

*Alternative Solution (Impedance Method)*

$$Z_1 = \frac{3V^2}{VA} = \frac{3 \times 220^2}{50{,}000\angle -45°} = 2.9\angle +45° = 2.05 + j2.05$$

$$Z_{\|\phi} = \frac{Z_1 Z_2}{Z_1 + Z_2} = \frac{(2.9\angle +45°)(25\angle -53.13°)}{2.05 + j2.05 + 15 - j20} = \frac{72.6\angle -8.13°}{24.76\angle -46.47°}$$
$$= 2.93\angle 38.3°; \quad \text{PF} = 0.78 \text{ lagging}$$

$$I_\phi = \frac{220}{2.93\angle 38.34°} = 75.03\angle -38.3° \text{ amps}$$

$$I_{line} = \sqrt{3}I_\phi = 130 \text{ amps}$$

The correct answer is b.

*Answer Rationale*

Incorrect solution a is the result of not having the minus sign for the angle of 38.3 in calculating the PF. Incorrect solution c is the result of incorrectly using the sine of the angle instead of the cosine of the angle 38.3 in calculating the PF. Incorrect solution d is the result of not having the $\sqrt{3}$ in the denominator in calculating the line current.

If your answers are correct, go on to Problem 3.3.

If your answers are not correct, review the subject of complex algebra in chapter 1, and polyphase power in this chapter. For more details, see chapter 3 in reference 1 and chapter 9 in reference 2.

## PROBLEM 3.3  Motor Input Current

Determine the line current to a 10-horsepower, 60-Hz, three-phase induction motor when it is operating at full load. Motor efficiency is 85% and power factor is 0.9. Line-to-line input voltage is 480 volts rms.

 a. $I_{line} = 11.73$ amps
 b. $I_{line} = 11.71$ amps
 c. $I_{line} = 0.1173$ amps
 d. $I_{line} = 20.3$ amps

### Solution

$$\text{Input power to motor} = \frac{10 \text{ HP}}{0.85} \times \frac{746 \text{ watts}}{\text{HP}} = 8776.5 \text{ watts}$$

$$P = \sqrt{3}\, V_L I_L \cos\theta$$

$$I_L = \frac{P}{\sqrt{3}V_L \cos\theta} = \frac{8776.5}{\sqrt{3} \times 480 \times 0.9} = 11.73 \text{ amps}$$

The correct answer is a.

### Answer Rationale

Incorrect solution b is the result of incorrectly using the factor 745 in the numerator when converting input power from HP to watts. Incorrect solution c is the result of incorrectly using the value of 85 instead of 0.85 in the denominator when converting input power from HP to watts. Incorrect solution d is the result of not having the $\sqrt{3}$ in the denominator in calculating the line current.

If your answer is correct, go on to Problem 3.4.

If your answer is not correct, review the subject of polyphase power in this chapter. For more details, see chapter 2 in reference 1 and chapter 13 in reference 3.

## PROBLEM 3.4  Phase Sequence (Lamp Test)

For the unbalanced load circuit shown in Exhibit 3.4, calculate the voltage across the lamps for phase sequences *abc* and *acb*. Line voltage and impedances are

$$V_{line} = 100 \text{ V}$$
$$Z_{an} = Z_{cn} = 100\angle 0° \, \Omega \text{ (resistance)}$$
$$Z_{bn} = 100\angle 90° \text{ (inductance)}$$

**Exhibit 3.4**

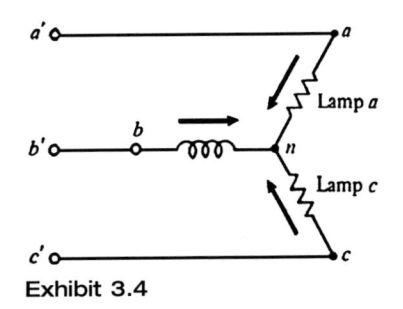

### Solution

Since the system is unbalanced, the neutral point is not at zero volts. For phase sequence *abc*, the voltage phasor diagram is shown in Exhibit 3.4a.

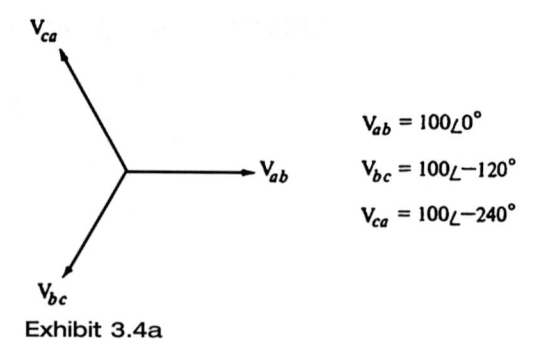

Exhibit 3.4a

$V_{ab} = 100\angle 0°$
$V_{bc} = 100\angle -120°$
$V_{ca} = 100\angle -240°$

Using Kirchhoff's laws, the following equations may be written:

$$I_{an} + I_{bn} + I_{cn} = 0$$
$$V_{ab} = V_{an} - V_{bn} = I_{an}Z_{an} - I_{bn}Z_{bn}$$
$$V_{bc} = V_{bn} - V_{cn} = I_{bn}Z_{bn} - I_{cn}Z_{cn}$$

Through a series of calculations, it can be shown that

$$V_{an} = Z_{an}\left[\frac{V_{ab}(Z_{bn}+Z_{cn})+V_{bc}Z_{bn}}{Z_{an}(Z_{bn}+Z_{cn})+Z_{cn}Z_{bn}}\right] \quad \text{(lamp } a\text{)}$$

$$V_{cn} = V_{ca} + V_{an} \quad \text{(lamp } c\text{)}$$

Substituting in the given values yields

$$V_{an} = 100\left[\frac{100(j100+100)+(100\angle -120°)(100\angle 90°)}{100(j100+100)+100(j100)}\right]$$

$$= 100\left[\frac{10{,}000+j10{,}000+8667-j5000}{10{,}000+j10{,}000+j10{,}000}\right]$$

$$= 100\left[\frac{18{,}667+j5000}{10{,}000+j20{,}000}\right] = 100\left[\frac{19{,}325\angle 15°}{22{,}361\angle 63.43°}\right] = 86.4\angle -48.43°$$

$$= 86.4\angle -48.43° \text{ volts across lamp } a$$

$$V_{cn} = 100\angle -240° + 86.4\angle -48.43° = 23.15\angle 71.55° \text{ across lamp } c$$

Thus, for sequence $abc$, lamp $a$ is brighter than lamp $c$.

For phase sequence $acb$, the voltage phasor diagram becomes as shown in Exhibit 3.4b.

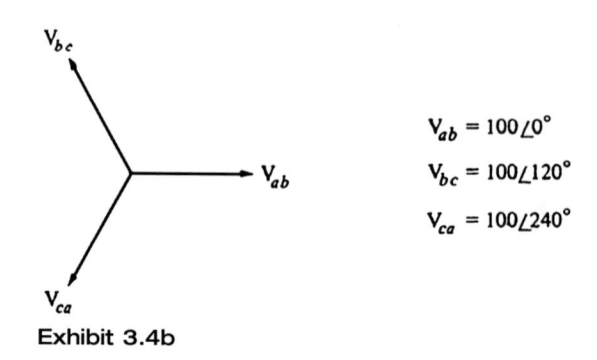

$V_{ab} = 100\angle 0°$
$V_{bc} = 100\angle 120°$
$V_{ca} = 100\angle 240°$

Exhibit 3.4b

Using the same equations to evaluate the lamp voltages,

$$V_{an} = 100 \left[ \frac{100(j100+100) + (100\angle 120°)(100\angle 90°)}{22{,}361\angle 63.43°} \right]$$

$$= 100 \left[ \frac{10{,}000 + j10{,}000 - 8667 - j5000}{22{,}361\angle 63.43°} \right]$$

$$= 100 \left[ \frac{5174.64\angle 75.07°}{22{,}361\angle 63.43°} \right] = 23.14\angle 11.64°$$

$$V_{cn} = 100\angle 240° + 23.14\angle 11.64° = 86.4\angle -108.45°$$

Thus, it is seen that by reversing the phase sequence the voltages across the two lamps exchange their magnitudes.

If your answers are correct, go on to Problem 3.5.

If your answers are not correct, review the subject of Kirchhoff's laws in chapter 1, and polyphase power in this chapter. For more details, see chapter 3 in reference 1 and chapter 8 in reference 3.

### PROBLEM 3.5  Unbalanced Load

A three-phase delta-connected load is supplied by a 480-volt power line as shown in Exhibit 3.5.

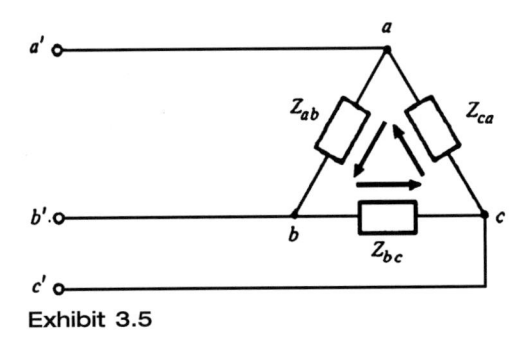

**Exhibit 3.5**

The load impedances are

$$Z_{ab} = 5 + j5$$
$$Z_{bc} = 3 + j4$$
$$Z_{ca} = 4 - j3$$

Determine the three-phase currents and the current in line $a$. Show these currents on a phasor diagram.

a. $I_{ab} = 67.88\angle -45°$ amps, $I_{bc} = \dfrac{V_{bc}}{Z_{bc}} = 96\angle -67°$ amps,

$I_{ca} = 96\angle 156.87°$ amps, $I_{a'a} = 160.99\angle -32.17°$ amps

b. $I_{ab} = 67.88\angle -45°$ amps, $I_{bc} = \dfrac{V_{bc}}{Z_{bc}} = 96\angle -173.13°$ amps,

$I_{ca} = 96\angle 156.87°$ amps, $I_{a'a} = 136.7\angle 4.3°$ amps

c. $I_{ab} = 67.88 \angle -45°$ amps, $I_{bc} = \dfrac{V_{bc}}{Z_{bc}} = 96 \angle -173.13°$ amps,

$I_{ca} = 96 \angle 156.87°$ amps, $I_{a'a} = 41.6 \angle 4.34°$ amps

d. $I_{ab} = 67.88 \angle -45°$ amps, $I_{bc} = \dfrac{V_{bc}}{Z_{bc}} = 96 \angle -173.13°$ amps,

$I_{ca} = 96 \angle 156.87°$ amps, $I_{a'a} = 160.99 \angle -32.17°$ amps

*Solution*

Assume phase sequence *abc*.

$$I_{ab} = \dfrac{V_{ab}}{Z_{ab}} = \dfrac{480 \angle 0°}{5+j5} = \dfrac{480 \angle 0°}{7.07 \angle 45°} = 67.88 \angle -45° \text{ amps}$$

$$I_{bc} = \dfrac{V_{bc}}{Z_{bc}} = \dfrac{480 \angle -120°}{3+j4} = \dfrac{480 \angle -120°}{5 \angle 53.13°} = 96 \angle -173.13° \text{ amps}$$

$$I_{ca} = \dfrac{V_{ca}}{Z_{ca}} = \dfrac{480 \angle +120°}{4-j3} = \dfrac{480 \angle 120°}{5 \angle -36.87°} = 96 \angle 156.87° \text{ amps}$$

$$I_{a'a} = I_{ab} - I_{ca} = 67.88 \angle -45° - 96 \angle 156.87° = 160.99 \angle -32.17° \text{ amps}$$

Exhibit 3.5a shows the phasor diagram.

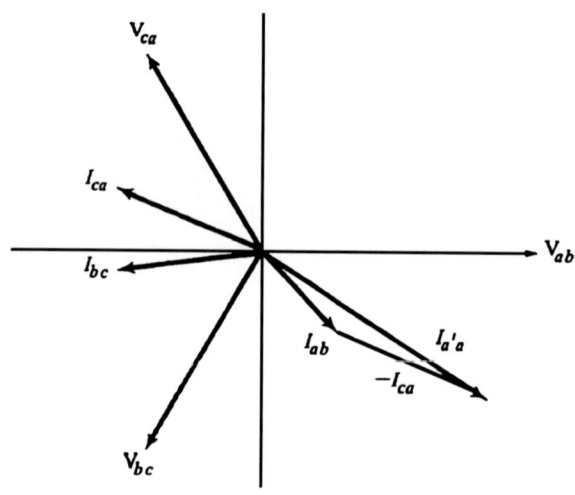

**Exhibit 3.5a**

The correct answer is d.

*Answer Rationale*

Incorrect solution a is the result of incorrectly adding the angle of 53.13 from the denominator to the angle −120 in the numerator in calculating $I_{bc}$. Incorrect solution b is the result of incorrectly using the plus sign for the angle 45 of the current $I_{ab}$ when calculating $I_{a'a}$. Incorrect solution c is the result of incorrectly adding the currents instead of subtracting in calculating $I_{a'a}$.

If your answers are correct, go on the next section.

If your answers are not correct, review the subjects of Ohm's law and complex algebra in chapter 1, and polyphase power in this chapter. For more details, see chapter 3 in reference 1 and chapter 8 in reference 3.

# POWER FACTOR CORRECTION

Most industrial loads contain many devices, such as induction motors, that have a lagging, or inductive, power factor. This power factor ranges from 0.9 to 0.7 in most cases. The effect of a low power factor is to require more current for the same power. Another undesirable effect is that it causes transmission lines to have larger regulation. To correct this situation, capacitance is added in parallel with the load as shown in Figure 3.7.

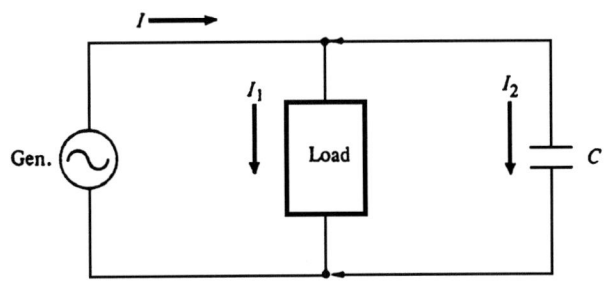

**Figure 3.7** Adding capacitance in parallel with load to correct power factor

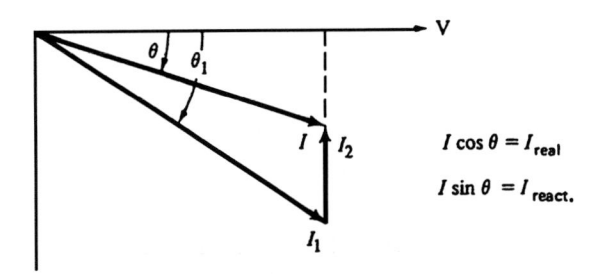

Figure 3.8 shows a phasor diagram of power factor correction in which $\theta_1$ is improved to $\theta$.

Power factor correction presents certain problems. For example, if the load changes, the resultant power factor will probably change, so it may be necessary to regulate the size of $I_2$, perhaps with automatic equipment, to keep the power factor at a steady value (not necessarily unity). This assures that the voltage is constant at the load. Sometimes each motor in a plant is equipped with its own capacitor, which is connected to the line only when the motor is connected.

Two types of devices are used in power factor correction, capacitors and synchronous motors. When a capacitor is used, it must be able to withstand the peak voltage ($\sqrt{2}$ times the RMS voltage) twice each cycle.

The following problem illustrates a power factor correction calculation.

$I \cos \theta = I_{real}$

$I \sin \theta = I_{react.}$

**Figure 3.8** Phasor diagram of power factor correction

## PROBLEM 3.6  Power Factor Correction (Capacitor)

A three-phase feeder line supplies a lighting load of 500 kW at unit power factor and a motor load of 1500 kVA at 0.7 power factor. Calculate the kVAR of shunt capacitance required to correct the power factor to 0.9.

*Solution*

Draw the uncorrected power factor phasor diagram (Exhibit 3.6).

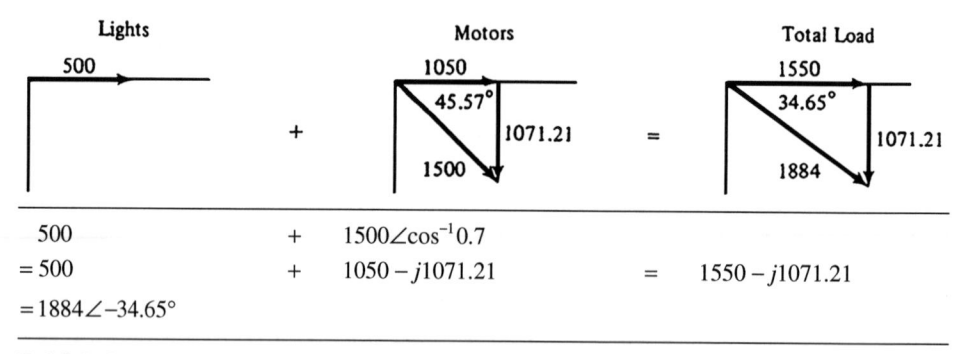

```
      Lights                    Motors                      Total Load
       500                       1050                         1550
                                 45.57°                       34.65°
                       +                    1071.21    =                1071.21
                                 1500                         1884
```

| | | | |
|---|---|---|---|
| 500 | + | $1500\angle\cos^{-1}0.7$ | |
| = 500 | + | $1050 - j1071.21$ | = $1550 - j1071.21$ |
| $= 1884\angle-34.65°$ | | | |

**Exhibit 3.6**

We would like the corrected load to be as shown in Exhibit 3.6a.

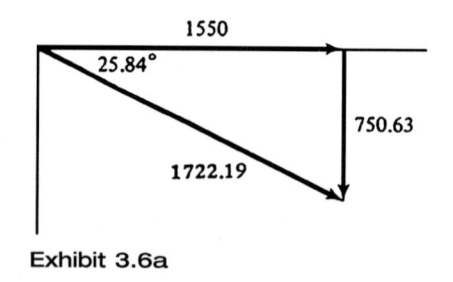

```
             1550
      25.84°
                        750.63
           1722.19
```

**Exhibit 3.6a**

The difference between corrected and uncorrected kVARs is

$$1071.21 - 750.63 = 320.6 \text{ kVAR}$$

This is the required kVAR amount of shunt capacitance.
  If your answer is correct, go on to Problem 3.7.
  If your answer is not correct, review the subjects of single-phase power, polyphase power, and power factor correction in this chapter. For more details, see chapter 3 in reference 1 and chapter 19 in reference 2.

## PROBLEM 3.7  Power Factor Correction (Synchronous Motor)

An overexcited synchronous motor is to be used to improve the power factor of a 2000-kW load having a lagging power factor of 0.83. The power factor of the motor will be adjusted to 0.8. If the resultant power factor is to be 0.94 lagging, what must be the kVA rating of the synchronous motor?

## Solution

Draw the power triangles (Exhibit 3.7).

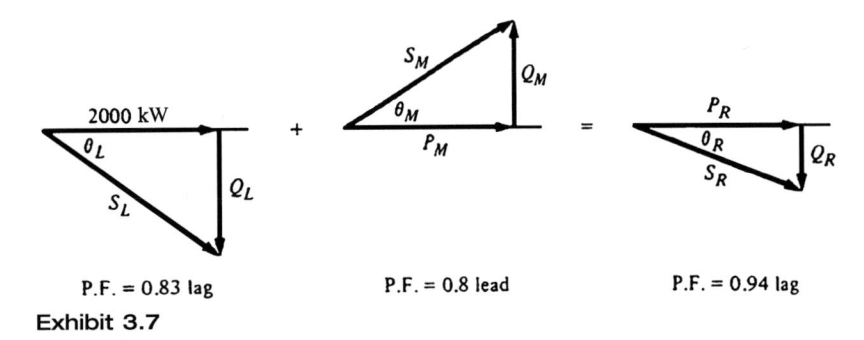

**Exhibit 3.7**

P.F. = 0.83 lag    P.F. = 0.8 lead    P.F. = 0.94 lag

The problem is to calculate $S_M$, the motor apparent power. Write two equations, one for reals and one for imaginaries:

1. $2000 + P_M = P_R$
2. $Q_L - Q_M = Q_R$

Calculate the phase angle, $S_L$ and $Q_L$:

$$\theta_L = \arccos 0.83 = 33.9°$$
$$S_L = \frac{2000}{0.83} = 2409.64 \text{ kVA}$$
$$Q_L = 2409.64 \sin 33.9° = 1344 \text{ kVAR}$$

Calculate the motor and resultant phase angles:

$$\theta_M = \arccos 0.8 = 36.87°$$
$$\theta_R = \arccos 0.94 = 19.95°$$

$$\tan \theta_R = \frac{Q_R}{P_R}, \quad Q_R = P_R \tan 19.95° = 0.36 P_R$$
$$\tan \theta_M = \frac{Q_M}{P_M}, \quad Q_M = P_M \tan 36.87° = 0.75 P_M$$

Substituting into equations (1) and (2) yields

1. $2000 + P_M = P_R$
2. $1344 - 0.75 P_M = 0.36 P_R$

Multiplying (1) by 0.75 and adding to (2) yields

$$\begin{array}{r} 1500 + 0.75 P_M = 0.75 P_R \\ 1344 - 0.75 P_M = 0.36 P_R \\ \hline 2844 \phantom{xxxxxxx} = 1.11 P_R \end{array}$$

$$P_R = \frac{2844}{1.11} = 2562 \text{ kW}$$

From (1),

$$P_M = P_R - 2000 = 2562 - 2000 = 562 \text{ kW}$$

Solving for $Q_M$ and $Q_R$ yields

$$Q_M = 0.75 P_M = 422 \text{ kVAR}$$
$$Q_M = 0.36 P_R = 922 \text{ kVAR}$$

Solving for $S_R$ and $S_M$ yields

$$S_R = \frac{P_R}{\cos\theta_R} = \frac{2562}{0.94} = 2726 \text{ kVA}$$
$$S_M = \frac{P_M}{\cos\theta_M} = \frac{562}{0.8} = 703 \text{ kVA}$$

If your answer is correct, go on to Problem 3.8.

If your answer is not correct, review the subjects of single-phase power and polyphase power in this chapter. For more details, see chapter 3 in reference 1 and chapter 17 in reference 3.

## PROBLEM 3.8  Power Factor (Adjusting Existing Load)

A small industrial company is being charged a premium rate for electricity because of poor power factor. Its electrical load of 500 kW has a lagging power factor of 0.75. However, the company is operating three 100 HP synchronous motors at unity power factor for maximum efficiency.

If the fields were to be overexcited, the motors could be made to operate at a leading power factor (at some small loss in efficiency), which would improve overall power factor. If the power factor could be improved to greater than 0.85, the premium electricity rate would be eliminated. Normally, the motors operate at 80% of capacity. However, the fields can be overexcited so that each motor is drawing rated current while still delivering 80% of rated capacity.

If the motors are operated in this manner, the new power factor is:

a. $PF_T = 0.87$  b. $PF_T = 0.91$

c. $PF_T = 0.64$  d. $PF_T = 0.49$

*Solution*

Assume the motors' efficiency is 85%. Maximum motor load is

$$3 \times \frac{100 \text{ HP}}{0.85} \times \frac{0.746 \text{ kW}}{\text{HP}} = 263.3 \text{ kVA}$$

The motors are presently drawing 263.3 × 0.8 = 210.64 kW. Draw the power vectors (Exhibit 3.8).

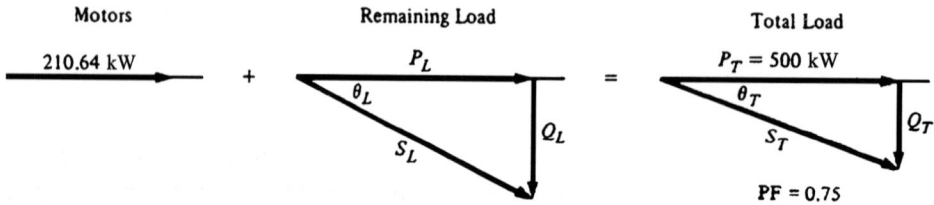

Exhibit 3.8

$$P_L = 500 - 210.64 = 289.36 \text{ kW}$$
$$\theta_T = \arccos 0.75 = 41.41°$$
$$S_T = \frac{500}{.75} = 666.67 \text{ kVA}$$
$$Q_T = 666.67 \sin 41.41° = 441 \text{ kVAR}$$
$$Q_L = Q_T = 441 \text{ kVAR}$$
$$\theta_L = \arctan \frac{441}{289.36} = 56.73°$$

Calculating the power triangle parameters of the synchronous motors when they are made to draw full load line current and output 80% of rated capacity results in Exhibit 3.8a.

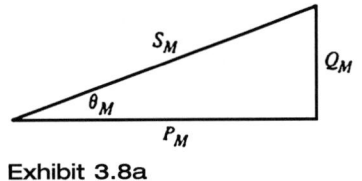

**Exhibit 3.8a**

$$S_M = 263.3 \text{ kVA}$$
$$P_M = 210.64 \text{ kW}$$
$$\theta_M = \arccos \frac{210.64}{263.3} = 36.87°$$
$$Q_M = 263.3 \sin 36.87 = 157.98 \text{ kVAR}$$
$$Q_T = Q_L - Q_M = 441 - 157.98 = 283.02$$
$$S_T = P_T + jQ_T = 500 + j283.02 = 574.54 \angle 29.51°$$
$$\text{PF}_T = \cos 29.51° = 0.87$$

This is above the 0.85 minimum, so normal rates will apply.

The correct answer is a.

*Answer Rationale*

Incorrect solution b is the result of incorrectly using the cosine of the angle 36.87 in calculating $Q_m$. Incorrect solution c is the result of incorrectly adding the values of $Q_L$ and $Q_m$ instead of subtracting them when calculating $Q_T$. Incorrect solution d is the result of incorrectly using the sine of the angle 29.51 in calculating the power factor.

If your answer is correct, go on to the next section.

If your answer is not correct, review the subjects of single-phase power and polyphase power in this chapter. For more details, see chapter 3 in reference 1 and chapters 8 and 17 in reference 3.

# TRANSMISSION LINE CALCULATIONS

A transmission line consists of the equivalent of two or more electrical conductors for the purpose of transmitting electrical energy. For single-phase transmission the line may consist of a single conductor with a ground return, or of two ordinary wires.

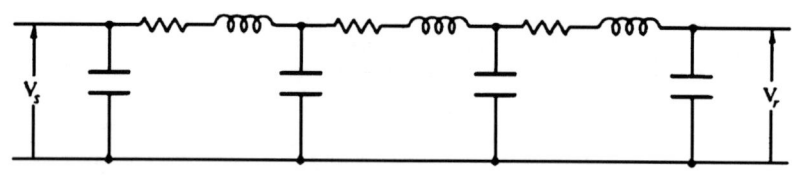

**Figure 3.9** Lumped-parameter transmission line

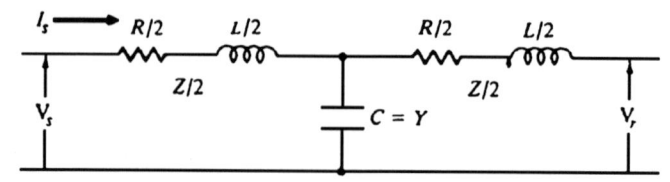

**Figure 3.10** T-line transmission line

For three-phase transmission, three wires are generally used, although in some installations a neutral wire or its equivalent is employed. Figure 3.9 shows a *lumped-parameter* transmission line, which is a good approximation to a real transmission line. $V_s$ and $I_s$ denote the sending-end voltage and current, and $V_r$ and $I_r$ represent receiver-end voltage and current.

Two popular arrangements for calculation are the T-line and the Π-line. Figure 3.10 shows the T-line representation of a transmission line.

When all the shunted capacitance $C$ of the line is concentrated in one capacitor and one-half of the total series impedance $Z$ is placed in each arm as indicated, the circuit is known as the nominal T-line. Calling $Y$ the admittance due to shunted capacitance, formulas for $V_s$ and $I_s$ can be derived to give the following results:

$$V_s = AV_r + BI_r$$
$$I_s = CV_r + DI_r$$

where

$$A = 1 + \frac{YZ}{2}$$
$$B = Z + \frac{YZ^2}{4}$$
$$C = Y$$
$$D = A$$

Figure 3.11 shows the Π-line representation of a transmission line.

In this case, one-half of the total line capacitance is concentrated at each end of the line, and all the series resistance and reactance are concentrated at the center. Formulas for $V_s$ and $I_s$ in this case are

$$V_s = AV_r + BI_r$$
$$I_s = CV_r + DI_r$$

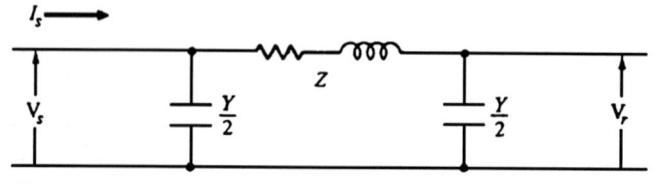

**Figure 3.11** Π-line transmission line

where

$$A = 1 + \frac{ZY}{2}$$
$$B = Z$$
$$C = Y\left(1 + \frac{ZY}{4}\right)$$
$$D = A$$

## PROBLEM 3.9 Transmission Line

A 60-Hz, three-phase line 200 miles long has a shunt capacitance to neutral per mile of $150 \times 10^{-4}$ $\mu$F, an inductive reactance of 0.75 ohm per wire per mile, and a resistance of 0.5 ohm per wire per mile. The receiver voltage is 120 kilovolts between lines. Find the sending-end voltage and current for a 0.8 power factor lagging load requiring 70 amps per line at the receiver.

*Solution*

Arbitrarily select the Π-line configuration (Exhibit 3.9).

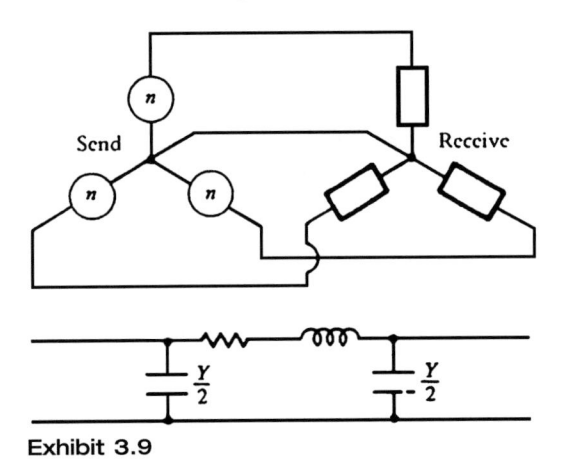

**Exhibit 3.9**

$$V_s = AV_r + BI_r \qquad A = 1 + \frac{Zy}{2}$$
$$I_s = CV_r + DI_r \qquad B = Z$$
$$C = Y\left(1 + \frac{ZY}{4}\right)$$
$$D = A$$
$$Z = (R + j\omega L)d = (R + jX_L)d$$
$$Y = j\omega Cd = j(1/X_C)d$$

Given (per phase):

$$R = 0.5 \, \Omega/\text{mile}$$
$$C = 150 \times 10^{-4} \, \mu\text{F/mile}$$
$$X_L = 0.75 \, \Omega/\text{mile}$$
$$d = 200 \text{ miles}$$
$$V_r = 120 \text{ kV line-to-line}$$
$$\text{PF} = 0.8 \text{ lag}$$
$$I_r = 70 \text{ amps/line}$$
$$f = 60 \text{ Hz}$$

$$Z = (0.5 + j0.75)200 = 0.9\angle 56.31° \times 200 = 180.28\angle 56.31°$$
$$Y = j2\pi \times 60 \times 150 \times 10^{-10} \times 200 = 1.131 \times 10^{-3}\angle 90°$$
$$ZY = 180.28\angle 56.31° \times 1.131 \times 10^{-3}\angle 90°$$
$$= 0.20389\angle 146.31° = -0.1696 + j0.1131$$

$$A = 1 - 0.0848 + j0.05655 = 0.9152 + j0.05655 = 0.9169\angle 3.536°$$
$$B = 180.28\angle 56.31°$$
$$C = (1.131 \times 10^{-3}\angle 90°)(1 - 0.0424 + j0.0283)$$
$$= (1.131 \times 10^{-3}\angle 90°)(0.9576 + j0.0283)$$
$$= 0.0010835\angle 91.69°$$

$$V_{s\phi} = 0.9169\angle 3.536° \times \frac{120 \times 10^3}{\sqrt{3}} + 180.28\angle 56.31° \times 70\angle -36.9°$$
$$= 6.353 \times 10^4 \angle 3.54° + 1.262 \times 10^4 \angle 19.44°$$
$$= 75{,}746\angle 6.16° \text{ volts line-to-neutral}$$

Converting to line voltage,

$$V_s = \sqrt{3}\, 75{,}746\angle 6.16° = 131{,}196\angle 6.16° \text{ volts line-to-line}$$

$$I_s = 0.0010835\angle 91.69° \times \frac{120 \times 10^3}{\sqrt{3}} + 0.9169\angle 3.536° \times 70\angle -36.9°$$
$$= 75.067\angle 91.69° + 64.183\angle -33.334°$$
$$= 64.994\angle 37.72° \text{ amps}$$

If the problem had been worked using the T-line configuration, the results would have been

$$V_s = 130{,}142\angle 6.32° \text{ volts} \quad \text{and} \quad I_s = 68.79\angle 38.8° \text{ amps.}$$

The difference is due to the fact that neither method is exact.

### Alternate Solution

The following solution is applicable to lines of 150 miles or less. Even though this line is 200 miles long, it may be interesting to compare the results. The circuit may be represented by Exhibit 3.9a.

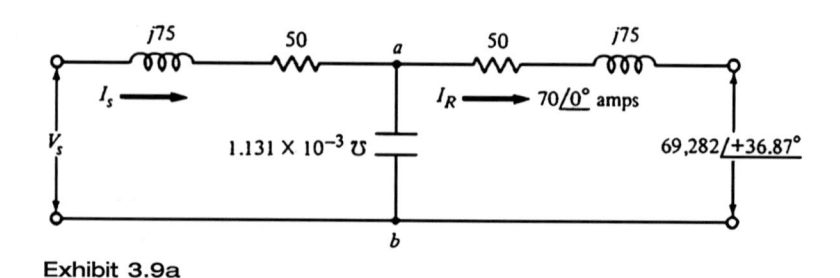

**Exhibit 3.9a**

$I_R$ is referenced to 0°.

$$V_{ab} = V_R + I_R Z = 69{,}282\angle{+36.87°} + 70(50+j75)$$
$$= 55{,}426 + j41{,}569 + 3500 + j5250$$
$$= 58{,}926 + j46{,}819 = 75{,}261.5\angle 38.47°$$

$$I_{ab} = V_{ab}Y_{ab} = (75{,}261.5\angle 38.47°)(1.131\times 10^{-3}\angle 90°)$$
$$= 85.12\angle 128.47° = -52.95 + j66.64$$

$$I_s = I_R + I_{ab} = 70 - 52.95 + j66.64 = 17.05 + j66.64$$
$$= 68.79\angle 75.65°$$

$$V_{s\phi} = V_{ab} + I_s Z = 75{,}261.5\angle 38.47 + (68.79\angle 75.65)(50+j75)$$
$$= 58{,}924.8 + j46{,}820.5 - 4145.8 - j4610.9$$
$$= 54{,}779 + j51{,}431.4 = 75{,}139.4\angle 43.19°$$

$$V_s = \sqrt{3}V_{s\phi} = 130{,}145 \text{ volts}$$

This method yields nearly the same results as the T-line configuration.

If your answers are correct, go on to Problem 3.10.

If your answers are not correct, review the subject of transmission line calculations in this chapter. For more details, see chapter 3 in reference 1 and chapter 25 in reference 3.

## PROBLEM 3.10 Transmission Line Regulation

A balanced three-phase load is supplied by a three-wire transmission line having a series impedance of 3 + j4 ohms in each line. The line voltage is 10 kV at the load. Power dissipated by the load is 3000 kW at 75% lagging power factor. The voltage regulation of the line is:

    a. 34.1%
    b. 92.8%
    c. 21.6%
    d. 19.65%

*Solution*

Regulation is defined as the difference between the full-load and no-load voltage at the load terminals and is expressed as a percentage of the full-load voltage as follows:

$$\%\text{ Regulation} = \frac{|V_{nl}| - |V_{fl}|}{|V_{fl}|} \times 100 = \frac{|V_s| - |V_r|}{|V_r|} \times 100$$

This formula is commonly used in calculating transformer regulation, as will be discussed in chapter 4. In this case, no-load voltage is the same as the sending-end voltage, and full-load voltage is the sending-end voltage minus the drop in the line. The circuit may be represented by Exhibit 3.10.

$$P = \sqrt{3}V_l I_l \cos\theta, \quad V_r = 10 \text{ KV}, \quad P = 3000 \text{ kW}, \quad \cos\theta = 0.75 \text{ lag}$$

$$I_l = \frac{P}{\sqrt{3}V_l \cos\theta} = \frac{3000\times 10^3}{\sqrt{3}\times 10^4 \times 0.75} = 231\angle -41.4° \text{ amps}$$

$$V_{s\phi} = \frac{V_r}{\sqrt{3}} + I_r Z_{\text{line}} = \frac{10{,}000}{\sqrt{3}} + 231(0.75 - j0.66)(3+j4)$$
$$= 5773.5 + 1154.7\angle 11.73° = 6908\angle 1.95° \text{ volts line-to-neutral}$$

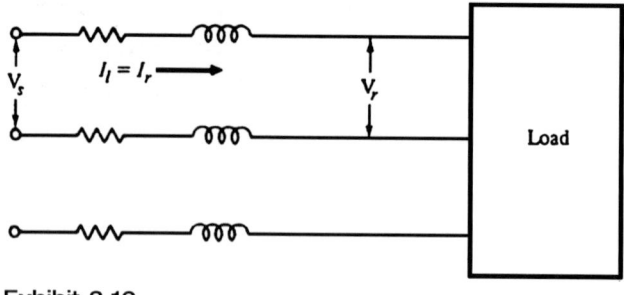

Converting to line voltage,

$$V_s = \sqrt{3}V_{s\phi} = \sqrt{3} \times 6908 \angle 1.95° = 11{,}965 \angle 1.95° \text{ volts line-to-line}$$

$$\% \text{ Regulation} = \frac{11{,}965 - 10{,}000}{10{,}000} \times 100 = \frac{1965}{10{,}000} \times 10 = 19.65\%$$

This is an example of very poor regulation.

The correct answer is d.

*Answer Rationale*

Incorrect solution a is the result of not having $\sqrt{3}$ in the denominator in calculating $I_l$. Incorrect solution b is the result of not having $\sqrt{3}$ in the denominator in calculating $V_{s\phi}$. Incorrect solution c is the result of incorrectly using the plus sign in the complex number (0.75 − j0.66) instead of the minus sign when calculating $V_{s\phi}$.

If your answer is correct, go on to the next section.

If your answer is not correct, review the subject of voltage regulation and Problem 4.13 in chapter 4. For more details, see chapter 3 in reference 1 and chapter 25 in reference 3.

# WATTMETER MEASUREMENTS

A wattmeter gives a reading proportional to the product of the current through its current coil, the voltage across its potential coil, and the cosine of the angle between the voltage and current. Since the total power in a three-phase circuit is the sum of the powers of the separate phases, the total power could be measured by placing a wattmeter in each phase. Since this method is undesirable unless the individual phase powers are required, another method making use of only two wattmeters is generally employed in making three-phase power measurements, as shown in Figure 3.12.

The power measured by each wattmeter is given by the following formulas:

$$W_a = V_{ac}I_{an} \cos(\theta - 30°)$$
$$W_b = V_{bc}I_{bn} \cos(\theta + 30°)$$
$$W_a + W_b = VI \cos(\theta - 30°) + VI \cos(\theta + 30°) = \sqrt{3}V_l I_l \cos\theta$$

Hence, $W_a + W_b$ correctly measures the power in a balanced three-phase system of any power factor. Furthermore, the algebraic sum of the readings of two wattmeters will give the correct value for power under any conditions of unbalance, waveform, or power factor.

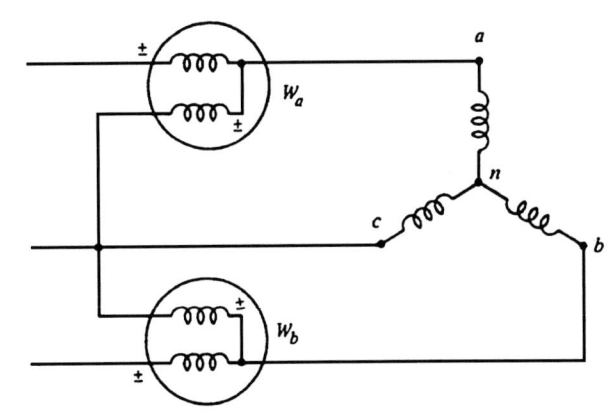

**Figure 3.12** Using two wattmeters for three-phase power measurements

The direction of rotation of polyphase induction motors is dependent upon the phase sequence of the applied voltages. Similarly, two wattmeters in the two-wattmeter method of measuring three-phase power interchange their readings when subjected to a reversal of phase sequence, even though the system is balanced. In general, $n - 1$ wattmeters can be used to measure $n$-phase power. Note that the return lines of all voltmeter windings must be connected to the same point.

### PROBLEM 3.11 Wattmeter

Two single-phase wattmeters are connected in a balanced three-phase circuit, as shown in Exhibit 3.11. The three load impedances are $Z = 4 + j3$, and the line voltage is 440 volts. Calculate the wattmeter readings for a phase sequence of *abc*.

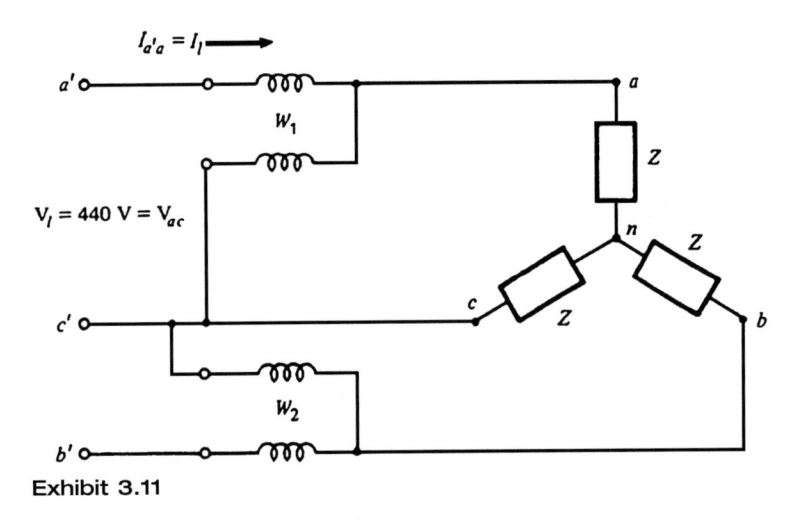

**Exhibit 3.11**

**Exhibit 3.11a**

*Solution*

In Exhibit 3.11a, the following relationships apply:

$$V_{an} = \frac{440}{\sqrt{3}} = 254 \text{ volts} = V_\phi$$

$$I_l = I_\phi = \frac{V_\phi}{Z} = \frac{254}{5\angle 36.87°} = 50.81 \text{ amps}$$

$$P_\phi = I^2 R = 50.81^2 \times 4 = 10{,}325 \text{ watts}$$

$$P_{tot} = 3P_\phi = 3 \times 10{,}326.6 = 30{,}976 \text{ watts}$$

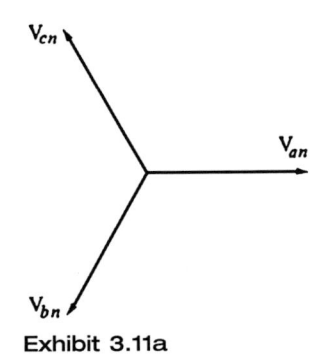

or

$$P_{tot} = \sqrt{3}V_l I_l \cos\theta = \sqrt{3} \times 440 \times 50.81 \times \cos 36.87° = 30{,}976 \text{ watts}$$

This represents the algebraic sum of the two wattmeter readings. Now, calculating the wattmeter readings:

$$W_1 = W_{ac-a'a} = V_{ac}I_{a'a}\cos\theta\Big|_{I_{a'a}}^{V_{ac}}$$

The notation $\cos\theta\Big|_{I_{a'a}}^{V_{ac}}$ represents the angle between line voltage $V_{ac}$ and line current $I_{a'a}$.

$$V_{ac} = V_{an} + V_{nc} = 254\angle 0° + 254\angle -60° = 440\angle -30° \text{ volts}$$
$$I_{a'a} = 50.81\angle -36.87° \text{ amps}$$
$$W_1 = 440 \times 50.81 \cos 6.87° = 22{,}194.5 \text{ watts}$$
$$W_2 = W_{bc-b'b} = V_{bc}I_{b'b}\cos\theta\Big|_{I_{b'b}}^{V_{bc}}$$
$$V_{bc} = V_{bn} + V_{nc} = 254\angle -120° + 254\angle -60° = 440\angle -90° \text{ volts}$$
$$I_{b'b} = \frac{V_{bn}}{Z_{bn}} = \frac{254\angle -120°}{5\angle 36.87°} = 50.81\angle -156.87°$$
$$W_2 = 440 \times 50.81\cos 66.87° = 8781.5 \text{ watts}$$

Check:

$$P_{tot} = W_1 + W_2 = 30{,}976 \text{ watts}$$

This agrees with the initial calculation.

An alternative solution is to use the equation given in the preceding text:

$$W_1 = V_{ac}I_{an}\cos(\theta - 30°)$$
$$= 440 \times 50.81\cos(36.87 - 30°) = 22{,}194.5 \text{ watts}$$
$$W_2 = V_{bc}I_{bn}\cos(36.87 + 30°) = 8781.5 \text{ watts}$$

These are identical to the previous calculations; the only difference is the manner in which the cosines of the angles are obtained.

If your answers are correct, go on to Problem 3.12.

If your answers are not correct, review the subject of wattmeter measurements in this chapter. For more details, see chapter 19 in reference 2 and chapter 8 in reference 3.

### PROBLEM 3.12 Wattmeter (Unbalanced Load)

Determine the reading of the wattmeter shown connected to the unbalanced three-phase load in Exhibit 3.12. Line voltage is 208 $V_{rms}$, and the phase sequence is *c-b-a*.

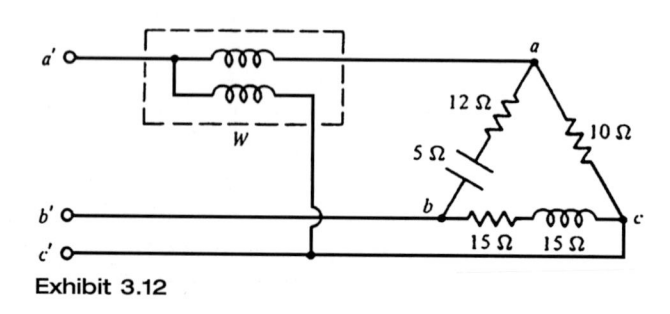

**Exhibit 3.12**

Exhibit 3.12a

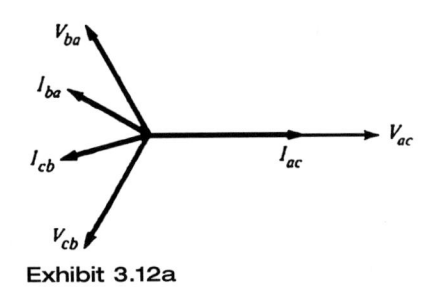

*Solution*

Exhibit 3.12a shows the phasor diagram of the three line voltages referenced to $V_{ac}$.

$$P = V_{ac}I_{a'a}\cos\theta\Big|_{I_{a'a}}^{V_{ac}}$$

Evaluating all the currents for future use in checking the solution:

$$I_{ab} = \frac{V_{ab}}{Z_{ab}} = \frac{208\angle -60°}{13\angle -22.62°} = 16\angle -37.38° = 12.71 - j9.71 \text{ amps}$$

$$I_{bc} = \frac{V_{bc}}{Z_{bc}} = \frac{208\angle +60°}{21.21\angle +45°} = 9.81\angle 15° = 9.48 + j2.54 \text{ amps}$$

$$I_{ca} = \frac{V_{ca}}{Z_{ca}} = \frac{208\angle -180°}{10} = 20.8\angle -180° = -20.8 \text{ amps}$$

$$I_{a'a} = I_{ab} + I_{ac} = 12.71 - j9.71 + 20.8 = 33.51 - j9.71 = 34.89\angle -16.16°$$
$$I_{b'b} = I_{ba} + I_{bc} = 12.71 + j9.71 + 9.48 + j2.54 = -3.23 + j12.25 = 12.67\angle 104.77°$$
$$I_{c'c} = I_{cb} + I_{ca} = -9.48 - j2.54 - 20.8 = -30.28 - j2.54 = 30.39\angle -175.21°$$

Calculating the individual phase power dissipations and summing them yields

$$P_{ab} = V_{ab}I_{ab}\cos\theta = (208\angle -60°)(16\angle -37.38°)\cos 22.62 = 3072 \text{ watts}$$
$$P_{bc} = V_{bc}I_{bc}\cos\theta = (208\angle +60°)(9.81\angle 15°)\cos 45° = 1443 \text{ watts}$$
$$P_{ca} = V_{ca}I_{ca}\cos\theta = (208\angle -180°)(20.8\angle -180°)\cos 0° = 4326 \text{ watts}$$
$$P_T = 8844 \text{ watts}$$

Now, returning to the calculation of the wattmeter reading,

$$P = V_{ac}I_{a'a}\cos\theta\Big|_{I_{a'a}}^{V_{ac}} = (208°\angle 0°)(34.89\angle 16.16°)\cos 16.16°$$
$$= 6970 \text{ watts}$$

Check:

Calculate the reading of a second wattmeter connected to phase *b* with its voltage winding return connected to the same line as first wattmeter, namely line phase *c*.

$$P = V_{bc}I_{b'b}\cos\theta\Big|_{I_{b'b}}^{V_{bc}} = (208\angle +60°)(12.67\angle +104.71°)\cos 44.71°$$
$$= 1873 \text{ watts}$$

Adding the two wattmeter readings yields the same result as the three-phase calculation of 8844 watts.

If your answer is correct, go on to the next section.

If your answer is not correct, review the subjects of single-phase power and polyphase power in this chapter. For more details, see chapter 19 in reference 2 and chapter 8 in reference 3.

# SHORT CIRCUIT CALCULATIONS

Power systems are susceptible to three kinds of short circuits:

- *Three-phase*—all three lines of a three-phase system electrically connected
- *Line-to-line*—two lines electrically connected
- *Line-to-ground*—a single wire electrically connected to ground

Short circuits are called *faults*. By use of a relay-operated circuit breaker, a distribution system may be protected from a faulty section. Circuit breakers are designed to trip under the lowest current fault condition, yet handle worst-case (greatest) fault currents.

A distribution network consists of many lines which may be connected by transformers and which generally operate at different voltages. To establish a simple network for purposes of calculation, the impedances of all lines and transformers are expressed in ohms referred to a common voltage base, or in percentage (or per unit) referred to a common kVA base. The latter method is preferred by most power engineers.

*Percentage reactance* is defined as the percentage of rated voltage that is consumed in a reactance drop when rated current flows. Expressed algebraically,

$$\% \text{ reactance} = \frac{I_{\text{rated}} \times \text{ohms}}{V_{\text{rated}}} \times 100$$

*Percent resistance* is similarly defined. Percentage values are manipulated like ohmic values.

By way of example, consider the per-phase equivalent circuit of a three-phase system in Figure 3.13.

Figure 3.14 shows the corresponding one-line diagram in which a fault is indicated on the secondary.

Referring the secondary to the primary, the equivalent current in Figure 3.15 results, assuming a transformer impedance per phase referred to the primary of $1 + j2$ ohms.

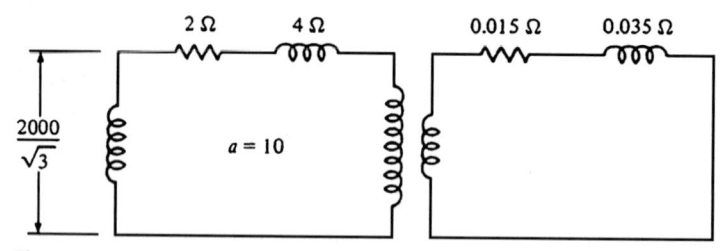

**Figure 3.13** Per-phase equivalent circuit of a three-phase system

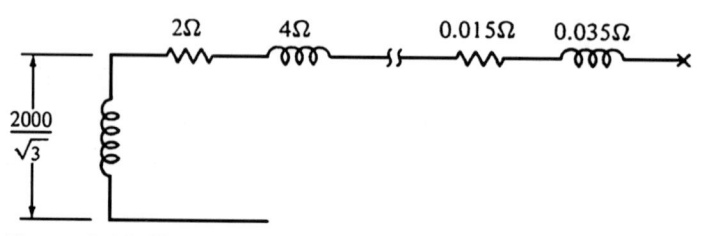

**Figure 3.14** The one-line diagram corresponding to Figure 3.13

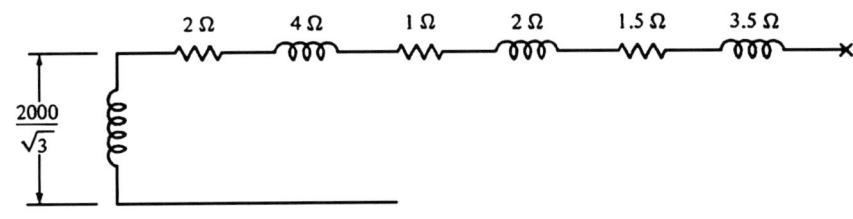

**Figure 3.15** Equivalent current

Converting from ohmic values to percentage values (assume a 10,000 kVA primary base and a 100 kVA secondary base):

$$\text{base current } I = \frac{10,000,000}{\sqrt{3} \times 2000} = 2887 \text{ amps}$$

$$\% IX \text{ drop due to base current} = 100 \times \frac{2887 \times 4}{2000/\sqrt{3}} = 1000$$

$$\% IR \text{ drop due to base current} = 100 \times \frac{2887 \times 2}{2000/\sqrt{3}} = 500$$

Transformer impedance on 10,000 kVA base is

$$\% IR \text{ drop} = 100 \times \frac{2887 \times 1}{2000/\sqrt{3}} = 250$$

$$\% IR \text{ drop} = 100 \times \frac{2887 \times 2}{2000/\sqrt{3}} = 500$$

Secondary line impedance based on 100 kVA is

$$\text{secondary voltage} = \frac{2000/\sqrt{3}}{10} = 200/\sqrt{3} \text{ volts}$$

$$\text{base current } I = \frac{100,000}{\sqrt{3} \times 200} = 288.7 \text{ amps}$$

$$\% IX \text{ drop} = 100 \times \frac{288.7 \times 0.035}{200/\sqrt{3}} = 8.75$$

$$\% IX \text{ drop} = 100 \times \frac{288.7 \times 0.015}{200/\sqrt{3}} = 3.75$$

Figure 3.16 shows a one-line diagram with parameters expressed on a percentage basis.

Now choose a common kVA base to which all constants may be referred. For discussion, arbitrarily select 1000 kVA. Since impedance varies directly with the kVA base, Figure 3.17 results.

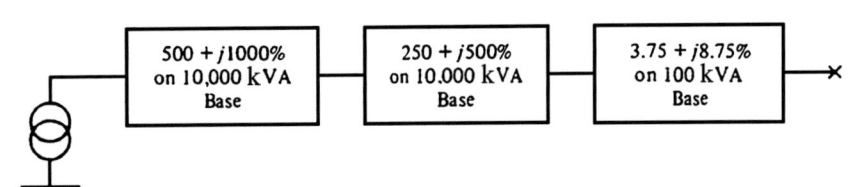

**Figure 3.16** Parameters on a percentage basis

[Figure 3.17: Generator connected through three series impedance blocks: 50 + j100%, 25 + j50%, 37.5 + j87.5, ending at fault point ×]

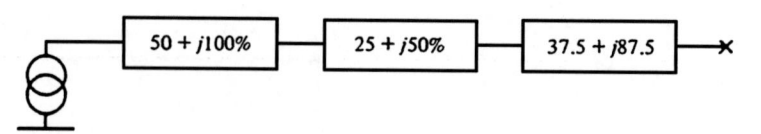

**Figure 3.17** Constants referred to a common kVA base

The combined impedance to the fault is

$$Z = 112.5 + j237.5\% = 263\angle 65°\%$$

Thus, 263% of rated voltage is required to cause 1000 kVA to be delivered by the generator. Since only rated voltage, or 100% voltage, is available, the total short circuit kVA must be (100/263) × 1000 = 380.5 kVA. Fault current at the actual voltage of the faulty line is calculated as follows:

$$I_{fault} = \frac{380.5 \times 1000}{200\sqrt{3}} = 1098.4 \text{ amps}$$

# CHAPTER 4

# Machinery

**OUTLINE**

dc MACHINES  99

ac MACHINES  108
Synchronous Generators ■ Synchronous Motors ■ Asynchronous Machines ■ Polyphase Induction Motor ■ Induction Generator ■ Single-Phase Induction Motors ■ Two-Winding Transformers ■ Autotransformer

MAGNETIC DEVICES  128

Electrical machinery may be divided into two categories: dc and ac. dc machinery consists, broadly, of motors and generators, whereas ac machinery consists of motors, generators, and transformers.

Figure 4.1 categorizes basic electric motors by type. Detailed characteristics are summarized in handbooks available from many motor manufacturers. Transformers also may be classified by phase (single phase or polyphase) and type (two-winding or autotransformer). The problems in this chapter illustrate the features and parameters of rotating machinery and transformers.

## dc MACHINES

dc motors and generators consist of an **armature,** a *field*, a **commutator, and poles.** The typical dc generator uses stationary electromagnets for producing the fields. Conductors for the generation of emf are carried on a rotating element called the armature.

The instantaneous emf induced in a conductor of length $l$ moving with a velocity $v$ within and perpendicular to a magnetic field of density $B$ is

$$e = Blv$$

Also, the force on a conductor of length $l$ when carrying a current $i$ within and perpendicular to a magnetic field of density $B$ is

| | | |
|---|---|---|
| $f = Bil$ | $\text{newtons} = \dfrac{\text{webers}}{\text{m}^2} \times \text{amps} \times \text{m}$ | (SI) |
| $f = 0.1\, Bil$ | $\text{dynes} = 10\,\text{gauss} \times \text{amps} \times \text{cm}$ | (CGS) |
| $f = 0.885 \times 10^{-7}\, Bil$ | $\text{pounds} = \dfrac{\text{lines}}{\text{m}^2} \times \text{amps} \times \dfrac{\text{in}}{0.885 \times 10^{-7}}$ | (English) |

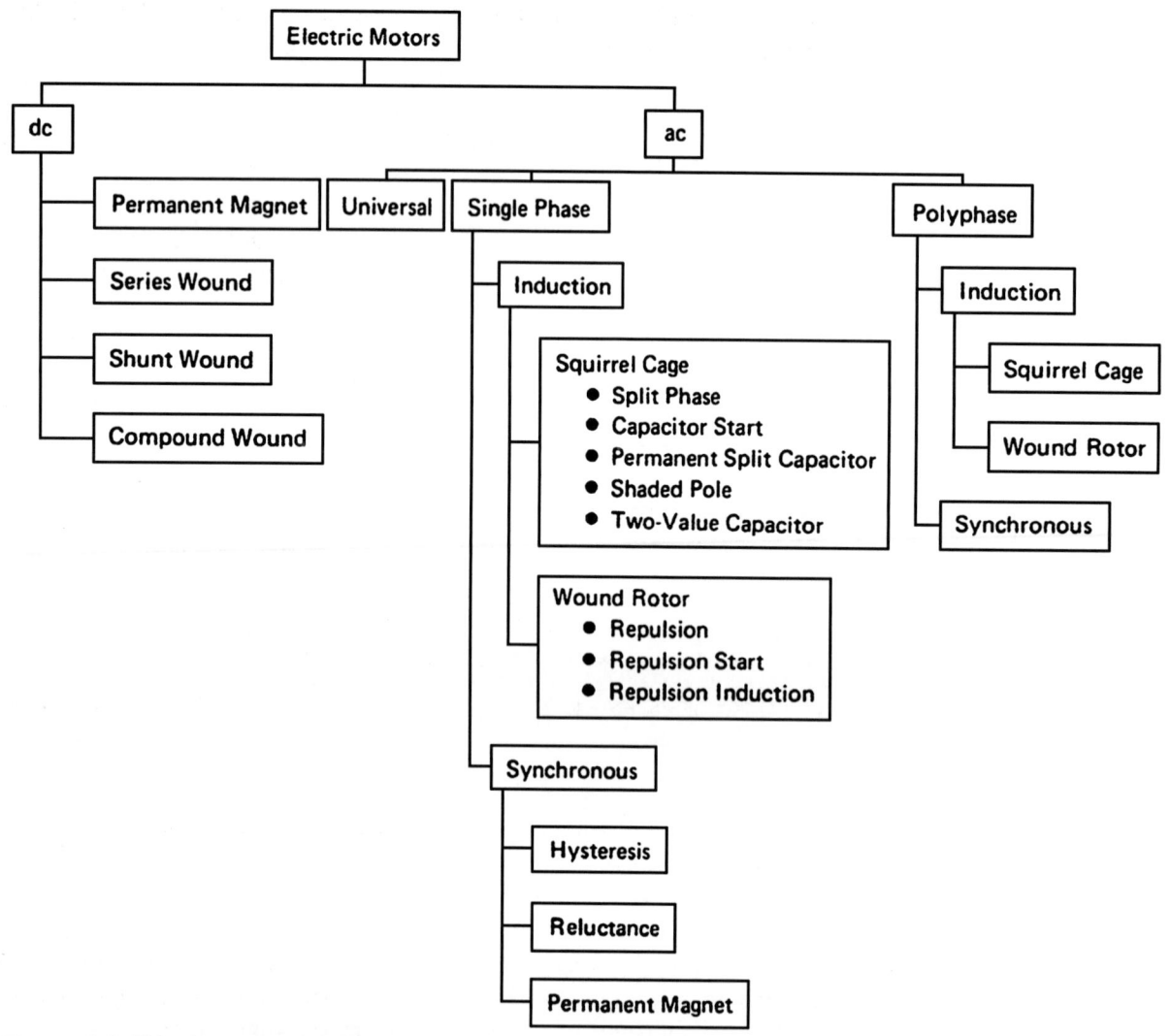

**Figure 4.1** Electric motor classification

Dividing $e = Blv$ by $f = Bil$ and rearranging slightly yields

$$ei = fv$$

or

electrical power = mechanical power

Because of the reversible property of the energy conversion, it is possible for the same machine to act as either a generator or a motor.

A conductor of $N$ turns moving across a magnetic field has an emf generated in it equal to:

$$E = BlvN \times 10^{-8} \text{ volts}$$

where
$B$ = in lines/in.$^2$
$l$ = in in.
$v$ = in in./sec
$N$ = number of turns

If the flux linking a **coil** of wire changes, an emf will be induced in that coil in accordance with the equation:

$$E = -N \frac{d\phi}{dt} \times 10^{-8}$$

The coil and field poles representing motor action are identical to those representing generator action. This fact gives rise to the term **back** or **counter emf** for the generated emf when a machine is acting as a motor. A change from motor to generator action for the same direction of rotation and polarity of the field is due only to a reversal of armature current. In the generator, the current flows in the direction of the induced emf, while in the motor it is opposite.

When considering the difference between electrical degrees and mechanical degrees, the number of poles must be taken into consideration. In ac machinery this concept is important because frequency enters in. A conductor must pass two poles to produce one cycle. Hence, the number of electrical degrees per revolution of the armature shaft is:

$$\frac{\text{electrical degree}}{\text{revolution}} = \frac{\text{no. of poles}}{2} \times 360$$

The terminal voltage of a generator tends to drop when the machine is loaded. The term **voltage regulation** is applied specifically to this situation and is defined by the following equation:

$$\% \text{ voltage regulation} = \frac{\text{(no-load voltage)} - \text{(full-load voltage)}}{\text{full-load voltage}} \times 100$$

The terminal voltage of a generator under load differs from the internal generated voltage by the amount of potential drop in the armature series circuit. The resistance drop is due to the total resistance encountered by the load current as it flows between the − and + terminals of the machine. This total resistance includes that of the brushes, brush contact, armature, series field, interpole field, and compensating windings. Thus, the terminal potential $V_t$ is always equal to the generated voltage $E$ minus the potential drop $I_a R_a$ in the armature series circuit as expressed by the equation:

$$V_t = E - I_a R_a$$

where the machine is operating as a **generator**. For **motor action** the sign of the $IR$ drop changes (since the direction of $I$ changes) and $E$ becomes the back emf.

A motor develops torque because of the emf exerted on a conductor when it is carrying current in a magnetic field. The torque can be expressed by the equations:

$$T = \frac{7.045}{S} E I_a = \frac{33{,}000 \text{ HP}}{2 \pi S}$$

where
  HP = gross mechanical hp of motor = $E I_a / 746$
  $S$ = speed of motor in rpm
  $T$ = in pound-feet

Speed regulation of a motor when operating at constant terminal potential is defined by the expression:

$$100 \times \frac{\text{(no-load speed)} - \text{(full-load speed)}}{\text{full-load speed}} \%$$

Efficiency is defined as output divided by input. For a generator,

$$\eta_G = \frac{\text{output}}{\text{output} + \text{losses}} = 1 - \frac{\text{losses}}{\text{output} + \text{losses}}$$

For a motor,

$$\eta_M = \frac{\text{input} - \text{losses}}{\text{input}} = 1 - \frac{\text{losses}}{\text{input}}$$

Losses are grouped as follows:

I. Electrical losses

   A. Ohmic losses

      1. $I^2 R_f$ loss in shunt field winding
      2. $I^2 R_a$ loss in armature winding
      3. $I^2 R_s$ loss in series field winding
      4. $I^2 R_{\text{rheo}}$ loss in rheostat

   B. Brush contact loss ($V_{\text{brush}} I_a$)

II. Rotational losses

   A. Mechanical

      1. Friction and windinge
      2. Brush friction

   B. Core loss (iron loss, hysteresis, and eddy current)
   C. Ventilating loss
   D. Stray-load loss

The all-day efficiency of a machine is defined as

$$\frac{\text{output}}{\text{output} + \text{constant losses} + \text{variable losses}}$$

Efficiency is high under full load and low at light loads, and this loading usually varies during a 24-hour period.

## PROBLEM 4.1 Series-Wound dc Motor

A series-wound dc motor is supplied from a source of $E$ volts having an internal resistance of $R_g$. The motor generates a back emf $E_m$ (proportional to its speed) in series with its armature and field windings represented by resistor $R_m$ as shown in the circuit in Exhibit 4.1a.

Exhibit 4.1a

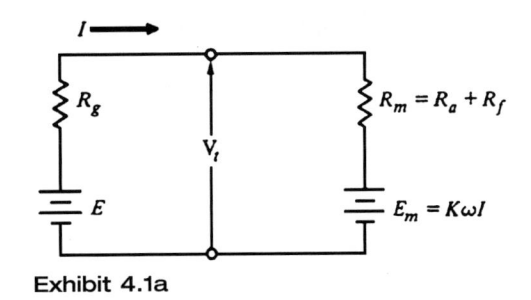

Determine the maximum average power that can be converted to mechanical form and the motor speed at which this occurs.

### Solution

Voltage at the motor terminals is given by the equation:

$$V_t = IR_m + E_m = E - IR_g$$

According to **Jacobi's law**, maximum motor power and torque occur when $E_m = IR_m$, or when the back emf equals one-half the motor terminal voltage. Mechanical power output is equal to the product of armature current and back emf, or

$$P_{mech} = IE_m$$

Therefore:

$$P_{mech\ max} = IV_t/2$$

In order to express power in terms of source voltage $E$, redraw the circuit so that $V_t \cong E$ (Exhibit 4.1b).

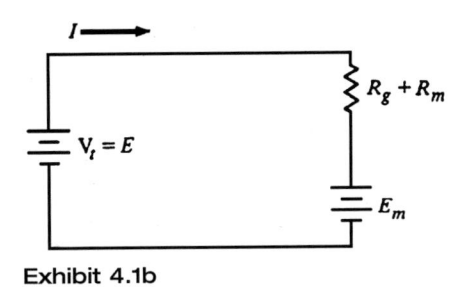

Exhibit 4.1b

Now,

$$P_{mech\ max} = IE/2$$

$$E_m = I(R_g + R_m) = \frac{E}{2}, \qquad I = \frac{E}{2(R_g + R_m)}$$

Finally,

$$P_{mech\ max} = \frac{E^2}{4(R_g + R_m)}$$

Determining motor speed:

$$E_m = \frac{E}{2} = I(R_g + R_m) = K\omega I$$

$$\omega = \frac{(R_g + R_m)}{K}$$

If your answers are correct, go on to Problem 4.2.

If your answers are not correct, review above solution. For more details, see chapter 2 in reference 1 and chapter 5 in reference 3.

## PROBLEM 4.2  Shunt-Wound dc Motor

For the shunt-wound dc motor having the circuit parameters shown in Exhibit 4.2, determine the armature current and field current when its speed drops to 1100 rpm under load. No-load speed is nearly 1200 rpm at an armature current of 1 amp.

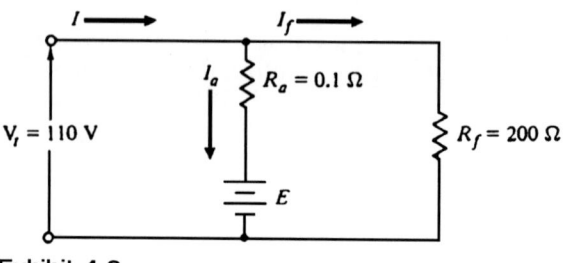

**Exhibit 4.2**

*Solution*

$$V_t = I_a R_a + E$$
$$E = V_t - I_a R_a = 110 - 0.1 I_a$$

Since $E$ is proportional to speed,

$$\frac{E_l}{E_{nl}} = \frac{1100}{1200} = \frac{110 - 0.1 I_{al}}{110 - 0.1 I_{anl}}$$

$$I_{al} = -\frac{1100}{1200}\left[\frac{110 - 0.1 I_{anl}}{0.1}\right] + \frac{110}{0.1}, \quad I_{anl} = 1 \text{ amp}$$

Therefore

$$I_{al} = 92.58 \text{ amps}$$

$$I_f = \frac{V_t}{R_f} = \frac{110}{200} = 0.55 \text{ amp}$$

If your answers are correct, go on to Problem 4.3.

If your answers are not correct, review the subject of dc machines in this chapter. For more details, see chapter 2 in reference 1 and chapter 5 in reference 3.

## PROBLEM 4.3 Compound (Series-Shunt)-Wound dc Motor

For the series-shunt-wound dc motor having the circuit parameters shown in the Exhibit 4.3, find the line current and gross mechanical hp output when the armature current $I_a = 25$ amps.

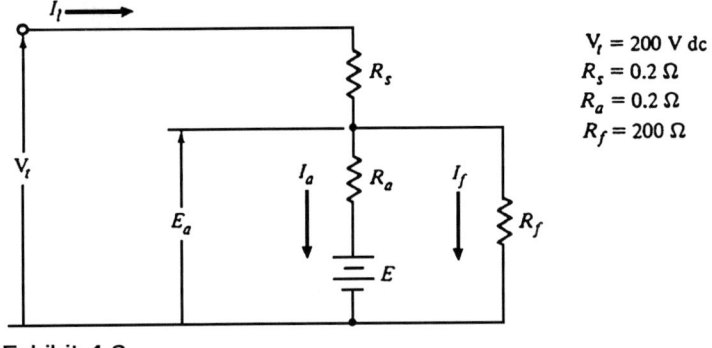

**Exhibit 4.3**

$V_t = 200$ V dc
$R_s = 0.2\ \Omega$
$R_a = 0.2\ \Omega$
$R_f = 200\ \Omega$

a. $I_l = 25.974$ amps, and HP = 6.361 hp
b. $I_l = 26$ amps, and HP = 6.361 hp
c. $I_l = 25.974$ amps, and HP = 6.53 hp
d. $I_l = 25.974$ amps, and HP = 6.61 hp

*Solution*

$$I_f = \frac{E_a}{R_f} = \frac{200 - I_l R_s}{R_f} = \frac{200 - 0.2 I_l}{200} = 1 - 0.001 I_l$$

$$I_l = I_a + I_f = 25 + I_f = 25 + 1 - 0.001 I_l = 26 - 0.001 I_l$$

$$I_l = \frac{26}{1.001} = 25.974 \text{ amps}$$

Gross mechanical hp = $I_a E / 746$

$$E_a = I_a R_a + E = V_t - I_l R_s = 200 - 25.97 \times 0.2 = 194.81 \text{ volts}$$
$$E = 194.81 - I_a R_a = 194.81 - 25 \times 0.2 = 189.81 \text{ volts}$$

$$\text{HP} = \frac{25 \times 189.81}{746} = 6.361 \text{ hp}$$

The correct answer is a.

*Answer Rationale*

Incorrect solution b is the result of incorrectly dividing by exactly 1 in the last step in calculating $I_l$. Incorrect solution c is the result of incorrectly using the value of $E_a$ instead of $E$ in the last step of calculating the HP. Incorrect solution d is the result of incorrectly using the value of $I_l$ instead of $I_a$ in the last step of calculating the HP.

If your answers are correct, go on to Problem 4.4.

If your answers are not correct, review the subject of dc machines in this chapter. For more details, see chapter 2 in reference 1 and chapter 5 in reference 3.

## PROBLEM 4.4 Shunt Motor Torque

A dc shunt wound motor, shown in Exhibit 4.4, has the following characteristics:

$$R_a = 0.5 \; \Omega \quad \text{no-load speed} = 1800 \text{ rpm}$$
$$R_f = 200 \; \Omega \quad \text{HP} = 7.5$$

The no-load current is 3.5 amps when the applied voltage is 230 volts. If the motor is loaded such that the speed drops to 1700 rpm, the line current and developed torque are:

a. $I_l = 3.4998$ amps, and $T_d = 2.23$ lb-ft
b. $I_l = 3.5$ amps, and $T_d = 2.1$ lb-ft
c. $I_l = 28.925$ amps, and $T_d = 25.89$ lb-ft
d. $I_l = 28.925$ amps, and $T_d = 24.86$ lb-ft

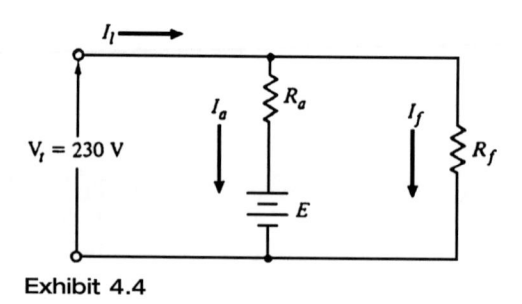

Exhibit 4.4

### Solution

$$I_f = \frac{V_{lt}}{R_f} = \frac{230}{200} = 1.15 \text{ amps}$$

$$I_a = I_l - I_f = 3.5 - 1.15 = 2.35 \text{ amps} @ 1800 \text{ rpm}$$

Speed in rpm:

$$S = \frac{V_a - I_a R_a}{K} = \frac{E}{K}$$

where
$K$ = a constant
$V_a$ = the voltage across the armature

Solving for $K$ yields

$$K = \frac{230 - 2.35 \times 0.5}{1800} = 0.127125$$

Solving the preceding equation for $I_a$ yields

$$I_a = \frac{V_a - KS}{R_a}$$

At 1700 rpm,

$$I_a = \frac{230 - 0.127125 \times 1700}{0.5} = 27.775 \text{ amps}$$

$$I_l = I_a + I_f = 28.925 \text{ amps}$$

Developed torque is defined by the formula

$$T_d = 7.04\, KI_a = 7.04 \times 0.127125 \times 27.78$$
$$= 24.86 \text{ lb-ft}$$

Solving for torque by an alternative method yields

$$HP = \frac{ST}{5252.12}$$

Rewriting results in the following:

$$T = \frac{5252.12\, HP}{S}$$

where

$$HP = \frac{I_a E}{746}$$
$$E = 230 - 27.78 \times 0.5 = 216.11$$

$$\therefore T = \frac{5252.12}{1700}\left[\frac{27.78 \times 216.11}{746}\right] = 24.86 \text{ lb-ft}$$

The correct answer is d.

*Answer Rationale*

Incorrect solution a is the result of incorrectly dividing by 1700 in calculating the constant $K$. Incorrect solution b is the result of incorrectly using the value of 1800 for the constant $K$ instead of 1700 in calculating $I_a$. Incorrect solution c is the result of incorrectly using the value of $I_l$ instead of $I_a$ in the last step of calculating the developed torque.

If your answers are correct, go to Problem 4.5.

If your answers are not correct, review the subject of dc machines in this chapter. For more details, see chapter 2 in reference 1 and chapter 5 in reference 3.

## PROBLEM 4.5  Separately-Excited dc Generator

A separately-excited 220 volt dc generator has a 5300 watt rating at 3000 rpm for an input power of 8 HP. Armature resistance is 0.4 ohms. It is proposed to increase the output voltage of this machine to 260 $V_{dc}$ at a speed of 3600 rpm.

Determine the machine ratings at the new operating conditions. Also determine the field current if the generator voltage is $E_g = 100\, I_f - I_f^2$.

*Solution*

The circuit diagram is shown in Exhibit 4.5. Generated voltage and power are proportional to speed.

Exhibit 4.5

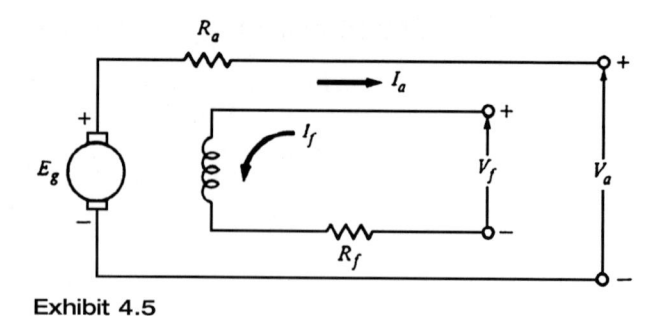

Therefore

$$V_a @ 3600 \text{ rpm} = 220 \times \frac{3600}{3000} = 264 \ V_{dc}$$

$$P_{out} @ 3600 \text{ rpm} = 5.3 \text{ kW} \times \frac{3600}{3000} = 6.36 \text{ kW}$$

Input horsepower also increases with speed, but not in exact proportion because part of the input power must supply $I_a^2 R_a$ losses, which do not vary with speed.

$$I_{\text{full load}} = \frac{5.3 \text{ kW}}{220 \text{ volts}} = 24.09 \text{ amps}$$

$$I_a^2 R_a = 24.09^2 \times 0.4 = 232.15 \text{ watts or } 0.31 \text{ HP}$$

Assuming the remainder of the 8 HP input is due to torque-speed, the speed increase will raise this portion of the input to

$$(8 - 0.31) \times \frac{3600}{3000} = 9.23 \text{ HP}$$

The total input HP, then, is

$$9.23 + 0.31 = 9.54 \text{ HP}$$

At full load for a terminal voltage of 260 $V_{dc}$ and a full load current of 24.09 amps through an armature resistance of 0.4 ohms, the generated voltage must be

$$E_g = 260 + 24.09 \times 0.4 = 269.64$$

Calculating the required field current:

$$269.64 = 100 \ I_f - I_f^2$$
$$I_f^2 - 100 \ I_f + 269.64 = 0$$
$$I_f = 2.77 \text{ amps}$$

If your answers are correct, go on to the next section.

If your answers are not correct, review the subject of dc machines in this chapter as well as problem 4.1. For more details, see chapter 2 in reference 1 and chapter 4 in reference 3.

# ac MACHINES

You may wish to return to Figure 4.1 and review the classification chart of electric motors. There are many more variables to consider when dealing with ac machines.

## Synchronous Generators

Synchronous generators do not differ in principle from dc generators. In fact, any dc generator is a synchronous generator in which the alternating voltage set up in the armature inductors is rectified by means of a commutator. Although any dc generator may be used as a synchronous generator by the addition of **slip rings** electrically connected to suitable points of its armature winding, it is more satisfactory to interchange the moving and fixed parts when only ac currents are to be generated. The only moving parts required are those necessary for the field excitation, which is carried at low potential.

The frequency of any synchronous generator is given by the formula:

$$f = \frac{p}{2} \times \frac{n}{60} = \frac{pn}{120}$$

where
  $p/2$ = the number of pole pairs, and
  $n$ = speed in rpm.

The voltage induced in a dc generator is equal to the average voltage induced in the coils in series between the brushes. In a synchronous generator, the induced voltage is the effective, or RMS, voltage. The instantaneous voltage induced in any coil on the armature of an alternator is:

$$e = -N\frac{d\phi}{dt}$$

where
  $N$ = the number of turns in the coil, and
  $\phi$ = the flux enclosed by the coil at any instant.

It can be shown that the rms voltage per coil is:

$$E = 4.44 N f \phi_m \text{ volts}$$

when flux distribution in the air gap is such as to produce a sine wave in the armature coils.

The **rating** or maximum output of any synchronous generator is limited by its mechanical strength, by the temperature rise of its parts produced by losses, and by the permissible increase in field excitation necessary to maintain rated voltage at some specified load and power factor. The maximum voltage at a given frequency depends upon the permissible pole flux and field heating. Synchronous generators are rated in kVA at a given frequency or in kW at a stated power factor, frequency, and voltage.

The **regulation** of a synchronous generator at a given power factor is the percentage rise in voltage under conditions of constant excitation and frequency when the rated kVA load at a given power factor is removed. Regulation is positive for an inductive load, since the voltage rises when the load is removed; conversely, regulation is negative for a capacitive load if the load angle is sufficiently great for the voltage to fall.

Efficiency of a synchronous generator is defined as

$$\frac{\text{output}}{\text{output} + \text{losses}}$$

that is, the same as for dc generators. The standard conditions under which efficiency is determined are rated voltage, frequency, power factor, and rated load, at 75°C. Efficiency is usually determined by output and losses. Losses are copper losses in the armature and field windings, core losses, and mechanical losses.

When two or more generators are operated in parallel, they must run at exactly the same frequency, phase, and voltage. When synchronous generators are paralleled, it is necessary to synchronize them before closing the switch. Once connected, the natural reactions resulting from departure from synchronism tend to re-establish synchronism. Unless mechanically coupled, synchronous generators cannot ordinarily operate in series. They are stable only in parallel.

## Synchronous Motors

There is no essential difference in construction between synchronous motors and generators. Any synchronous generator can operate as a synchronous motor and vice versa. As a general rule, a synchronous motor is designed with a more effective damping device to prevent hunting, and its armature reaction is larger than is desirable for a generator. A synchronous motor operates at one speed, the synchronous speed, which depends solely upon the number of poles and the frequency of the line voltage. Speed is independent of load. A load change is accompanied by a change in phase and in instantaneous speed but not by a change in average speed. If, because of excessive load or any other cause, the average speed differs from synchronous speed, the average torque developed is zero, and the motor comes to rest. A synchronous motor, as such, has no starting torque.

The power factor of a synchronous motor operating from a constant potential is fixed by its field excitation and load. At any given load the power factor can be varied over wide limits by altering the field excitation. A motor may be overexcited, underexcited, on normally excited. Normal excitation produces unity power factor. Overexcitation produces capacitive action and causes the motor to take leading current. Underexcitation causes the motor to take lagging current. The field current that produces normal excitation depends upon the load and, in general, it increases with the load.

A synchronous motor operating with constant impressed voltage and frequency can operate at constant speed only under constant load conditions. The speed in rpm is given by the formula:

$$n = \frac{2f}{p} \times 60$$

This is another version of the formula given previously for the synchronous generator. As mentioned above, if a load is increased, the motor begins to slow down and continues to do so until sufficient change in phase has been produced. If the motor is not properly damped, it may overrun and develop too much power; it then speeds up and may again overrun in the reverse direction, developing too little power. Repeated action of this sort is called **hunting.**

A synchronous motor is not inherently self-starting. The average synchronous motor torque is zero at rest and until synchronous speed is reached. Some auxiliary device is necessary to bring a synchronous motor up to speed.

There are many uses for synchronous motors. A synchronous motor is frequently used for one unit of a motor-generator set when the other unit is a dc generator or a synchronous generator. When the other unit is a synchronous generator, the M-G set acts as a frequency converter (e.g., 50 to 60 Hz or vice versa). When operated at unity PF, a synchronous motor weights and costs less than an equivalent induction motor.

Synchronous motors are often operated without load on a power transmission system to control power factor and to improve voltage regulation. A polyphase synchronous motor floated on a circuit carrying an unbalanced load tends to restore balanced conditions in regard to both current and voltage. If the system is badly out of balance, the synchronous motor may take power from the phases at high voltage and deliver power to the phase, or phases, at low voltage.

## Asynchronous Machines

The previous paragraphs considered machines that operate at synchronous speeds. There is another class known as asynchronous machines that do not operate at synchronous speed. Their speed varies with the load.

A commercial synchronous machine requires fields excited by direct current. An asynchronous machine requires no dc excitation. Both parts of an asynchronous machine (armature and field) carry alternating current and are either connected in series, as in the series motor, or are inductively related, as in the induction motor. The induction motor and generator, the series motor, the repulsion motor, and all forms of ac commutator motors are included in the general class known as asynchronous machines. The induction motor is the most important and most widely used type of asynchronous motor. It has essentially the same speed and torque characteristics as a dc shunt motor and is suitable for the same type of work.

## Polyphase Induction Motor

The polyphase induction motor is equivalent to a static transformer operating on a noninductive load whose magnetic circuit is separated by an air gap into two portions that rotate with respect to each other. In most cases the primary is the stator and the secondary is the rotor. The stator winding is similar to the armature winding of a synchronous machine. The polyphase currents in the stator winding produce a rotating magnetic field corresponding to the armature reaction in a synchronous generator. The fundamental of this field revolves at synchronous speed with respect to the stator in the same way the armature reaction of a synchronous machine revolves with respect to the armature. With respect to the rotor, the field revolves at a speed that is the difference between synchronous speed and the speed of the rotor. This difference is known as **slip.**

Slip is defined by the formula

$$S = \frac{n_1 - n_2}{n_1}$$

where
  $n_1$ = the speed of the revolving magnetic field and also the synchronous speed in rpm, and
  $n_2$ = the actual speed of the rotor in rpm.

The formula for $n_1$ is

$$n_1 = \frac{120 f_1}{p}$$

## Induction Generator

An induction generator does not differ in construction from an induction motor. Whether an induction machine acts as a generator or as a motor depends solely upon its slip. Below synchronous speed it can operate only as a motor; above synchronous speed it operates as a generator.

The power factor at which an induction generator operates is fixed by its slip and its constants and not by the load.

An induction generator is free from hunting since it does not operate at synchronous speed.

## Single-Phase Induction Motors

A single-phase induction motor possesses no starting torque. It is heavier, has lower efficiency and lower power factor than a polyphase motor for the same speed and output. These characteristics are true for any single-phase motor or generator.

The single-phase induction motor must be started by some form of auxiliary device and must attain considerable speed before it develops sufficient torque to overcome its own friction and windage. The direction of rotation depends merely upon the direction in which it is started. Once started, it operates as well in one direction as the other.

### PROBLEM 4.6 Induction Motor Speed

A three-phase induction motor has the following nameplate data:

$$15 \text{ hp, } 230 \text{ V, } 60 \text{ Hz, } 1150 \text{ rpm}$$

Determine the number of poles, slip and rotor current frequency at full load. Also, estimate the speed at half of full load.

*Solution*

Exhibit 4.6 shows a typical speed-torque motor curve.

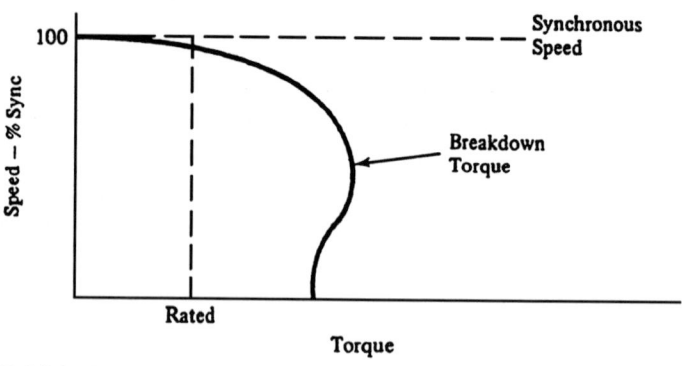

Exhibit 4.6

An induction motor operating at rated load runs at a speed somewhat lower than synchronous speed. In this case, synchronous speed is most likely 1200 rpm. Based on this assumption, the number of poles may be calculated as follows:

$$f = \frac{p}{2} \times \frac{n}{60}$$

where

$f$ = frequency in Hz
$p/2$ = number of pole pairs
$n$ = synchronous speed in rpm

Thus,

$$p = \frac{120 f_1}{n_1} = \frac{120 \times 60}{1200} = 6 \text{ poles}$$

Slip is defined by the formula

$$S = \frac{n_1 - n_2}{n_1}$$

where

$S$ = slip
$n_1$ = synchronous speed in rpm
$n_2$ = actual rotor speed in rpm

Thus,

$$S = \frac{1200 - 1150}{1200} = 0.0417$$

Rotor current frequency is given by the formula

$$f_s = S \times f = 0.0417 \times 60 = 2.5 \text{ Hz}$$

As a point of information, rotor currents at frequency $f_s$ produce a rotating **magnetomotive force** on the rotor. This mmf revolves at a speed $n_s$ with respect to the rotor. This relative speed is the difference between synchronous speed and actual speed, and is given by the formula

$$n_s = n_1 - n_2 = \frac{120 f_1 S}{p}$$

Assuming the speed-torque curve is linear over the range of operation, speed varies directly with load. Thus, the speed at half of full load is calculated as follows:

$$\frac{7.5 \text{ hp}}{15 \text{ hp}} = \frac{1200 - n_2}{1200 - 1150} = \frac{1200 - n_2}{50} = \frac{1}{2}$$

$$n_2 = 1200 - 25 = 1175 \text{ rpm}$$

If your answers are correct, go to Problem 4.7.

If your answers are not correct, review the subject of ac machines in this chapter. For more details, see chapter 2 in reference 1 and chapter 13 in reference 3.

## PROBLEM 4.7 Induction Motor Efficiency

A six-pole, 220-volt, three-phase induction motor has the following measured losses:

- stator copper loss = 0.6 kW

- friction, no load core and windage losses = 1 kW

The motor runs at 1150 rpm when drawing 25 kW from the line. The motor efficiency and output horsepower under these conditions are:

a. $\eta = 0.895\%$, and $P_{out} = 30$ hp
b. $\eta = 89.5\%$, and $P_{out} = 30$ hp
c. $\eta = 89.5\%$, and $P_{out} = 33.5$ hp
d. $\eta = 89.5\%$, and $P_{out} = 22.4$ hp

*Solution*

Exhibit 4.7 shows the equivalent circuit of a polyphase induction motor. It has been simplified so that there is a nearly constant error of 2–3% in induced voltages $E_1$ and $E_2$ between no load and full load. Everything in the equivalent circuit is referred to the primary.

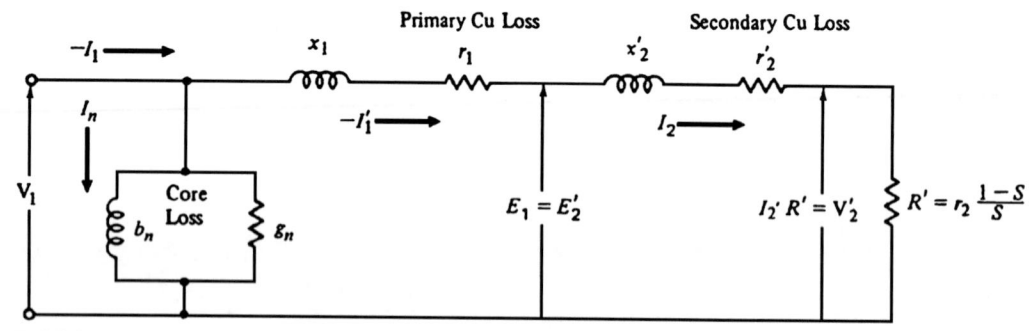

**Exhibit 4.7** Losses in an induction motor include the following:

- primary copper loss;
- secondary copper loss;
- core loss;
- friction and windage loss;
- stray-load loss.

When it is possible to load the motor and measure its input and slip but not possible to measure the output directly, the efficiency is obtained from the formula

$$\eta = \frac{\text{input} - \text{losses}}{\text{input}}$$

Primary copper loss is usually obtained by measuring the ohmic resistance per phase with dc and multiplying by the square of the primary phase current and the number of phases. In this problem, this loss is given as

$$0.6 \text{ kW} = 3I_1^2 r_1$$

Whenever slip is accurately measurable, the secondary copper loss may be taken as the product of the measured primary input minus the primary copper and core losses and the slip. In this case,

$$\text{slip} = \frac{1200 - 1150}{1200} = 0.0417 = \frac{\text{sec. Cu loss}}{P_{in} - (\text{prime. Cu loss} + \text{core loss})}$$

Thus, rotor (secondary) copper loss = (25 − 0.6) 0.00417 = 1.017 kW.

Sometimes it is difficult to separate core loss from friction and windage losses. In this case, core loss may be excluded from the primary copper losses with little effect on the final result.

The friction and windage loss plus the core loss may be obtained by running the motor at no load with normal frequency and voltage applied. The input is then equal to the no-load primary copper loss, friction and windage, core loss and the no-load secondary copper loss (which is small and usually neglected). In this problem, these losses are given as 1 kW.

Stray-load loss includes all losses not otherwise accounted for. This is obtained from two tests: a dc excitation test and a test to find the blocked rotor torque with balanced polyphase voltages at rated frequency. In this problem, this loss will be assumed to be combined with all other losses.

Total losses, then, equal

$$P_{total\ loss} = 0.6 + 1.017 + 1 = 2.617\ kW$$

Efficiency is

$$\eta = \frac{25 - 2.617}{25} = 0.895 \quad \text{or} \quad 89.5\%$$

Output horsepower is

$$P_{out} = \frac{25 - 2.617}{0.746} = 30\ hp$$

The answer is b.

*Answer Rationale*

Incorrect solution a is the result of not multiplying by 100 to show the efficiency as a percentage. Incorrect solution c is the result of not subtracting the total loss value of 2.617 kW in the numerator when calculating $P_{out}$. Incorrect solution d is the result of not dividing by the factor 0.746 to convert the units of $P_{out}$ from watts to hp.

If your answers are correct, go to Problem 4.8.

If your answers are not correct, review the topics of efficiency and losses at the beginning of this chapter. For more details, see chapter 2 in reference 1 and chapter 13 in reference 3.

Note: The term "copper loss" is no longer accurate since many machines make use of aluminum rather than copper; "winding loss" is a better term.

## PROBLEM 4.8 Induction Motor Losses

A 4-hp, four-pole, 60-Hz, three-phase, wound-rotor motor draws 3670 watts from the input power line. The following losses have been measured:

- primary copper (stator) loss: 350 watts
- core loss: 200 watts.
- friction, windage, stray-load loss: 50 watts

What is the speed of the motor when delivering 3.95 hp?

*Solution*

$$n_1 = \frac{120f}{4} = \frac{120 \times 60}{4} = 1800 \text{ rpm}$$

$$n_2 = n_1(1 - S) = 1800(1 - S)$$

$$S = \frac{\text{secondary copper loss}}{P_{in} - (\text{primary copper loss} + \text{core loss})} = \frac{\text{second copper loss}}{P_{em}}$$

$$= \frac{\text{secondary copper loss}}{3670 - 550}$$

We need to find secondary copper loss.

$$P_{\text{total losses}} = P_{in} - P_{out} = 3670 - 3.95 \times 746$$
$$= 3670 - 2946 = 723.3 \text{ watts}$$

$$\text{Secondary copper losses} = P_{\text{total loss}} - \text{all other losses}$$
$$= 723.3 - (350 + 200 + 50) = 123.3 \text{ watts}$$

Now we can calculate slip from the above formula (assuming core loss charged to stator):

$$S = \frac{123.3}{3670 - 550} = 0.0395$$

Finally, the speed is

$$n_2 = 1800(1 - S) = 1800(1 - 0.0395) = 1729 \text{ rpm}$$

If core loss is charged to rotor,

$$S = \frac{123.3}{3670 - 350} = 0.0371$$

$$n_2 = 1800(1 - 0.0371) = 1733 \text{ rpm}$$

Another way to calculate slip is from the formula

$$P_{out} = \text{rotor loss} \times \left(\frac{1-S}{S}\right)$$

$$2946.7 = 123.3\left(\frac{1-S}{S}\right), \quad \frac{1-S}{S} = 23.9$$

$$1 - S = 23.9S, \quad 1 = 24.9S, \quad S = \frac{1}{24.9} = 0.0402$$

If your answers are correct, go to Problem 4.9.

If your answers are not correct, review the topics of efficiency and losses at the beginning of this chapter as well as Problem 4.6. For more details, see chapter 2 in reference 1 and chapters 13 and 14 in reference 3.

### PROBLEM 4.9 Induction Motor Speed

Find the speed of a three-phase, 60 Hz, 12 HP, six-pole induction motor, given the following information:

- $P_{in} = 11{,}000$ watts
- $P_{out} = 11.9$ HP

- $P_{core} = 540$ watts
- $P_{stator\ copper\ loss} = 770$ watts
- $P_{rotational} = 175$ watts

The speed is:
- a. $n_r = 1142$ rpm
- b. $n_r = 1200$ rpm
- c. $n_r = 560$ rpm
- d. $n_r = 1121$ rpm

*Solution*

$$P_{in} - P_c - P_s = 11{,}000 - 540 - 770 = 9690 \text{ watts}$$
$$I_R^2 R_R = 9690\, S$$
$$P_{out} + P_{rot} = 11.9 \times 746 + 175 = 9052 \text{ watts}$$

$$I_R^2 R_R \frac{(1-S)}{S} = 9052 = 9690\,(1-S)$$

$$1 - S = \frac{9052}{9690} = 0.934$$

$$n_s = \frac{60f}{P/2} = \frac{3600}{6/2} = 1200 \text{ rpm}$$
$$n_r = n_s\,(1 - S) = 1121 \text{ rpm}$$

The correct answer is d.

*Answer Rationale*

Incorrect solution a is the result of incorrectly subtracting the rotational osses values of 175 from the input power in the first step. Incorrect solution b is the result of not multiplying the speed $n_s$ by $(1 - S)$ to obtain $n_r$. Incorrect solution c is the result of not dividing the number of poles P by 2 in the denominator when calculating $n_s$.

If your answer is correct, go on to Problem 4.10.

If your answer is not correct, review the previous two problems. For more details, see chapter 2 in reference 1 and chapter 13 in reference 3.

## PROBLEM 4.10   Motor Starting-Line Voltage Drop

The impedance of a three-phase power distribution system is $0.06 + j0.3$. Open circuit line voltage is 2400 volts. The percent voltage drop when a 400 hp induction motor is connected to the line and started is:
- a. 14.1%
- b. 20.5%
- c. 11.85%
- d. 2.0%

*Solution*

$$P_{out} = 400 \text{ hp} \times 0.746\,\frac{\text{kW}}{\text{hp}} = 298.4 \text{ kW}$$

Exhibit 4.10

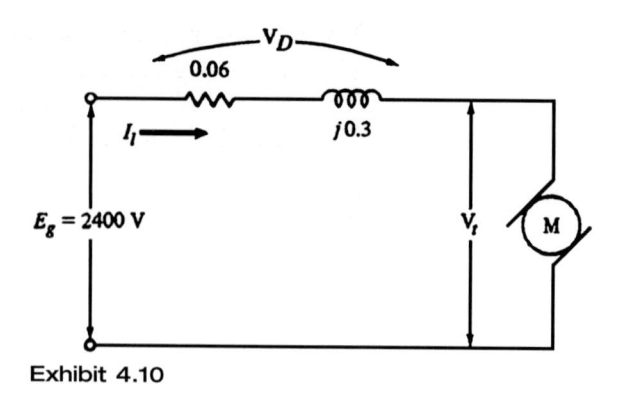

Before this problem can be solved, certain required information not given must be assumed about the motor:

$$\eta = 89\%$$
$$PF_{FL} = 0.9 \text{ lag}$$
$$I_{start} = 6 I_l$$
$$PF_{start} = 0.25 \text{ lag}$$

$$P_{out} = \sqrt{3} V_l I_l \eta \cos\theta$$

$$I_l = \frac{P}{\sqrt{3} V_l \eta \cos\theta} = \frac{298.4}{\sqrt{3}(2.4)(0.89)(0.9)} = 89.62 \text{ amps}$$

$$I_{start} = 6 I_l = 537.7 \text{ amps}$$

Using $I$ as reference:

$$V_t = E_g - V_D = \frac{2400}{\sqrt{3}}(0.25 + j0.968) - 537.7(0.06 + j0.3)$$
$$= 1385.64\angle 75.5° - 164.5\angle 78.7° = 1221.4\angle 75.1°$$

$$\% \text{ drop} = \frac{1385.64 - 1221.4}{1385.64} \times 100 = 11.85\%$$

An alternative way to calculate voltage drop of the distribution line is to use the following formula:

$$V_{drop} = IR \cos\theta + IX \sin\theta$$
$$= 537.7 \times 0.06 \times 0.25 + 537.7 \times 0.3 \times 0.968 = 164.22 \text{ volts}$$

Thus

$$\% \text{ drop} = \frac{164.22}{1385.66} \times 100 = 11.85\%$$

The correct answer is c.

*Answer Rationale*

Incorrect solution a is the result of incorrectly using the value of $P = 400$ kW without converting from hp to kW in calculating $I_l$. Incorrect solution b is the result of not multiplying by $\sqrt{3}$ in the denominator when calculating $I_l$. Incorrect solution d is the result of incorrectly using the value of $I_l$ instead of $I_{start}$ in calculating $V_t$.

If your answers are correct, go to Problem 4.11.

If your answers are not correct, review the subject of polyphase power in chapter 3 as well as Problem 3.10. For more details, see chapter 3 in reference 1 and chapter 26 in reference 3.

## PROBLEM 4.11  Induction Motor Connections

The following test data was obtained on a 60-hp, 900-rpm, three-phase, 60-Hz, 440-volt, delta-connected induction motor:

|  | Voltage | Current | Power | Torque |
| --- | --- | --- | --- | --- |
| No load | 440 V | 31 A | 2.6 kW | 0 |
| Locked rotor | 230 V | 201 A | 30.0 kW | 144 lb-ft |

If the machine is reconnected in wye and 440 volts line-to-line is applied, the starting torque and current will be:
- a.  $T_{ST} = 176$ lb-ft, and $I_{ST} = 129$ amps
- b.  $T_{ST} = 527$ lb-ft, and $I_{ST} = 223$ amps
- c.  $T_{ST} = 176$ lb-ft, and $I_{ST} = 223.4$ amps
- d.  $T_{ST} = 159$ lb-ft, and $I_{ST} = 129$ amps

*Solution*

If the line voltage remains at 440 and the windings are connected in wye, the new phase voltage is

$$V_\phi = \frac{440}{\sqrt{3}} = 254 \text{ volts}$$

Phase current in the delta-connected configuration with a locked rotor line current of 202 amps was

$$I_\phi = \frac{202}{\sqrt{3}} = 116.6 \text{ amps}$$

$$T = \frac{7.045}{n_1} EI_a$$

Torque is proportional to the square of the voltage as seen from the preceding equation, because voltage is proportional to current ($V = IR$). Calculating new starting torque based on the original locked rotor phase voltage versus the new phase voltage of 254 volts yields

$$T_{ST} = \left(\frac{254}{230}\right)^2 \times 144 = 176 \text{ lb-ft}$$

Since current is directly proportional to voltage, the new starting current per phase is

$$I_{ST} = \frac{254}{230} \times 116.6 = 129 \text{ amps}$$

The correct answer is a.

*Answer Rationale*

Incorrect solution b is the result of not dividing by $\sqrt{3}$ in calculating $V_\phi$. Incorrect solution c is the result of not dividing by $\sqrt{3}$ in calculating $I_\phi$. Incorrect solution d is the result of not taking the square of the ratio 254/230 when calculating $T_{ST}$.

If your answers are correct, go on to the next section.

If your answers are not correct, review the topics of efficiency and losses at the beginning of this chapter. For more details, see chapter 2 in reference 1 and chapters 13 and 14 in reference 3.

## Two-Winding Transformers

A two-winding transformer consists of two or more insulated coils coupled by mutual induction. Its action depends upon the self-inductances of its coils and the mutual induction between them. In its simplest form, the transformer consists of two coils linked by a common magnetic circuit. When power is supplied to one coil at a definite frequency and voltage, power can be taken from the other at the same frequency, but, in general, at a different voltage.

Transformers may be of the air-core types, but for power purposes they are always of the iron-core type.

The voltage induced in any winding depends only on the number of turns in the winding and the rate of change of the flux linking it. If $N_1$ is the number of turns in the coil, the voltage rise induced in the coil at any instant by the flux $\phi$ is

$$e = -N_1 \frac{d\phi}{dt} = -\omega N_1 \phi_m \cos \omega t$$

The maximum voltage is

$$e_{max} = \omega N_1 \phi_{max} = 2\pi f N_1 \phi_{max}$$

The effective RMS voltage in volts, when $\phi_{max}$ is expressed in webers, is

$$E_1 = \frac{2\pi f}{\sqrt{2}} N_1 \phi_{max} = 4.44 f N_1 \phi_{max}$$

If the voltage is not a sine wave, then

$$E_1 = 4 \times (\text{form factor})^* \times f N_1 \phi_{max}$$

The ratio of transformation of a transformer is given by the formula

$$\frac{E_1}{E_2} = \frac{N_1}{N_2} = a$$

Efficiency of a transformer is defined as

$$\eta = \frac{\text{full load output (watts) at rated PF}}{\text{full load output} + \text{losses}}$$

Losses include core loss and copper loss. These losses may be determined from open circuit (core loss) and short circuit (copper loss) tests.

---

[*] Where the **form factor** is the ratio of the rms value of the waveform to its half-period average value, the form factor for a sine wave is 1.11.

## PROBLEM 4.12 Transformer Efficiency and Regulation

Short circuit and open circuit tests were run on a single-phase transformer to determine its characteristics. The 2400/240 volt, 60 Hz transformer is rated at 100 kVA. Test data is shown below.

1. High-voltage winding open; rated voltage applied to the low-voltage terminals:

$$I = 32 \text{ amps}$$
$$P = 600 \text{ watts}$$
$$V = 240 \text{ volts}$$

2. High-voltage winding short-circuited; voltage applied to the low-voltage winding such that rated current flows in the windings:

$$I = 417 \text{ amps}$$
$$P = 800 \text{ watts}$$
$$V = 15 \text{ volts}$$

Determine efficiency $\eta$ and full load regulation at PF = 0.8 lag. Also, determine efficiency at 1.25 times rated load.
    a.  $\eta = 98.28\%$, Regulation = 4.26%, and at 1.25 rated load $\eta = 98.18\%$
    b.  $\eta = 98.4\%$, Regulation = 4.46%, and at 1.25 rated load $\eta = 98.18\%$
    c.  $\eta = 98.28\%$, Regulation = 4.46%, and at 1.25 rated load $\eta = 98.77\%$
    d.  $\eta = 98.28\%$, Regulation = 4.46%, and at 1.25 rated load $\eta = 98.18\%$

*Solution*

Exhibit 4.12 shows the equivalent circuit of a two-winding transformer.

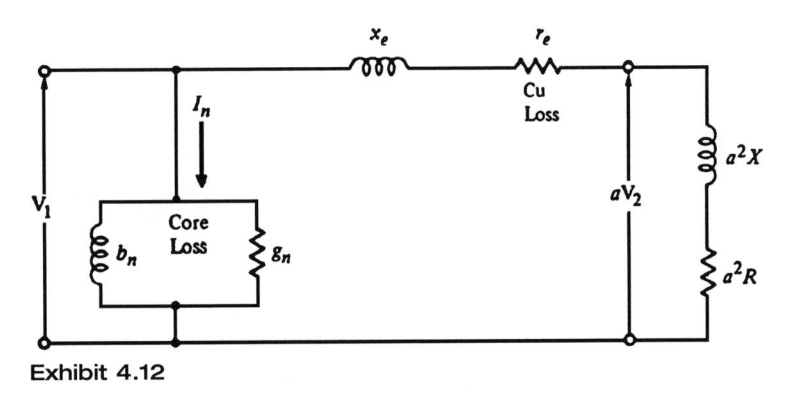

Exhibit 4.12

$$\left.\begin{array}{l} x_e = x_1 + a^2 x_2 \\ r_e = r_1 + a^2 r_2 \end{array}\right\} \text{ referred to primary}$$

In this figure, $x_e$ and $r_e$ are the equivalent reactance and equivalent resistance referred to the primary side and must be used with the secondary current referred to the primary side. The open-circuit test (test 1) gives core loss data, and the short-circuit test (test 2) gives copper loss data.

Efficiency is given by the formula

$$\eta = \frac{\text{full load output (watts) at rated PF}}{\text{full load output (watts) at rated PF} + \text{core loss} + \text{copper loss}}$$

Thus,

$$\eta = \frac{100{,}000 \times 0.8}{100{,}000 \times 0.8 + 600 + 800} = 0.9828 \quad \text{or} \quad 98.28\%$$

$$\text{Regulation} = \frac{|V_{\text{no load}}| - |V_{\text{full load}}|}{|V_{\text{full load}}|} = \frac{|V_{\text{in}}| - |V_{\text{out}}|}{|V_{\text{out}}|}$$

When referred to secondary

$$\frac{V_1}{a} = V_{\text{in}} = V_2 + I_2(\cos\theta \pm j\sin\theta)(r_e + jX_e)$$

$$\begin{cases} + \text{ for leading PF} \\ - \text{ for lagging PF} \end{cases}$$

The equivalent impedance is obtained from the short-circuit test

$$P = I^2 r_e, \quad r_e = \frac{P}{I^2} = \frac{800}{417^2} = 0.00461 \,\Omega$$

$$Z_e = \frac{V}{I} = \frac{15}{417} = 0.036 = r_e + jx_e = 0.00461 + j0.0357$$
$$= 0.036 \angle 82.64°$$

Calculate $V_{\text{in}}$:

$$V_{\text{in}} = 240 + 417(0.8 - j0.6)(0.036 \angle 82.64°) = 250.7 \angle 2.46°$$

Regulation may now be determined:

$$\text{Regulation} = \frac{250.7}{240} = 0.0446 \quad \text{or} \quad 4.46\%$$

Efficiency at 1.25 of full load is calculated as follows: copper loss varies as the square of the load; thus,

$$P_{\text{in}} = 800 \times 1.25^2 = 1250 \text{ watts}$$

At 125% full load,

$$P_{\text{out}} = 100{,}000 \times 0.8 \times 1.25 = 100{,}000 \text{ watts}$$

$$\eta = \frac{100{,}000}{100{,}000 + 600 + 1250} = 0.9818 \quad \text{or} \quad 98.18\%$$

The correct answer is d.

*Answer Rationale*

Incorrect solution a is the result of incorrectly dividing by $V_{\text{in}} = 250.7$ instead of 240 in calculating regulation. Incorrect solution b is the result of not taking the square of the ratio 1.25 when calculating $P_{\text{in}}$. Incorrect solution c is the result of not adding the value of 600 watts in the denominator when calculating $\eta$ at 125% full load.

If your answers are correct, go to Problem 4.13.

If your answers are not correct, review the subject of polyphase power as well as Problem 3.10 in chapter 3, the topics of efficiency and losses at the beginning of this chapter, and problem 4.10. For more details, see chapter 2 in reference 1 and chapter 10 in reference 3.

## PROBLEM 4.13  Regulation Improvement

A 750-kVA, three-phase, balanced load of 60% power factor is supplied from a 1000-kVA, three-phase transformer and distribution line as shown in Exhibit 4.13a. The line impedance relative to the 1000 kVA is 3% resistance and 5% reactance. What will be the voltage regulation improvement if the power factor is correct to 85%?

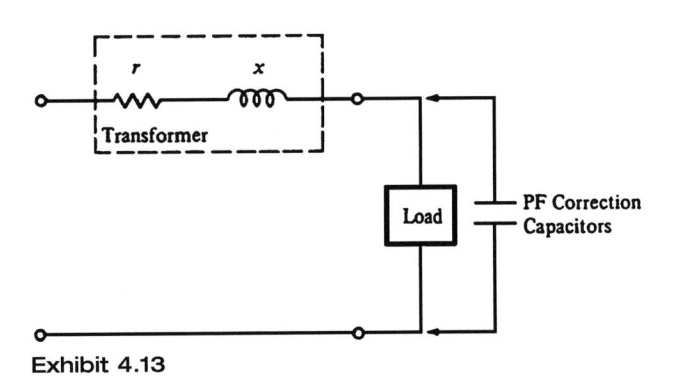

Exhibit 4.13

*Solution*

The problem is easily solved using the equation

$$\% \text{ regulation} = 100\left[mr + nx + \frac{(mx-nr)^2}{2}\right]\left[\frac{\text{actual kVA load}}{\text{rated kVA load}}\right]$$

where
  $m$ = PF of load = $\cos\theta$
  $n$ = $\pm\sin\theta$ (+ for lag, − for lead)
  $r$ = resistance factor = $\dfrac{\text{resistance loss in kW}}{\text{rated kVA of transformer}}$
  $x$ = reactance factor = $\sqrt{Z^2 - r^2}$
  $Z$ = $\dfrac{\text{impedance in kVA}}{\text{rated kVA of transformer}}$

The terms $r$ and $x$, shown in Exhibit 4.13, are on a per-unit basis. Thus, if resistance and reactance factors are 3% and 5%, then $r = 0.03$ and $x = 0.05$, respectively.

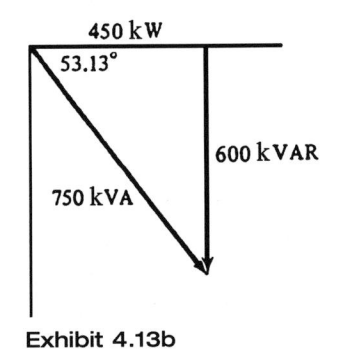

Exhibit 4.13b shows the phasor diagram before PF correction.

$$\% \text{ regulation} = 100\left[0.6 \times 0.03 + 0.8 \times 0.05 + \frac{(0.6 \times 0.05 - 0.8 \times 0.03)^2}{2}\right]\frac{750}{1000}$$

$$= 100\,[0.018 + 0.4 + 0.000018]\,0.75 = 4.35\%$$

Exhibit 4.13b

**Exhibit 4.13c**

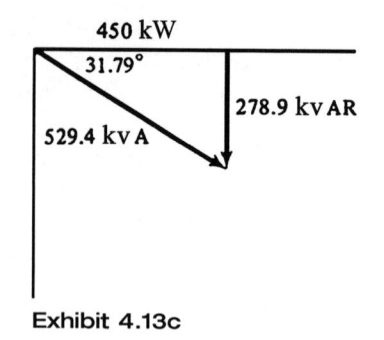

Exhibit 4.13c shows the phasor diagram after PF correction.

% regulation
$$= 100\left[0.85 \times 0.03 + 0.527 \times 0.05 + \frac{(0.85 \times 0.05 - 0.527 \times 0.03)^2}{2}\right]\frac{529.4}{1000}$$
$$= 100[0.0255 + 0.02635 + 0.0004]0.5294 = 2.75\%$$
$$\text{Regulation improvement} = 4.35 - 2.75 = 1.6\%$$

The number of kVARs of capacitance to achieve this improvement is

$$600 - 279 = 321 \text{ kVARs}$$

If your answers are correct, go to Problem 4.14.

If your answers are not correct, review the subject of polyphase power in chapter 3. For more details, see chapter 2 in reference 1 and chapters 10 and 12 in reference 3.

## PROBLEM 4.14 Transformer Specifications

A new machine shop is being established and will have the following loads:

(1) 55 kVA, 240 volt, single-phase, unity power factor

(2) 14 kW, 240 volt, three-phase, 0.82 power factor

Primary power to the shop is 7200 volts, three-phase, 60 Hz. Design a power distribution system for these two loads.

### Solution

With the single-phase load considerably larger than the three-phase load, a V-V, or open delta, transformer configuration would be the most economical to install. Assuming an *a-b-c* phase sequence and the single-phase load across line *a-b* as follows, the ratings of the two transformers may be specified as shown in Exhibit 4.14.

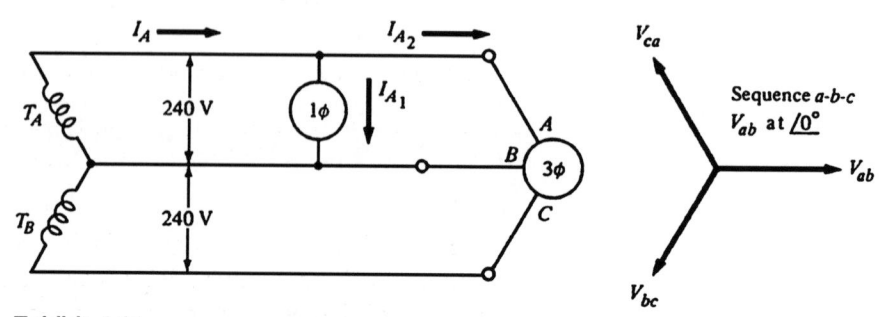

**Exhibit 4.14**

$$I_{A1} = I_{1\phi} = \frac{\text{KVA}}{V_{ab}} = \frac{55{,}000}{240\angle 0°} = 229.17\angle 0° \text{ amps}$$

$$I_{A2} = I_{3\phi} = \frac{P}{\sqrt{3}V_L \cos\theta}\angle-(\theta+30°) = \frac{14{,}000}{\sqrt{3}\times 240\times 0.82}\angle-64.92°$$

$$= 41.07\angle-64.92° = 19.41 - j36.19 \text{ amps}$$

$$I_A = I_{A1} + I_{A2} = 229.17 + 19.41 - j36.19 = 248.58 - j36.19 = 251.2\angle-8.28 \text{ amps}$$

kVA$_A$ = (kV) ($I_A$) = 0.24 × 251.2 = 60 kVA

kVA$_B$ = (kV) ($I_{3\phi}$) = 0.24 × 41.04 = 9.86 kVA

In summary, the two transformers to be installed have the following specifications:

$T_A$: 60 kVA, 7200-to-120/240 volts, 60 Hz, 1$\phi$
$T_B$: 10 kVA, 7200-to-120/240 volts, 60 HZ, 1$\phi$

If your answers are correct, go on the next section.

If your answers are not correct, review the subject of polyphase power in chapter 3. For more details, see chapter 2 in reference 1 and chapters 10 and 12 in reference 3.

## Autotransformer

In many cases either an autotransformer or a two-winding transformer may be used to accomplish the same transformation. In such a case, they would have the same primary and secondary voltages and equal kVA ratings, which would mean also the same rated values of primary and secondary currents. The size of wire used for the continuous winding is not the same throughout unless the ratio of transformation is such that its two parts carry equal currents. Since part of the winding serves for both primary and secondary, an autotransformer requires less material and is therefore cheaper than a two-winding transformer of the same output and efficiency. The saving, however, is substantial only when the ratio of transformation is small (less than 2).

The advantages of the autotransformer are better regulation and efficiency and lower cost for the same output. Its disadvantages are that its low-voltage winding is in electrical connection with the high-voltage winding and forms part of it, and that because of its reduced impedances larger currents result from a short-circuit on its secondary side.

For purposes of comparison, the following assumptions are made with respect to the autotransformer:

1. Its core is identical with that of the corresponding two-winding transformer.

2. The number of turns used in its winding equals that of the high-voltage winding transformer. The two transformers would then operate at the same flux density and their core losses would be equal.

3. The wire size for the series winding is the same as that used in the high voltage winding of a two-winding transformer. At rated load these windings carry equal currents.

4. The wire size for the common winding is so chosen that at rated load the current density in this winding equals the current density in the low voltage winding of the two-winding transformer at the same load.

A useful formula for determining the rating of either the series or common winding is

$$\text{Rating} = \text{Output}\left[1 - \frac{V_2}{V_1}\right]$$

## PROBLEM 4.15  Autotransformer Currents

An autotransformer as shown in Exhibit 4.15 is adjusted to provide 240 volts to a 5-kVA single-phase load from a-480 volt distribution line. The currents in the series and common winding of the autotransformer are:

a. $I_S = 10.42$ amps, and $I_C = 10.42$ amps
b. $I_S = 20.83$ amps, and $I_C = 0$ amps
c. $I_S = 10.42$ amps, and $I_C = 0$ amps
d. $I_S = 10.42$ amps, and $I_C = 31.25$ amps

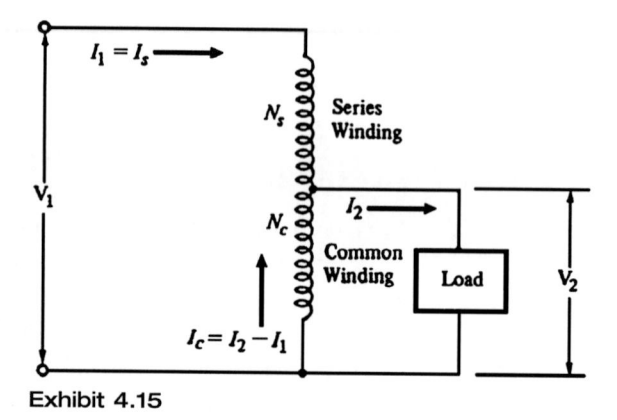

Exhibit 4.15

### Solution

The turns ratio is

$$a = \frac{V_1}{V_2} = \frac{N_1}{N_2} = \frac{480}{240} = 2$$

where
$N_1 = N_s + N_c$
$N_2 = N_c$

Assuming an ideal transformer (no losses),

$$I_1 = \frac{5000}{480} = 10.42 \text{ amps} = I_S$$

$$I_2 = \frac{5000}{240} = 20.83 \text{ amps}$$

$$I_c = I_2 - I_1 = 20.83 - 10.42 = 10.42 \text{ amps}$$

The correct answer is a.

*Answer Rationale*

Incorrect solution b is the result of incorrectly dividing by 240 instead of 480 in calculating $I_S$. Incorrect solution c is the result of incorrectly dividing by 480 instead of 240 in calculating $I_2$. Incorrect solution d is the result of incorrectly adding the current value of $I_1$ instead of subtracting it when calculating $I_c$ in the last step.

If your answers are correct, go to Problem 4.16.

If your answers are not correct, review the subject of autotransformer. For more details, see chapter 2 in reference 1 and chapter 11 in reference 3.

## PROBLEM 4.16 Autotransformer Rating

A 220 microfarad capacitor is connected across the primary of a 440/120 volt autotransformer as shown in Exhibit 4.16.

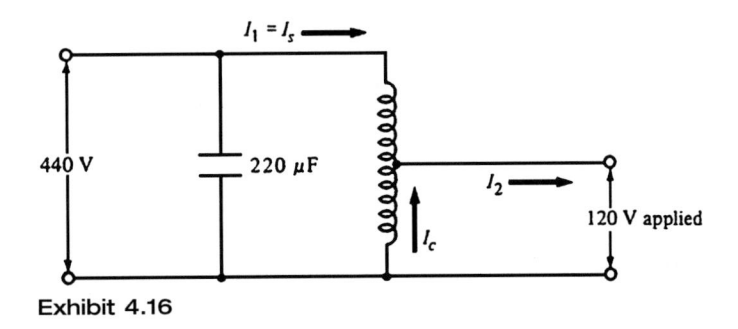

**Exhibit 4.16**

What is the equivalent capacitance when referred to the secondary? What are the kVA ratings of the common and series windings when 120 volts ac is impressed across the secondary winding?

*Solution*

The turns ratio is

$$a = \frac{440}{120} = 3.67$$

Calculating the equivalent secondary capacitance yields

$$X_{Cp} = a^2 X_{C_s}, \quad \frac{1}{\omega Cp} = \frac{a^2}{\omega Cs}$$

$$Cs = a^2 Cp = (3.67)^2 (220 \ \mu f) = 2958 \ \mu F$$

Calculating the winding kVAs:

$$X_{CP} = \frac{1}{\omega c \ Cp} = \frac{1}{377 \times 220 \times 10^{-6}} = 12.06 \ \Omega$$

$$I_S = \frac{440}{12.06} = 36.49 \text{ amps}$$

$$I_2 = aI_s = 3.67 \times 36.49 = 133.81 \text{ amps}$$

$$I_c = I_2 - I_s = 133.81 - 36.49 = 97.32 \text{ amps}$$

$$(\text{kVA})_{\text{series}} = (\text{kV})_{\text{series}} I_s = (0.440 - 0.120) \times 36.49 = 11.68 \text{ kVA}$$

$$(\text{kVA})_{\text{common}} = (\text{kV})_{\text{common}} I_c = (0.120) \times 97.32 = 11.68 \text{ kVA}$$

An alternate method of calculating the rating of either the common or series winding of an autotransformer (they are always the same) is to use this formula:

$$\text{rating} = \text{output}\left[1 - \frac{V_2}{V_1}\right]$$

$$\text{rating} = 440 \times 36.49 \left[1 - \frac{120}{440}\right] = 16.06 \times 10^3 \, [0.73] = 11.68 \text{ kVA}$$

If your answers are correct, go to the next section.

If your answers are not correct, review the subject of autotransformer. For more details, see chapter 2 in reference 1 and chapters 10 and 11 in reference 3.

## MAGNETIC DEVICES

The following are problems related to magnetic devices. The principles of magnetic circuits as discussed in Chapter 1 are applied to the solutions of these problems.

### PROBLEM 4.17   Reactor

A reactive device has 250 turns of wire on a closed iron core having an average length of 18 inches. The iron cross-sectional area is 8 square inches. The device is to have an air gap cut in the iron so that the coil emf will be 120 volts RMS when the magnetizing current is 2.5 amps at 60 Hz.

The required length of the air gap is:
   a. $l_{air} = 2.8 \times 10^{-3}$ m
   b. $l_{air} = 2.966 \times 10^{-3}$ m
   c. $l_{air} = 4.773 \times 10^{-3}$ m
   d. $l_{air} = 4.2 \times 10^{-3}$ m

*Solution*

In order to solve this problem certain assumptions must be made. The following are reasonable assumptions:

1. Effective cross-sectional air gap area including fringing = 12 square inches
2. Flux is uniformly distributed
3. Twelve percent of the total mmf is required to overcome the iron path reluctance
4. Winding factor = 1

$$E_{rms} = 4.44 \, fN\phi_{max}$$

$$\phi_{max} = \frac{E_{rms}}{4.44 \, Nf} = \frac{120}{4.44 \times 250 \times 60} = 1.802 \times 10^{-3} \text{ weber}$$

$$A_{air} = (12 \text{ in}^2)(6.45 \times 10^{-4} \text{ m}^2/\text{in}^2) = 7.74 \times 10^{-3} \text{ m}^2$$

$$B_{air} = \frac{\phi_{max}}{A_{air}} = \frac{1.802 \times 10^{-3} \text{ weber}}{7.74 \times 10^{-3} \text{ m}^2} = 0.233 \text{ weber/m}^2$$

$$l_{air} = \frac{\mu_0 \times 0.88(NI_{max})_{air}}{B_{air}} \text{ meter}$$

$$I_{max} = \sqrt{2} I_{rms} = \sqrt{2} \times 2.5 = 3.54 \text{ amps}$$

$$l_{air} = \frac{4\pi \times 10^{-7} \times 0.88 \times 250 \times 3.54}{0.233} = 4.2 \times 10^{-3} \text{ m}$$

The correct answer is d.

*Answer Rationale*

Incorrect solution a is the result of incorrectly using the value of 8 in$^2$ instead of 12 in$^2$ in calculating $A_{air}$. Incorrect solution b is the result of not multiplying by $\sqrt{2}$ in calculating $I_{max}$. Incorrect solution c is the result of not multiplying by the factor 0.88 in the numerator when calculating $l_{air}$ in the last step.

If your answer is correct, go to Problem 4.18.

If your answer is not correct, review the subject of magnetic circuit terms in chapter 1. For more details, see chapter 1 in reference 1 and chapter 2 in reference 3.

### PROBLEM 4.18 Magnetic Device

A magnetic device is designed to operate at 115 V, 60 Hz with an eddy-current loss of 4 watts and a hysteresis loss of 9 watts. The core loss of the device when it is operated at 100 V, 50 Hz is:

    a. Core loss$_{50}$ = 11.05 watts
    b. Core loss$_{50}$ = 5.94 watts
    c. Core loss$_{50}$ = 11.64 watts
    d. Core loss$_{50}$ = 10.82 watts

*Solution*

$$\text{Core loss} = P_e + P_b$$
$$P_e = Ke t^2 f^2 B_{max}^2$$
$$P_b = \eta f B_{max}^n \quad \text{(Steinmetz empirical equation)}$$

where
    $Ke$ = coefficient of eddy-current loss
    $t$ = thickness of laminations
    $f$ = frequency
    $B_{max}$ = maximum flux density
    $\eta$ = coefficient of hysteresis loss
    $n$ = hysteresis exponent

$$E = 4.44 f N \phi_{max}, \phi_{max}$$

is directly proportional to $B_{max}$. Thus, $B$ is directly proportional to the line voltage and inversely proportional to frequency

$$\frac{B_{50}}{B_{60}} = \left(\frac{100}{115}\right)\left(\frac{60}{50}\right) = 1.04$$

$$P_{e_{50}} = P_{e_{60}} \left(\frac{50}{60}\right)^2 (1.04)^2 = 4 \times 0.76 = 3.02 \text{ watts}$$

$$P_{b_{50}} = P_{b_{60}} \left(\frac{50}{60}\right)(1.04)^{1.6} = 9 \times 0.89 = 8.03 \text{ watts}$$

$$\text{Core loss}_{50} = 3.02 + 8.03 = 11.05 \text{ watts}$$

The correct answer is a.

*Answer Rationale*

Incorrect solution b is the result of incorrectly using the ratio of 50/60 instead of 60/50 in calculating $\frac{B_{50}}{B_{60}}$. Incorrect solution c is the result of not multiplying taking the square of the ratio 50/60 in calculating $P_{e_{50}}$. Incorrect solution d is the result of not raising the value 1.04 to the power of 1.6 when calculating $P_{b_{50}}$.

If your answer is correct, go to the next chapter.

If your answer is not correct, review above solution and the subject of magnetic circuit terms in chapter 1. For more details, see chapter 1 in reference 1 and chapter 2 in reference 3.

# CHAPTER 5

# Control Theory

**OUTLINE**

BASIC FEEDBACK SYSTEMS TERMS  132

SINGULARITY FUNCTIONS  133
Unit Step ■ Unit Impulse ■ Unit Ramp

SECOND-ORDER SYSTEMS  134

POLES AND ZEROS  136

PARTIAL FRACTIONS  136
Non-Repeated Poles ■ Repeated Poles

STABILITY  139
Example of a Stable System

TRANSFER FUNCTION  144

COMPENSATION  145
Lead Compensator ■ Lag Compensator ■ Lag-Lead Compensator ■ Simple Lag Compensator ■ System Compensation

BODE ANALYSIS  150

ROOT LOCUS  154

The field of control theory brings together a broad spectrum of engineering principles and specialized mathematical and graphical techniques. This chapter presents the minimal technical background needed to demonstrate the solutions to the types of problems often found in the PE Exam. The solutions require a knowledge of basic feedback system terms, block diagram algebra, singularity functions, second order systems, differential equations, convolution, translation, inversion, Laplace transforms, pole-zero maps, *RLC* circuit analysis, partial fractions, stability, errors, system type, transfer functions, compensation, Bode analysis, root locus analysis, and system analysis and design.

If you have sufficient background in these areas, you will recognize the problem-solving techniques. If you do *not* have sufficient background, you may wish to study a basic college text before working through this chapter.

**132** Chapter 5 Control Theory

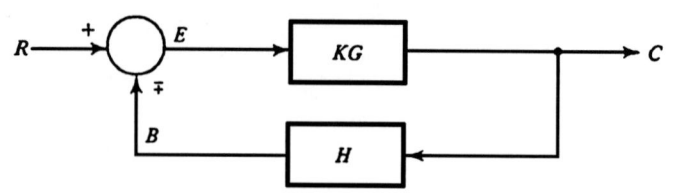

**Figure 5.1** A basic feedback system

## BASIC FEEDBACK SYSTEMS TERMS

Figure 5.1 shows a simple basic feedback system. The following discussion defines terms shown in Figure 5.1 as well as other fundamental terminology.

In the equations below, a minus sign denotes positive feedback; a plus sign denotes negative feedback.

$K$ = constant forward loop gain
$G$ = frequency-sensitive forward loop gain
$E$ = error signal = $R - B$
$R$ = reference or command input
$B = CH$ = primary feedback signal
$C$ = controlled variable output
$H$ = feedback gain ($H = 1$ for unity feedback)
$KG$ = forward loop transfer function = forward gain
$KGH$ = open loop transfer function

$$\frac{C}{R} = \frac{KG}{1 \pm KGH} = \text{system transfer function} = \text{closed loop gain}$$

$$\frac{B}{R} = \frac{KGH}{1 \pm KGH} = \text{primary feedback ratio}$$

$$\frac{E}{R} = \text{error ratio}$$

The *characteristic equation* of a feedback system is determined from $1 \pm KGH = 0$; alternatively, from the loop gain $KGH$, it is:

$$D_{GH} \pm N_{GH} = 0$$

where
$D_{GH}$ is the denominator of $KGH$
$N_{GH}$ is the numerator of $KGH$

Figure 5.2 illustrates unity negative feedback, where

$$\frac{C}{R} = \frac{V_o}{V_i} = \frac{KG}{1 + KG}$$

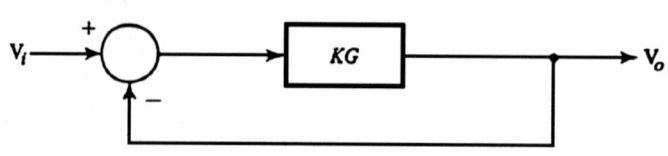

**Figure 5.2** Unity negative feedback

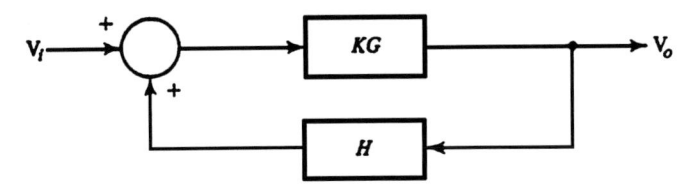

Figure 5.3 Positive feedback oscillator

Figure 5.3 illustrates positive feedback oscillator, where

$$\frac{C}{R} = \frac{V_o}{V_i} = \frac{KG}{1-KGH}$$

This type of feedback system is seldom used.

# SINGULARITY FUNCTIONS

Three *singularity functions* are used extensively in the study of control systems: the unit step, the unit impulse, and the unit ramp.

### Unit Step, u(t)

Response of a control system to the unit step input (Figure 5.4) is the output $y(t)$ when input $x(t) = u(t)$ and all initial conditions are zero. Figure 5.5 shows the responses of a second-order system to a unit step input for various damping factors $\zeta$. The integral of the unit impulse is the unit step.

### Unit Impulse, δ(t)

The quotient in brackets in Figure 5.5 represents a rectangle of height $1/\Delta t$ and width $\Delta t$. In the limit, the unit impulse is a rectangle of infinite height and zero width. The area under the curve is always equal to unity. The derivative of the unit step is the unit impulse.

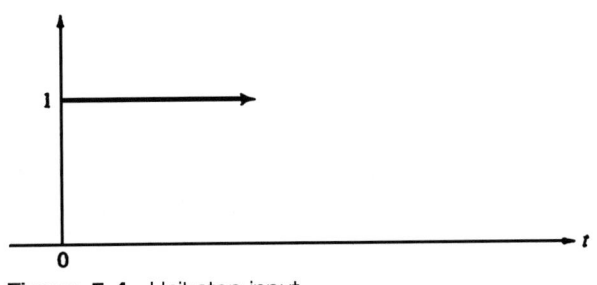

Figure 5.4 Unit step input

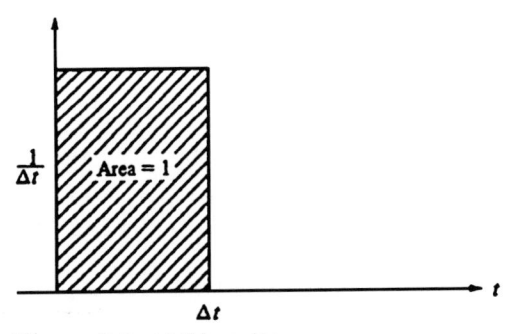

Figure 5.5 Unit impulse

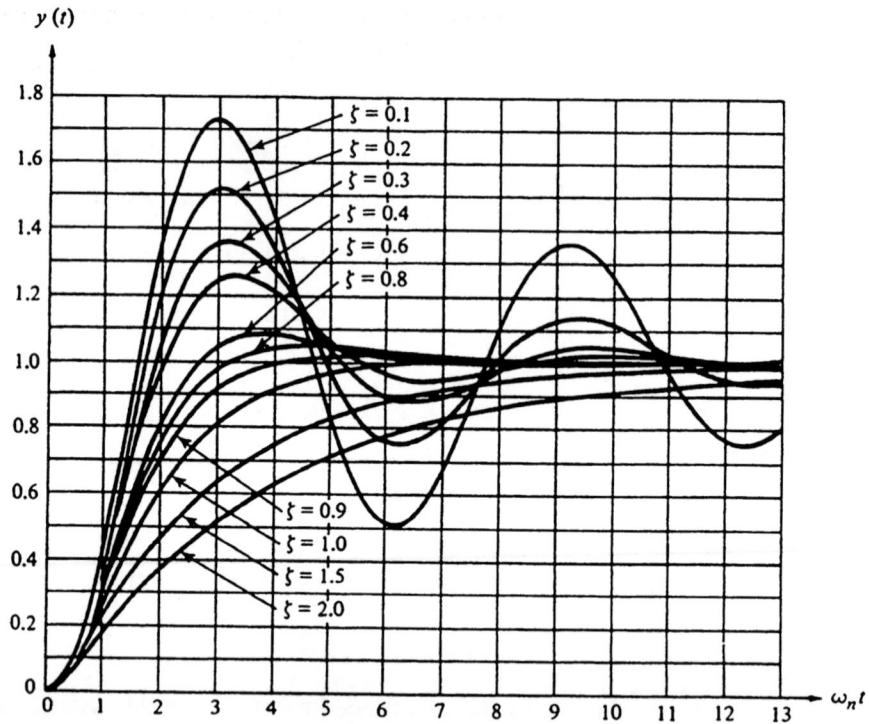

**Figure 5.6** Universal transient response curves for second-order systems subjected to a unit step input

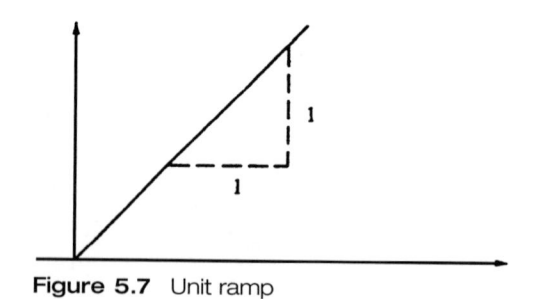

**Figure 5.7** Unit ramp

### Unit Ramp

The derivative of the unit ramp (Figure 5.7) is the unit step. The integral of the unit step is the unit ramp.

## SECOND-ORDER SYSTEMS

Second-order systems are important because most higher-order systems can be approximated by them. The general form of a linear, constant-coefficient, second-order differential equation is given below:

$$\frac{d^2y}{dt^2} + 2\zeta\omega_n\frac{dy}{dt} + \omega_n^2 y = \omega_n^2 x$$

where
  $\zeta$ is the *damping ratio (damping factor)*
  $\omega_n$ is the *undamped natural frequency* of the system

The characteristic equation for the above equation is

$$D^2 + 2\zeta\omega_n D + \omega_n^2$$

## PROBLEM 5.1  Second-Order System

The block diagram of a second-order servo system is shown in Exhibit 5.1.

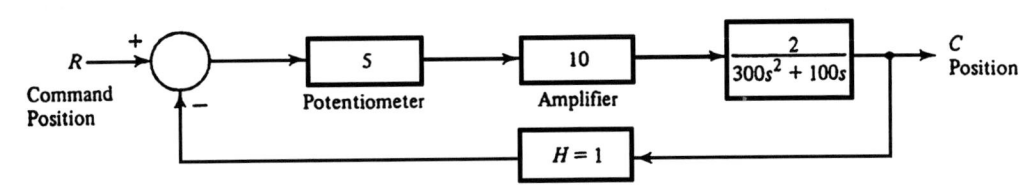

**Exhibit 5.1**

The natural frequency and the damping ratio are:

a. $\omega_n = \sqrt{0.333}$ rad/sec, and $\zeta = 0.29$

b. $\omega_n = 0.333$ rad/sec, and $\zeta = 0.5$

c. $\omega_n = \sqrt{0.333}$ rad/sec, and $\zeta = 0.5$

d. $\omega_n = \sqrt{0.333}$ rad/sec, and $\zeta = 0.58$

*Solution*

The system transfer function for $H = 1$ is

$$\frac{C}{R} = \frac{KG}{1+KG} = \frac{1}{1+\frac{1}{KG}} = \frac{1}{1+\frac{300s^2+100s}{5\times10\times2}}$$

$$= \frac{100}{300s^2+100s+100} = \frac{0.333}{s^2+0.333s+0.333} = \frac{\omega_n^2}{s^2+2\zeta\omega_n s+\omega_n^2}$$

$\omega_n = \sqrt{0.333}$ rad/sec

$\zeta = \dfrac{0.333}{2\sqrt{0.333}} = 0.29$ (underdamped)

The correct answer is a.

*Answer Rationale*

Incorrect solution b is the result of no taking the square root of 0.333 in calculating $\omega_n$. Incorrect solution c is the result of incorrectly using $\sqrt{0.333}$ in the numerator instead of 0.333 when calculating $\zeta$. Incorrect solution d is the result of not multiplying by 2 in the denominator when calculating $\zeta$.

If your answers are correct go to the next section.

If your answers are not correct, review the subjects of basic feedback systems and second-order systems in this chapter. For more details, see chapter 5 in reference 1 and chapter 4 in reference 4.

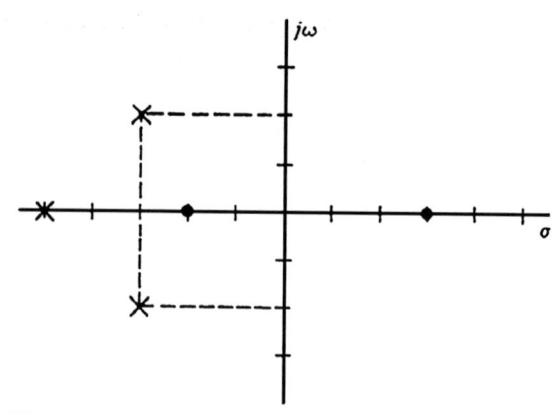

**Figure 5.8** A pole-zero map

## POLES AND ZEROS

Poles and zeros are complex numbers. A pole or a zero can be represented as a point in rectangular coordinates on a map called a *complex plane* or *s-plane*. The abscissa is called the $\sigma$-axis (real), and the ordinate is called the $j\omega$-axis (imaginary). A pole on an *s*-plane is denoted by an X, and a zero is denoted by a O. The *s*-plane showing the location of poles and zeros of $F(s)$ is called a *pole-zero map* of $F(s)$.

Figure 5.8 shows the pole-zero map of

$$F(s) = \frac{(s+2)(s-3)}{(s+5)(s+3-2j)(s+3+2j)}$$

Those values of $s$ of which $F(s) = 0$ are called zeros (numerator), and those values of $s$ for which $F(s) = \infty$ are called poles (denominator).

## PARTIAL FRACTIONS

In converting from the *s*-domain to the time domain, it is sometimes handy to make use of partial fraction expansion. Consider, for example, the following:

$$C(s) = \frac{R(s)G(s)}{1+G(s)H(s)}$$

If, by partial fraction expansion, the above equation can be rewritten as

$$C(s) = \frac{A}{s-P_1} + \frac{B}{S-P_2} + \frac{C}{s-P_3} + \cdots$$

then, converting to the time domain,

$$C(t) = Ae^{P_1 t} + Be^{P_2 t} + Ce^{P_3 t} + \cdots$$

A general algebra book will show several examples of partial fraction expansion for both non-repeated and repeated linear factors and also factors of higher degree.

In general,

$$\frac{N(X)}{D(X)} = \frac{N(X)}{G(X)H(X)L(X)} = \frac{A(X)}{G(X)} + \frac{B(X)}{H(X)} + \frac{C(X)}{L(X)}$$

where the numerator $N(X)$ must be of lower order than the denominator $D(X)$. To evaluate the coefficients, multiply both sides of the equation by $D(X)$ to clear fractions. The collect terms, equate like powers of $X$, and solve the resulting simultaneous equations for the unknown coefficients.

## Non-Repeated Poles

$$G(S) = \frac{2s+1}{s(s+1)(s+2)} = \frac{A}{s} + \frac{B}{s+1} + \frac{C}{s+2}$$

$$2s+1 = A(s+1)(s+2) + Bs(s+2) + Cs(s+1)$$

$$A = \frac{2s+1}{(s+1)(s+2)}\bigg|_{s=0} = \frac{1}{2}$$

$$B = \frac{2s+1}{s(s+2)}\bigg|_{s=-1} = \frac{-2+1}{-1(1)} = 1$$

$$C = \frac{2s+1}{s(s+1)}\bigg|_{s=-2} = \frac{-4+1}{-2(-1)} = \frac{-3}{2}$$

## Repeated Poles

$$G(S) = \frac{s^2+2}{s^3(s^2+2s-4)} = \frac{A}{s^3} + \frac{B}{s^2} + \frac{C}{s} + \frac{Ds+E}{s^2+2s-4}$$

A general expression for this type of equation is

$$\frac{N(X)}{X^m G(X)} = \frac{A_0}{X^m} + \frac{A_1}{X^{m-1}} + \cdots + \frac{A_{m-1}}{X} + \frac{F(X)}{G(X)}$$

where

$$F(X) = f_0 + f_1 X + f_2 X^2 + \cdots$$

$$G(X) = g_0 + g_1 X + g_2 X^2 + \cdots$$

$$A_0 = \frac{n_0}{g_0}, \quad A_1 = \frac{n_1 - A_0 g_1}{g_0}, \quad A_2 = \frac{n_2 - A_0 g_2 - A_1 g_1}{g_0}$$

$$A_k = \frac{1}{g_0}\left[n_k - \sum_{i=0}^{k-1} A_i g_{k-1}\right]$$

any $m: f_j = n_{m+j} - \sum_{i=0}^{m-1} A_i g_{m+j-i}$

$$N(X) = n_0 + n_1 X + n_2 X^2 + \cdots + n_i X^i$$

Evaluating coefficients of the example,

$$A_0 = A = \frac{n_0}{g_0} = \frac{2}{-4} = -\frac{1}{2}, \quad A_1 = B = \frac{n_1 - A_0 g_1}{g_0} = \frac{0 + \frac{1}{2}(2)}{-4} = -\frac{1}{4}$$

$$A_2 = C = \frac{n_2 - Ag_2 - Bg_1}{g_0} = \frac{1 - \left(-\frac{1}{2}\right)(1) - \left(-\frac{1}{4}\right)(2)}{-4} = -\frac{1}{2}$$

$$m = 3 \begin{cases} D = n_3 - A_0 g_4 - A_1 g_3 - A_2 g_2 = 0 - \left(-\frac{1}{2}\right)(0) - \left(-\frac{1}{4}\right)(0) - \left(-\frac{1}{2}\right)(1) = \frac{1}{2} \\ E = n_3 - A_0 g_3 - A_1 g_2 - A_2 g_1 = 0 - \left(-\frac{1}{2}\right)(0) - \left(-\frac{1}{4}\right)(1) - \left(-\frac{1}{2}\right)(2) = \frac{5}{4} \end{cases}$$

## PROBLEM 5.2   Partial Fractions

Given

$$C(s) = \frac{8s^2 + 3}{(s^2 + s + 1)(s - 2)} = \frac{As + B}{(s^2 + s + 1)} + \frac{C}{s - 2}$$

$$= \frac{As(s-2) + B(s-2) + C(s^2 + s + 1)}{(s^2 + s + 1)(s - 2)}$$

The values of $A$, $B$, and $C$ are:
   a.  $A = -2.75$, $B = -0.145$, and $C = 2.71$
   b.  $A = 8$, $B = 2$, and $C = 7$
   c.  $A = 3.42$, $B = 2.29$, and $C = 4.57$
   d.  $A = 3$, $B = 1$, and $C = 5$

### Solution

$$8s^2 + 3 = As(s - 2) + B(s - 2) + C(s^2 + s + 1)$$

For $s = 2$, $35 = 7C$, $C = 5$

For $s = 0$, $3 = -2B + C = -2B + 5$, $B = 1$

For $s = 1$, $11 = -A - B + 3C = -A + 14$, $A = 3$

The correct answer is d.

### Answer Rationale

Incorrect solution a is the result of not taking the square of $s$ in the first term of the equation. Incorrect solution b is the result of not taking the square of $s$ in the last term of the equation. Incorrect solution c is the result of not adding 3 to the term $8s^2$ in the equation.

If your answers are correct, go to Problem 5.3.

If your answers are not correct, review the subjects of poles and zeros and partial fractions in this chapter. For more details, see chapter 5 in reference 1 and chapter 2 in reference 4.

## PROBLEM 5.3  Inverse Laplace Transform

Find the inverse Laplace transform of

$$F(s) = \frac{10\omega^2}{s^2 - \omega^2}$$

a. $f(t) = 10\omega \cosh \omega t$
b. $f(t) = 10\omega \sinh \omega t$
c. $f(t) = 10\omega \sinh \omega$
d. $f(t) = 5\omega \sinh \omega t$

*Solution*

$$\frac{10\omega^2}{s^2 - \omega^2} = \frac{10\omega^2}{(s+\omega)(s-\omega)} = \frac{A}{(s+\omega)} + \frac{B}{(s-\omega)}$$

$$10\omega^2 = A(s-\omega) + B(s+\omega)$$

For $s = -\omega$, $10\omega^2 = -2\omega A$, $A = -5\omega$
For $s = \omega$, $10\omega^2 = 2B\omega$, $B = 5\omega$

$$F(s) = \frac{-5\omega}{s+\omega} + \frac{5\omega}{s-\omega} = 5\omega \left[ \frac{1}{s-\omega} - \frac{1}{s+\omega} \right]$$

$$f(t) = \mathcal{L}^{-1} F(s) = 5\omega(e^{\omega t} - e^{-\omega t}) = 10\omega \sinh \omega t$$

The final form is based upon the exponential definition of hyperbolic sine, as found in any book of math tables.

The correct answer is b.

*Answer Rationale*

Incorrect solution a is the result of incorrectly using the cosh function instead of the sinh function in the final answer. Incorrect solution c is the result of incorrectly using $\omega$ instead of $\omega t$ in the sinh function in the final answer. Incorrect solution d is the result of incorrectly multiplying by 5 instead of 10 times the sinh function for the final answer.

If your answer is correct, go to the next section.
If your answer is not correct, review p. 125. For more details, see chapter 5 in reference 1 and chapter 2 in reference 4.

## STABILITY

A system is stable if its impulse response approaches zero as time approaches infinity, or if every bounded input produces a bounded output.

For a system to be stable, the real parts of the roots of the characteristic equation must be negative. If they are zero, the system is *marginally stable*, meaning that the impulse response does not decay to zero although it is bounded. Additionally, certain inputs will produce unbounded outputs. Therefore, marginally stable systems are considered to be unstable.

## Example of a Stable System

Let

$$G(s) = K/s, \quad H(s) = H$$

$$\frac{C}{R}(s) = \frac{\frac{K}{s}}{1 + \frac{HK}{s}} = \frac{K}{s + HK} = K\left[\frac{1}{s + HK}\right]$$

Let $R(s) = 1/s$ (unit step input). Then

$$C(s) = K\left[\frac{1}{s + HK}\right]\left[\frac{1}{s}\right] = \frac{A}{s + HK} + \frac{B}{s} \quad \text{(partial fraction expansion)}$$

$$K = As + B(s + HK)$$

$$B = \left.\frac{K}{s + HK}\right|_{s=0} = \frac{1}{H}$$

$$A = \left.\frac{K}{s}\right|_{s=-HK} = -\frac{1}{H}$$

$$V_0(s) = \frac{1}{H}\left[\frac{1}{s} - \frac{1}{s + HK}\right]$$

$$V_0(t) = \frac{1}{H}[1 - e^{-HKt}]$$

This system is stable because as time increases, the exponential term approaches zero.

Stability may be determined by several methods. Some methods indicate only that the system is stable or not stable. Other methods are used to determine *how* stable (or unstable) the system.

Stability may be determined by obtaining the roots of the characteristic equation. Sometimes finding the roots of an *n*th order characteristic equation is not easy, in which case graphical methods can be used. Graphical methods also give a feel for *relative stability*; these methods include:

- Root locus    (time-domain technique)
- Bode Plot
- Nyquist diagram   (frequency-domain techniques)
- Nichols chart

Non-graphical methods indicate whether or not a system is stable, but not always to what extent. These methods include the following:

- Routh stability criterion
- Hurwitz stability criterion
- Continued fraction criterion

Graphical methods will be illustrated later.

The Routh stability criterion method of analysis is a simple technique and is defined as follows:

Consider an *n*th-order characteristic equation.

$$a_n s^n + a_{n-1} s^{n-1} + \cdots a_1 s + a_o = 0$$

Routh Table

| | | | | |
|---|---|---|---|---|
| $s^n$ | $a_n$ | $a_{n-2}$ | $a_{n-4}$ | ... |
| $s^{n-1}$ | $a_{n-1}$ | $a_{n-3}$ | $a_{n-5}$ | ... |
| $s^{n-2}$ | $b_1$ | $b_2$ | $b_3$ | ... |
| $s^{n-3}$ | $c_1$ | $c_2$ | $c_3$ | ... |
| . | ⋮ | ⋮ | ⋮ | ⋮ |

where $a_n, a_{n-1}, \ldots, a_o$ are coefficients of the characteristic equation and

$$b_1 = \frac{a_{n-1} a_{n-2} - a_n a_{n-3}}{a_{n-1}}$$

$$b_2 = \frac{a_{n-1} a_{n-4} - a_n a_{n-5}}{a_{n-1}}$$

$$c_1 = \frac{b_1 a_{n-3} - a_{n-1} b_2}{b_1}$$

$$c_2 = \frac{b_1 a_{n-s} - a_{n-1} b_3}{b_1}$$

All roots of this characteristic equation have negative real parts if, and only if, the elements of the first column have the same sign. Otherwise, the number of roots with positive real parts is equal to the number of sign changes. If a zero occurs in the first column, further investigation is required before stability is determined.

## PROBLEM 5.4  Routh Stability

Using the Routh stability criterion, determine the range of values of $K$ for which the control system shown in Exhibit 5.4 is stable.

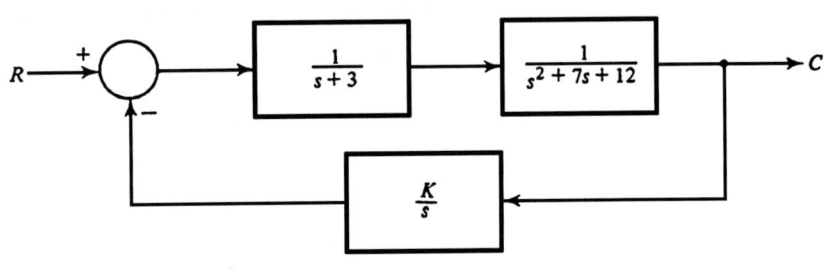

**Exhibit 5.4**

*Solution*

$$GH = \frac{K}{s(s+3)(s^2 + 7s + 12)}$$

The characteristic equation is

$$D_{GH} \pm N_{GH} = 0$$
$$s^4 + 10s^3 + 33s^2 + 36s + K = 0$$

| | | | |
|---|---|---|---|
| $s^4$ | 1 | 33 | K |
| $s^3$ | 10 | 36 | 0 |
| $s^2$ | 29.4 | K | 0 |
| $s^1$ | $36 - 0.34K$ | 0 | 0 |
| $s^0$ | K | | |

$a_n = 1,\ a_{n-1} = 10,\ a_{n-2} = 33,\ a_{n-3} = 36,\ a_{n-4} = K$

$$b_1 = \frac{a_{n-1}a_{n-2} - a_n a_{n-3}}{a_{n-1}} = \frac{330 - 36}{10} = \frac{294}{10} = 29.4$$

$$b_2 = \frac{a_{n-1}a_{n-4} - a_n a_{n-5}}{a_{n-1}} = \frac{10K - 1 \times 0}{10} = K$$

$$c_1 = \frac{b_1 a_{n-3} - a_{n-1} b_3}{b_1} = \frac{29.4 \times 36 - 10K}{29.4} = 36 - \frac{10}{29.4}K$$

All signs in the first column must be positive for stability. Therefore,

$$K > 0 \quad \text{and} \quad 36 - 0.34K > 0$$

$$K < \frac{36}{0.34} = 105.88$$

For stability, $0 < K < 105.88$.
If your answer is correct go to Problem 5.5.
If your answer is not correct, review the subject of stability in this chapter. For more details, see chapter 5 in reference 1 and chapter 6 in reference 4.

## PROBLEM 5.5  Error and Stability

For the control system in Exhibit 5.5, determine the steady-state error for a step input and the range of $K$ for which this system is stable.

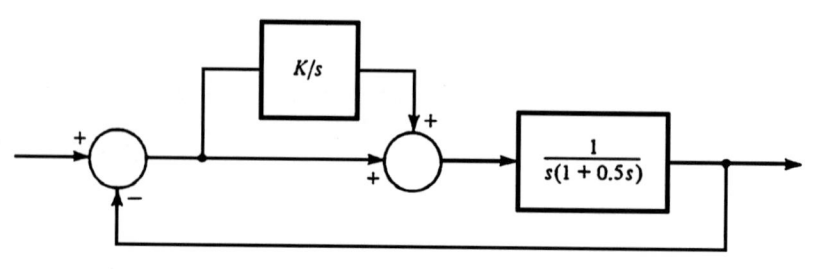

Exhibit 5.5

## Solution

Using block diagram algebra, simplify the diagram as shown in Exhibit 5.5a.

$$1 + \frac{K}{s} = \frac{s+K}{s}$$

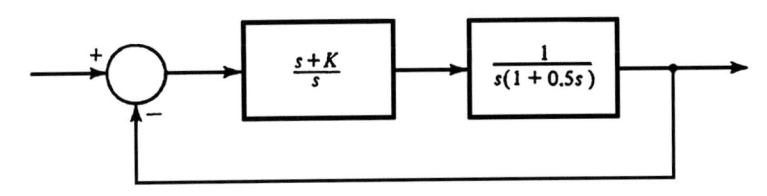

Exhibit 5.5a

$$GH = \frac{s+K}{s^2(1+0.5s)}$$

Now determine the system type. In general, for type $\ell$, $\ell = b - a$ for $b \geq a$, where

$$GH = \frac{Ks^a \prod_{i=1}^{m-a} (s+Z_i)}{s^b \prod_{i=1}^{n-b} (s+P_i)} = \frac{KB_1(s)}{s^\ell B_2(s)}$$

Therefore, a system with $\ell = 0$ is a type 0 system. If $\ell = 1$ or $\ell = 2$, the corresponding system is a type 1 or type 2 system, respectively. In this problem $a = 0$ and $b = 2$; therefore, $\ell = b - a = 2 - 0 = 2$, so this is a type 2 system. The steady-state error for a type 2 system is zero.

The characteristic equation is

$$D_{GH} \pm N_{GH} = s^2(1 + 0.5s) + s + K = 0$$
$$s^3 + 2s^2 + 2s + 2K = 0$$

Routh Table

| | | |
|---|---|---|
| $s^3$ | 1 | 2 |
| $s^2$ | 2 | $2K$ |
| $s^1$ | $2-K$ | 0 |
| $s^0$ | $2K$ | |

To assure the sign for each term in column 1 is the same (positive) for stability:

$$2 - K > 0, \ K < 2$$

Also,

$$2K > 0, \ K > 0$$

Thus, the range on $K$ is

$$0 < K < 2$$

If your answers are correct, go on to the next section.
   If your answers are not correct, review the subject of stability in this chapter. For more details, see chapter 6 in reference 1 and chapter 7 in reference 4.

# TRANSFER FUNCTION

The transfer function of a system is that factor $P(s)$ in the equation for $Y(s)$ multiplying the transform of the input $X(s)$.

$$Y(s) = P(s) X(s) + \underbrace{\text{(terms due to initial conditions)}}_{= 0 \text{ if system is at rest prior to application of the input}}$$

This equation separates the forced response excluding initial values (on the left) from the free response and the forced response initial value terms (on the right). If all initial values are zero, the equation becomes:

$$Y(s) = P(s)X(s)$$

and

$$Y(t) = \mathcal{L}^{-1}[P(s)X(s)]$$

$$P(s) = \frac{Y(s)}{X(s)}$$

Not all transfer functions are rational algebraic expressions. The transfer function of a system having time delays contains terms of the form $e^{-sT}$. The transfer function of pure time delay is $P(s) = e^{-sT}$, where $T$ is the time delay in units of time.

### PROBLEM 5.6   Transfer Function

A servomechanism for controlling angular position by means of differentially connected potentiometers consists of a servo-amplifier, a servomotor, and a feedback potentiometer (Exhibit 5.6) whose transfer functions are as follows.

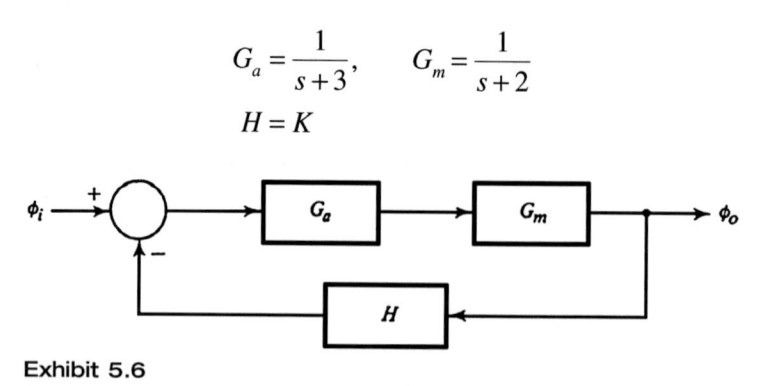

$$G_a = \frac{1}{s+3}, \quad G_m = \frac{1}{s+2}$$

$$H = K$$

**Exhibit 5.6**

Determine:

1. Open loop transfer function.
2. System transfer function.
3. Range of $K$ for a stable system.

## Solution

Exhibit 5.6a shows the system block diagram.

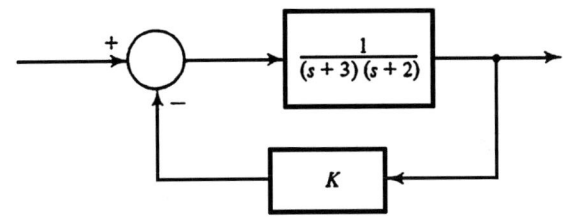

**Exhibit 5.6a**

### 1. Open Loop Transfer Function

$$\text{Open loop transfer function} = GH$$

$$= \frac{K}{(s+3)(s+2)}$$

### 2. System Transfer Function

$$\text{System transfer function} = \frac{G}{1+GH}$$

$$= \frac{\frac{1}{(s+3)(s+2)}}{1+\frac{K}{(s+3)(s+2)}} = \frac{1}{K+(s+3)(s+2)}$$

$$= \frac{1}{s^2 + 5s + (6+K)}$$

### 3. Range of K for Stable System

For the system to be stable there must be no negative factors in the denominator. Therefore, $(6 + K)$ must always be positive. Hence,

$$(6 + K) > 0$$
$$K > -6$$

If your answers are correct, go on to the next section.

If your answers are not correct, review the subject of basic feedback system in this chapter. For more details, see chapter 5 in reference 1 and chapter 2 in reference 4.

# COMPENSATION

Compensation networks are sometimes introduced into a control system to improve its operation. Compensators may be introduced into either the forward path or the feedback path in order to achieve the desired results. Compensation may be either passive or active.

Three commonly used control system compensators using passive $RC$ networks are illustrated in the following paragraphs.

## Lead Compensator

The transfer function of a lead compensator is

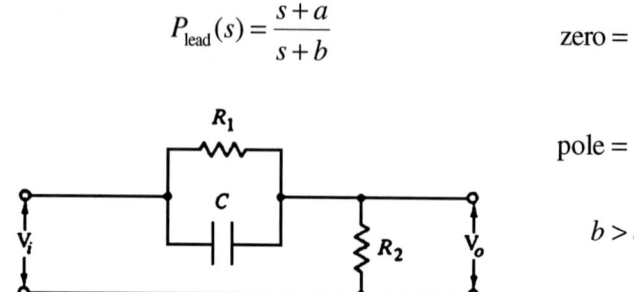

$$P_{lead}(s) = \frac{s+a}{s+b}$$

$$\text{zero} = -a = -\left[\frac{1}{R_1 C}\right]$$

$$\text{pole} = -b = -\left[\frac{1}{R_1 C} + \frac{1}{R_2 C}\right]$$

$$b > a$$

## Lag Compensator

The transfer function of a lag compensator is

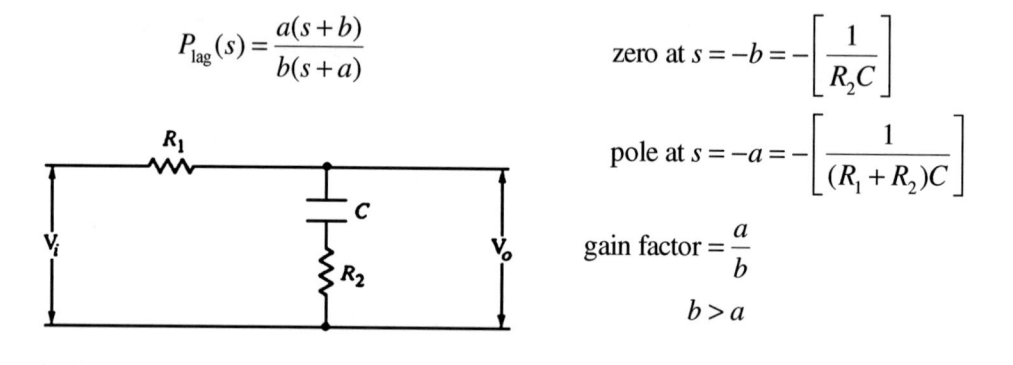

$$P_{lag}(s) = \frac{a(s+b)}{b(s+a)}$$

$$\text{zero at } s = -b = -\left[\frac{1}{R_2 C}\right]$$

$$\text{pole at } s = -a = -\left[\frac{1}{(R_1 + R_2)C}\right]$$

$$\text{gain factor} = \frac{a}{b}$$

$$b > a$$

## Lag-Lead Compensator

The transfer function of a lag-lead compensator is

$$P_{ll}(s) = \frac{(s+a_1)(s+b_2)}{(s+b_1)(s+a_2)}$$

zeros at $-a_1, -b_2$
poles at $-b_1, -a_2$
$a_1 b_2 = b_1 a_2$
$b_1 > a_1, b_2 > a_2$

**Figure 5.9** Block diagram of an unstable second order system

## Simple Lag Compensator

Shown below is a simple lag compensator with its transfer function:

$$P(s) = \frac{V_0(s)}{V_i(s)} = \frac{\frac{1}{C_S}}{R + \frac{1}{C_S}} = \frac{\frac{1}{RC}}{s + \frac{1}{RC}}$$

## System Compensation

Consider the unstable second order system shown in Figure 5.9.

$$\frac{C}{R}(s) = \frac{1/s^2}{1 + K/s^2} = \frac{1}{s^2 + K'} \quad s = \pm j\sqrt{K}$$

This system is characterized by the following:

- Crosses 0 db at −40 db/decade
- Phase margin = 0°
- Poles on $j\omega$ axis
- System oscillates

Now, add lead compensation to the forward path as shown in Figure 5.10. Or add compensation in the feedback loop as shown in Figure 5.11. The system is stable in either configuration.

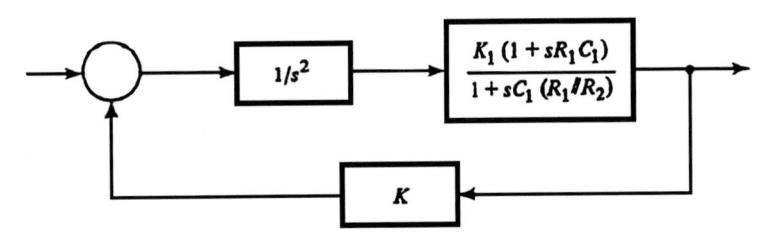

**Figure 5.10** Adding lead compensation to the forward path

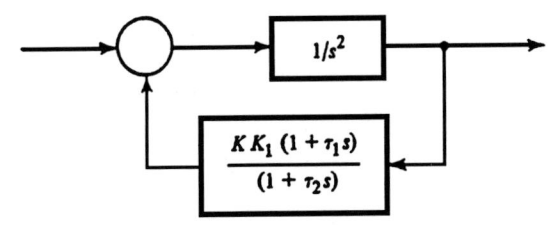

**Figure 5.11** Adding compensation in the feedback loop

## PROBLEM 5.7 Cancellation Compensation

Using cancellation compensation in the feedback control system of Exhibit 5.7, design a lead network that will yield a system having a damping ration, $\zeta$, of 0.7 and an undamped natural frequency, $\omega_n$, of 4. The resulting system gain will be

  a. System gain = 2
  b. System gain = 5.7
  c. System gain = 16
  d. System gain = 2.86

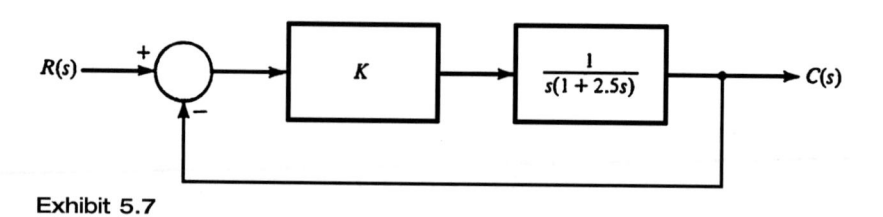

Exhibit 5.7

### Solution

$$\frac{C(s)}{R(s)} = \frac{\frac{0.4K}{s(s+0.4)}}{1+\frac{0.4K}{s(s+0.4)}} = \frac{0.4K}{s^2+0.4s+0.4K}$$

This is of the form

$$\frac{0.4K}{s^2+2\zeta\omega_n s+\omega_n^2}$$

Thus,

$$\omega_n = \sqrt{0.4K} = 4, \quad K = 40$$

$2\zeta\omega_n s = 0.4s, \quad \zeta = \dfrac{0.4}{2\omega_n} = 0.05$, which does not satisfy the requirement that $\zeta = 0.7$.

Therefore, add lead compensation to the forward path as shown in Exhibit 5.7a.

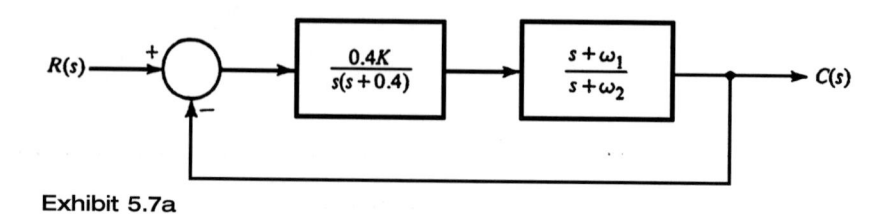

Exhibit 5.7a

For cancellation, $(s + \omega_1)$ cancels out $(s + 0.4)$, $\omega_1 = 0.4$, $\frac{s+\omega_1}{s+0.4} = 1$

$$\frac{C(s)}{R(s)} = \frac{\frac{(0.4K)(s+\omega_1)}{s(s+0.4)(s+\omega_2)}}{1+\frac{(0.4K)(s+\omega_1)}{s_s(s+0.4)(s+\omega_2)}} = \frac{0.4K(s+\omega_1)}{s(s+0.4)(s+\omega_2)+0.4K(s+\omega_1)}$$

$$= \frac{0.4K}{s(s+\omega_2)+0.4K} = \frac{0.4K}{s^2+\omega_2 s+0.4K}$$

$$= \frac{0.4K}{s^2+2\zeta\omega_n s+\omega_n^2}$$

$$\zeta = 0.7 = \frac{\omega_2}{2\omega_n}, \quad \omega_2 = 0.7 \times 2 \times 4 = 5.6 \text{ rad/sec}$$

Forward path lead compensation $= G_C(s) = \frac{s+0.4}{s+5.6}$

For the compensated system,

$$G(s) = \left(\frac{s+0.4}{s+5.6}\right)(16)\left(\frac{1}{s(s+0.4)}\right) = \frac{16}{s(s+5.6)} = \frac{16/5.6}{s(1+s/5.6)} = 2.86$$

System gain $= 2.86$

The correct answer is d.

### Answer Rationale

Incorrect solution a is the result of not multiplying by the factor 0.7 in calculating $\zeta$. Incorrect solution b is the result of not multiplying by 2 in calculating $\zeta$. Incorrect solution c is the result of not dividing by 5.6 in the numerator when calculating the system gain in the last step.

If your answer is correct, go on to problem 5.8.

If your answer is not correct, review the subject of compensation in this chapter. For more details, see chapter 5 in reference 1 and chapter 11 in reference 4.

### PROBLEM 5.8 Lead Compensation

The performance of an automatic control system is to be improved by use of a phase lead compensator. If the input to the network in Exhibit 5.8 is of the form $e_1 = E_m \sin\omega t$ and the desired output is of the form $e_2 = KE_m \sin(\omega t + \theta)$, determine the general relationship for the phase lead angle, $\theta$, and the amplifier gain $1/K$ in terms of $R_1$, $R_2$, and $C$. Assume the amplifier has a infinite input impedance.

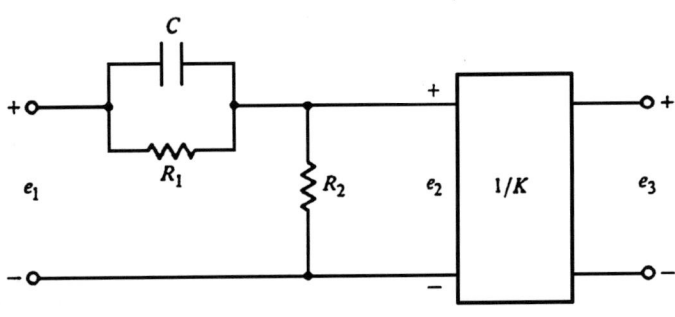

**Exhibit 5.8**

### Solution

Exhibit 5.8a shows the lead circuit.

$$e_1 = E_m \sin \omega t$$
$$e_2 = KE_m \sin(\omega t + \theta)$$
$$e_3 = E_m \sin(\omega t + \theta)$$

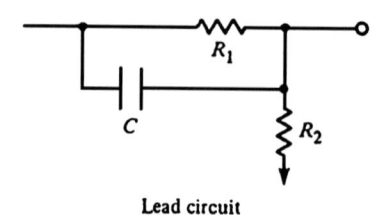

**Lead circuit**
**Exhibit 5.8a**

$$\text{Transfer function} = K \frac{(T_1 s + 1)}{(T_2 s + 1)}$$

$$\text{First break} = T_1 = R_1 C$$

$$\text{Second break} = T_2 = \frac{R_1 R_2 C}{R_1 + R_2}$$

$$K = \frac{R_2}{R_1 + R_2} \quad \text{(voltage divider)}$$

Exhibit 5.8b shows the Bode plot for this system.

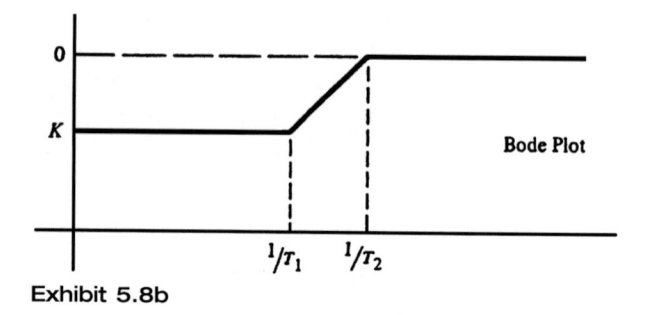

**Exhibit 5.8b**

$$\sin \theta = \frac{|K - 1|}{|K + 1|}$$

If your answer is correct, go on to the next section.

If your answer is not correct, review the subjects of compensation and Bode analysis in this chapter. For more details, see chapter 5 in reference 1 and chapter 11 in reference 4.

## BODE ANALYSIS

Bode analysis is a graphical technique which can be used to determine the relative stability of a system. Bode plots consist of two graphs: the magnitude of $GH(j\omega)$, and the phase angle of $GH(j\omega)$, both plotted as a function of frequency, $j\omega$.

Log scales are usually used for the frequency axes and for $|GH(j\omega)|$. Gain and phase margins are often defined in terms of Bode plots.

For example, given

$$G(s) = \frac{K_1}{(s+\omega_1)(s+\omega_2)}$$

$$H(s) = \frac{K_2(s+\omega_3)}{(s+\omega_4)(s+\omega_s)}$$

$$GH(s) = \frac{\frac{K_1 K_2 \omega_3 (1+\frac{s}{\omega_3})}{\omega_1 \omega_2 \omega_4 \omega_5}}{(1+\frac{\omega}{\omega_2})(1+\frac{s}{\omega_2})(1+\frac{s}{\omega_4})(1+\frac{s}{\omega_5})}$$

$$GH(j\omega) = \frac{K_3(1+j\frac{\omega}{\omega_3})}{(1+j\frac{\omega}{\omega_1})(1+j\frac{\omega}{\omega_2})(1+j\frac{\omega}{\omega_4})(1+j\frac{\omega}{\omega_5})}$$

where

$$K_3 = \frac{K_1 K_2 \omega_3}{\omega_1 \omega_2 \omega_4 \omega_5}$$

$\phi$ = phase shift

$$= -\arctan\frac{\omega}{\omega_1} - \arctan\frac{\omega}{\omega_2} + \arctan\frac{\omega}{\omega_3} - \arctan\frac{\omega}{\omega_4} - \arctan\frac{\omega}{\omega_5}$$

consider the Bode plot 20 log $|GH(j\omega)|$ vs. log $\omega$.

For $\omega \ll \omega_i$,

$$\text{gain of } \left(1+j\frac{\omega}{\omega_i}\right) = 1$$

For $\omega \gg \omega_i$,

$$\text{gain of } \left(1+j\frac{\omega}{\omega_i}\right) = +20 \text{ db/decade}$$

$$\text{gain of } \frac{1}{1+j\frac{\omega}{\omega_i}} = -20 \text{ db/decade}$$

Figure 5.12 shows the Bode plot for the system.

For the composite Bode plot as shown in Figure 5.13,

$$\text{gain total} = \Sigma \text{ individual gain plots}$$
$$= \log G_1 + \log G_2 - \log G_3 - \log G_4 - \log G_5$$

**152** Chapter 5 Control Theory

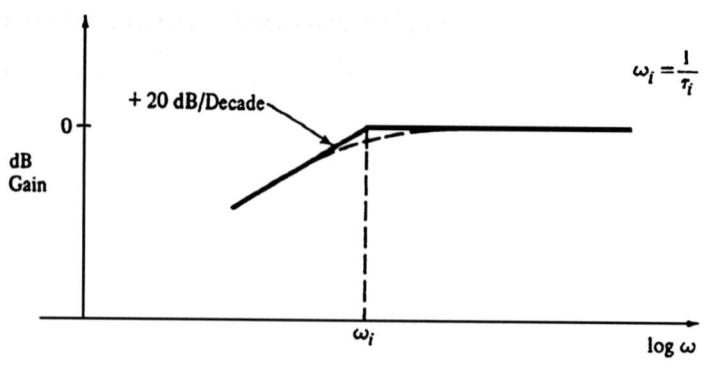

*Figure 5.12 (top): Numerator, or Lead Break — +20 dB/Decade, $\omega_i = \frac{1}{\tau_i}$*

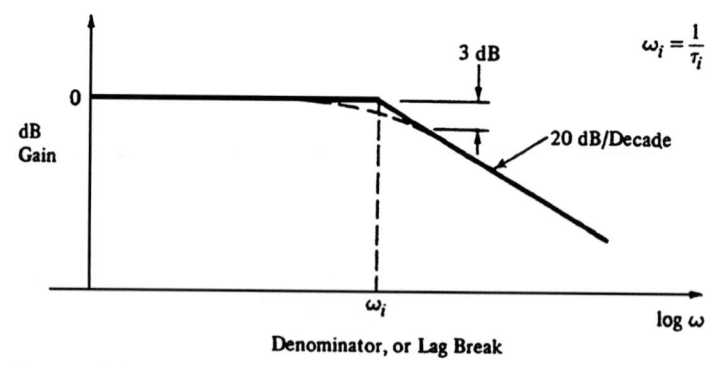

*Figure 5.12 (bottom): Denominator, or Lag Break — −20 dB/Decade, 3 dB, $\omega_i = \frac{1}{\tau_i}$*

**Figure 5.12**

The system is stable if the Bode plot is characterized as follows:

- Slope of gain plot is −20 db/decade at $\omega_{co}$ (db|GH| = 0 db)
- Total phase lag <180° at $\omega_{co}$

Thus, 1st order has −90° and is stable, while 2nd order has $-90° > \phi > -180°$.

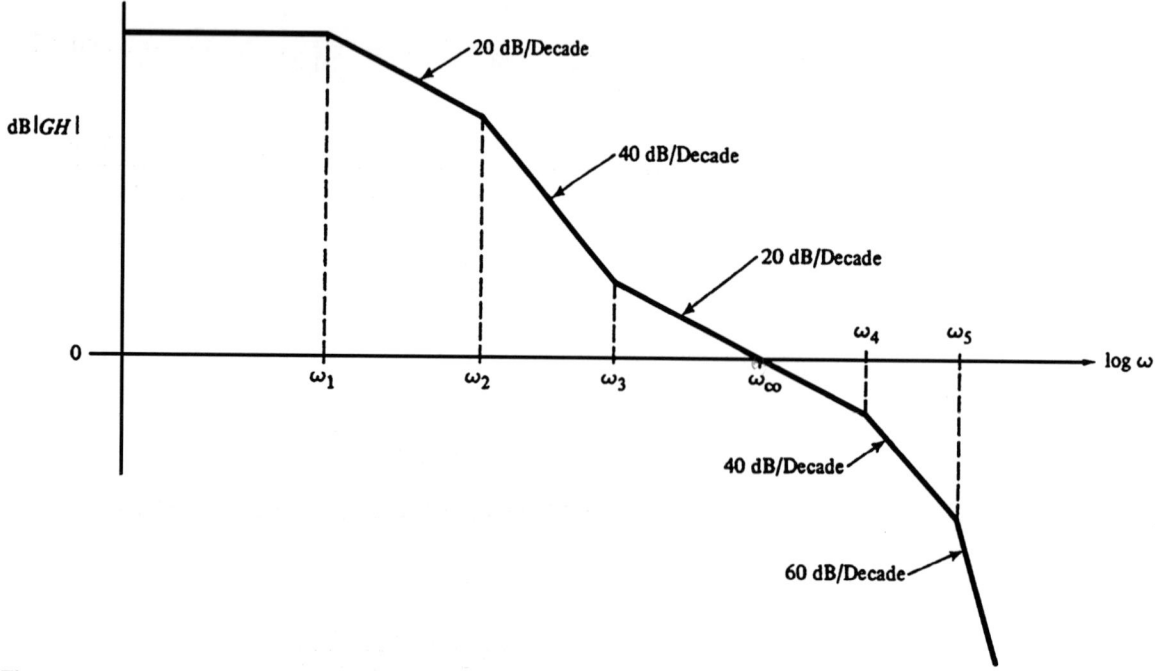

**Figure 5.13**

## PROBLEM 5.9  Bode Analysis

A control system has the block diagram shown in Exhibit 5.9.

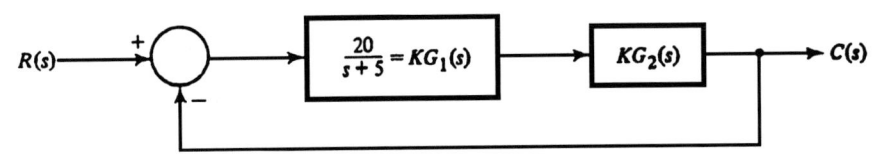

Exhibit 5.9

The frequency response of $KH_2(s)$ has been measured and plotted on the graph in Exhibit 5.9a.

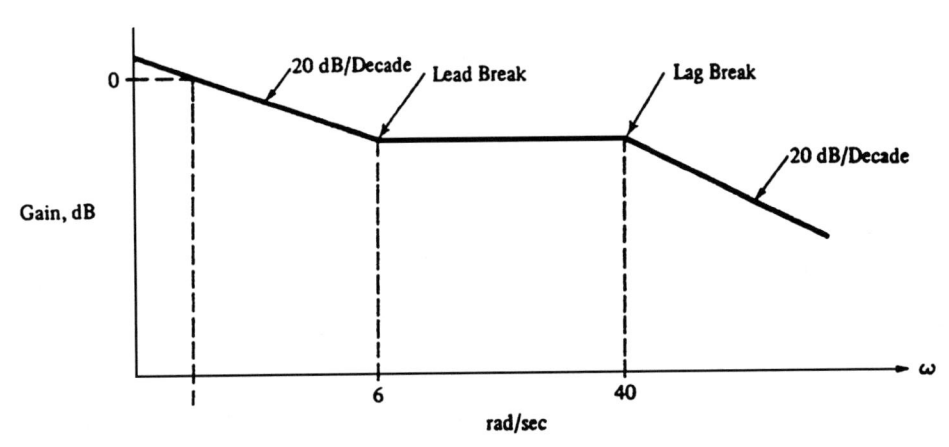

Exhibit 5.9a

Determine the system transfer function $\dfrac{C}{R}(s)$.

### Solution

From the Bode plot, it is seen that there is a lead break at $\omega = 6$ and a lag break at $\omega = 40$. Therefore, the transfer function for $KG(s)$ is

$$KG_2(s) = \frac{K(s+6)}{s(s+40)} = \frac{K(1+0.167s)6}{s(1+0.025s)40} \rightarrow \frac{0.15K(1+j0.167\omega)}{j\omega(1+j0.025\omega)}$$

$$= \frac{0.15K(1+j0.167\omega)}{-0.025\omega^2 + j\omega}$$

To find $K$, evaluate gain = 0 db at $\omega = 1$.

$$|KG_2(s)| = \frac{0.15K(1+j0.167)}{(-0.025+j1)} = \frac{0.15K(1.01)}{1} = 0.15K$$

$$db = 20 \log 0.15K \equiv 0, \ 0.15K = 1, \ \therefore K = 6.67$$

Thus,

$$KG_2(s) = \frac{6.67(s+6)}{s(s+40)}$$

Evaluating the system transfer function,

$$\frac{C}{R}(s) = \frac{KG}{1+KGH} = \frac{KG}{1+KG} \quad \text{for} \quad H=1$$

$$\frac{C}{R}(s) = \frac{1}{1+\frac{s(s+40)(s+5)}{6.67(s+6)20}} = \frac{133(s+6)}{133(s+6)+s(s+40)(s+5)}$$

$$= \frac{133(s+6)}{s^3+45s^2+333s+800}$$

If your answer is correct go to the next section.

If your answer is not correct, review the subject of Bode analysis in this chapter. For more details, see chapter 5 in reference 1 and chapter 10 in reference 4.

# ROOT LOCUS

Root locus analysis of a control system is an analytical method for displaying the location of the poles and zeros of the closed-loop transfer function as a function of the gain factor $K$ of the open loop transfer function. Root locus analysis yields accurate time-domain response characteristics while also providing frequency response information.

## PROBLEM 5.10  Root Locus

Given the feedback control system defined by the block diagram in Exhibit 5.10, determine using root locus analysis whether or not the system is stable.

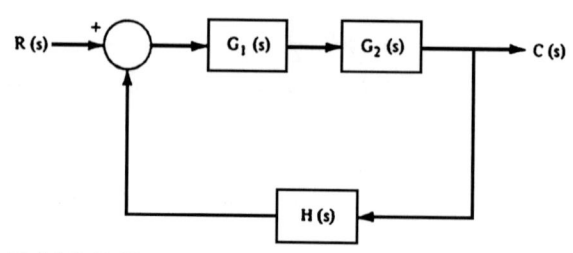

Exhibit 5.10

$$G_1(s) = \frac{50}{(s+2)(s+4)}, \quad G_2(s) = \frac{3}{s+3}, \quad H(s) = 1$$

*Solution*

The open loop transfer function is

$$GH = \frac{K}{(s+2)(s+4)(s+3)}$$

where $K = 150$.

The characteristic equation is

$$D_{GH} \pm N_{GH} = 0$$
$$150 + (s+2)(s+4)(s+3) = 0$$
$$150 + s^3 + 9s^2 + 26s + 24 = 0$$
$$s^3 + 9s^2 + 26s + 174 = 0$$

From the open loop transfer function, open loop poles (there are no zeros) are located as shown on the s-plane in Exhibit 5.10a.

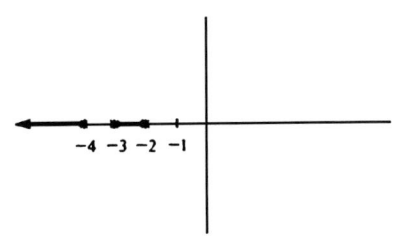

**Exhibit 5.10a**

Calculating the breakaway point, $\sigma_b$, yields

$$\frac{1}{(\sigma_b + 2)} + \frac{1}{(\sigma_b + 4)} + \frac{1}{(\sigma_b + 3)} = 0$$
$$(\sigma_b + 4)(\sigma_b + 3) + (\sigma_b + 2)(\sigma_b + 3) + (\sigma_b + 2)(\sigma_b + 4) = 0$$
$$3\sigma_b^2 + 18\sigma_b + 26 = 0$$
$$\sigma_b^2 + 6\sigma_b + \frac{26}{3} = 0$$

Using the quadratic formula*,

$$\sigma_b = \frac{-6 \pm \sqrt{36 - \frac{4 \times 26}{3}}}{2} = -3 \pm 0.58 = -3.58 \quad \text{or} \quad -2.43$$

Select −2.43 since it lies between the poles at −2 and −3.

Calculating the center of the asymptotes yields

$$\sigma_c = \frac{\sum_{i=1}^{n} P_i - \sum_{i=1}^{m} Z_i}{n - m}$$

where
$n$ = number of poles = 3
$m$ = number of zeros = 0

$$\therefore \sigma_c = \frac{9 - 0}{3} = 3$$

---

$$*x = \frac{-b \pm \sqrt{b^2 - 4ac}}{2a}$$

Calculating the angle between asymptotes and the real axis yields

$$\beta = \frac{(2N+1)180}{n-m} \text{ degrees for } K > 0$$

where $N$ is any arbitrary integer (in this case let $N = 0$).

$$\beta = \frac{180°}{3} = 60°$$

Determine the value of $K$ at the crossing of root locus on the imaginary axis. For the characteristic equation,

$$s^3 + 9s^2 + 26s + 24 + K = 0$$

let $s = j\omega$ and rewrite

$$-j\omega^3 - 9\omega^2 + 26j\omega + 24 + K = 0$$

Separating the real and imaginary parts and equating to zero,

$$-9\omega^2 + 24 + K = 0, \quad K = 9\omega^2 - 24$$
$$-\omega^3 + 26\omega = 0, \quad \omega^2 = 26$$

Thus, the root locus crosses the imaginary axis at $\omega = \sqrt{26} = 5.1$, and $K$ at this point is

$$K = 9 \times 26 - 24 = 210$$

For stability $K$ must be less than 210 in the LHP. Since $K = 150$, the system is stable.

Exhibit 5.10b shows the final plot of the root loci based on the preceding calculations.

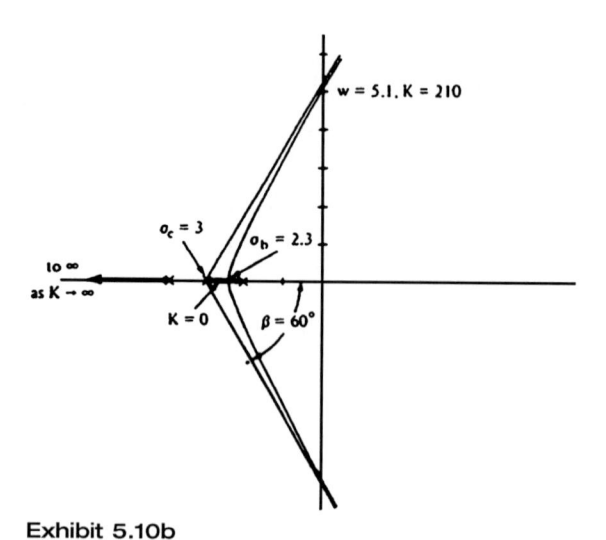

**Exhibit 5.10b**

### Alternate Solution

The system may be checked for stability using alternative methods. One quick method is by use of the Routh criteria:

$$s^3 + 9s^2 + 26s + 174 = 0$$

Routh Table

| | | | |
|---|---|---|---|
| $s^3$ | 1 | 26 | 0 |
| $s^2$ | 9 | 174 | |
| $s^1$ | 6.67 | 0 | |
| $s^0$ | 174 | | |

$$b_1 = \frac{(9)(26)-(1)(174)}{9} = 6.67$$

$$b_2 = \frac{(9)(0)-(1)(0)}{9} = 0$$

$$C_1 = \frac{(6.67)(174)-(9)(0)}{6.67} = 174$$

Since there are no sign changes in the first column, the system is stable.

If your answer is correct, go to the next chapter.

If your answer is not correct, review the subjects of stability and root locus in this chapter. For more details, see chapter 5 in reference 1 and chapter 8 in reference 4.

# CHAPTER 6

# Electronics

**OUTLINE**

DEPENDENT SOURCES 159

DIODES 159
Diode Symbol and Characteristics ■ Diode Circuit Model ■ Zener Diode ■ Zener Model

TRANSISTORS 162
Transistor Equivalent Circuits ■ Common Emitter Equivalent Circuits ■ Common Base Equivalent Circuits ■ Common Collector (Emitter-Follower) Equivalent Circuits

BIASING AND STABILITY 180

FIELD EFFECT TRANSISTORS 185

OPERATIONAL AMPLIFIERS 189
Special Cases

AMPLIFIER CLASS 193

POWER SUPPLY CIRCUITS 195

Only the most basic aspects of electronics are presented as they relate to the type of problems found in the PE Exam. Reference to a basic college text on electronics may be helpful in reviewing this subject.

## DEPENDENT SOURCES

Dependent sources are typically represented by a diamond shaped symbol instead of a circle. This is done to distinguish dependent sources, which depend on other voltages and currents, from independent sources. The figures in this chapter will reflect this symbol convention.

## DIODES

For a p-n junction diode, a simple explanation of the operation at the junction between the two types of semiconductor materials is that in each type there are many available charge carriers. These are referred to as holes (p, for positive) and electrons (n, for negative), respectively. If the diode is biased such that positive charges are forced near the boundary, a rapid recombination of the charges takes place. This represents the direction of easy current flow (forward bias). However, if the diode is reverse biased, the free charges are attracted toward their bias polarities. This creates a depletion region at the junction with very little recombination and almost no current flows.

The classical diode equation is

$$I = I_s(e^{qV/kT} - 1)$$

where
- $I_s$ = saturation current
- $q$ = charge on an electron = $1.602 \times 10^{-19}$ coulomb
- $k$ = Boltzmann's constant = $1.38 \times 10^{-16}$ erg/K
- $T$ = temperature in K (K = °C + 273)
- $V$ = forward voltage across the diode

If $V$ is about four times greater than $kT/q$ (say 0.1 volt), the diode equation simplifies to

$$I = I_s(e^{qV/kT})$$

and the dynamic forward resistance at a specific operating point is

$$R_f = \frac{kT}{qI} \Omega$$

At room temperature, $kT/q$ is approximately 26 mV; thus, a diode with 1 mA of forward current will exhibit 26 ohms of dynamic resistance.

The following are diode models, characteristics, and applications.

## Diode Symbol and Characteristics

The diode symbol used in circuit diagrams and the volt-amp characteristics of a diode are shown in Figure 6.1.

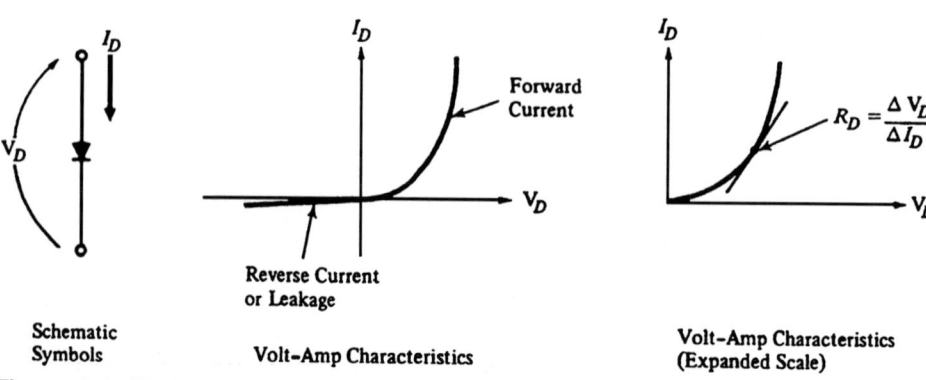

**Figure 6.1** Diode symbol and characteristics

## Diode Circuit Model

The diode circuit model is shown in Figure 6.2.

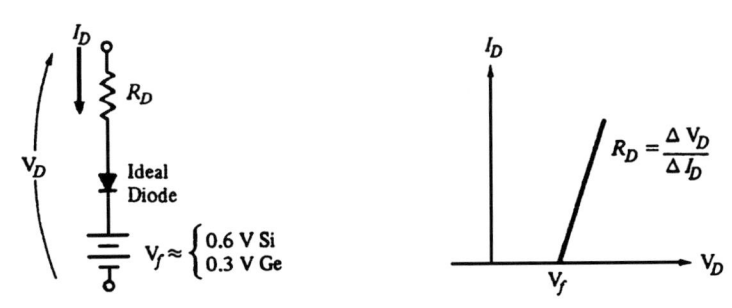

**Figure 6.2** Diode circuit model

## Zener Diode

The volt-amp characteristics of a zener diode are shown in Figure 6.3.

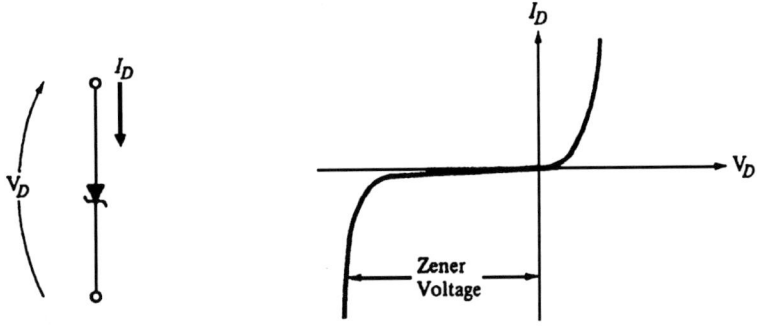

**Figure 6.3** Zener diode

## Zener Model

The zener model is shown in Figure 6.4.

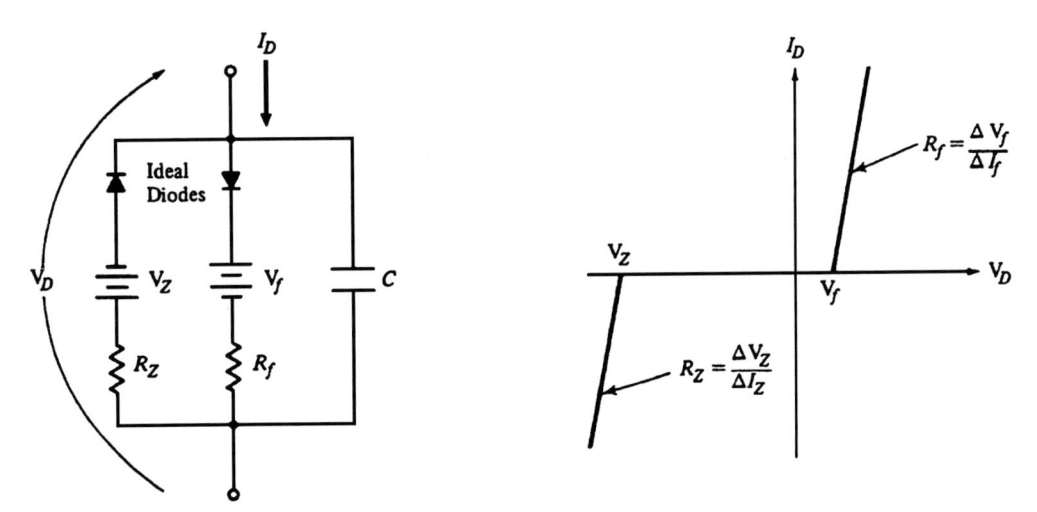

**Figure 6.4** Zener model

### PROBLEM 6.1 Diode Suppression

A relay coil is to be protected against excessively high induced voltage when it is open-circuited. Exhibit 6.1 shows the circuit. Show the correct connection of the diode and specify its minimum voltage and current ratings.

Exhibit 6.1

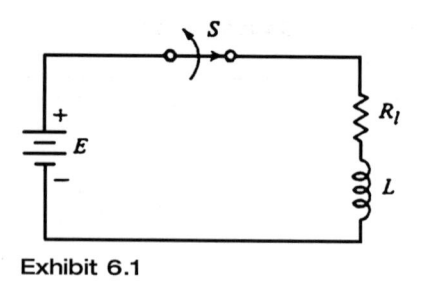

### Solution

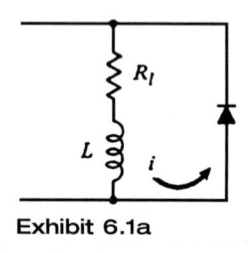

The diode is connected as shown in Exhibit 6.1a so that it will conduct when the switch is open, thus permitting the coil current to decay to zero while preventing arcing of the switch contacts.

Maximum voltage across the diode is $E$ volts, and maximum current through the diode is $E/R$ amps. To provide a conservative safety margin, a diode should be selected that can withstand twice these values.

If your answer is correct, continue to the next section.

If your answer is not correct, review the subject of diodes in this chapter. For more details, see chapter 4 in reference 1 and chapter 2 in reference 5.

Exhibit 6.1a

## TRANSISTORS

Bipolar transistors may be used in three configurations: common emitter, common base, and common collector (emitter-follower). Transistor models with their parameters expressed in different ways are delineated below. Several aspects of transistors must be considered in practical designs, including temperature variation, leakage current, gain-bandwidth, maximum operating frequency, operating load current, biasing, and stability.

Temperature variation can affect current gain, collector leakage, and power dissipation. As a rule of thumb, collector leakage $I_{co}$ doubles with every $\Delta t$ temperature increase, where $\Delta t = 10°C$ for germanium and $6°C$ for silicon. It also increases with increased collector voltage. The *gain-bandwidth product*, $f_T$, is a common emitter parameter; it is the frequency at which gain $h_{fe}$, drops to unity (0 dB). Although common emitter current gain is 0 dB at $f_T$, there may still be considerable power gain at $f_T$ due to different input and output impedance levels. Thus, $f_T$ is not necessarily the maximum useful operating frequency. In a practical circuit, the minimum load current should be no less than 10 times the leakage current. The maximum load current should not exceed the maximum power dissipation voltage. The operating load current should be midway between these two extremes. When the operating load current is flowing at the selected no-signal point, the collector load resistance should drop the collector voltage to one-half the supply voltage, for class A operation.

### Transistor Equivalent Circuits

Transistor equivalent circuits in the common emitter, common base, and common collector configurations are shown below. Direction of current flow depends upon whether the transistor is PNP or NPN. Table 6.1 delineates the small signal characteristics; it provides for conversion between *T*-parameters and *h*-parameters in either direction for the three circuit configurations, and it gives typical actual parameter values. Table 6.2 summarizes the *h*-parameters for the three circuit configurations.

**Table 6.1** Transistor Small Signal Characteristics
(Numerical values are for a typical transistor operating under standard conditions)

| Symbols | | | Common Emitter | Common Base | Common Collector | T-Equivalent |
|---|---|---|---|---|---|---|
| $h_{11e}$ | $h_{ie}$ | $r_i$ | 1500 ohms | $\dfrac{h_{ib}}{1+h_{fb}}$ | $h_{ic}$ | $r_b + \dfrac{r_e}{1-\alpha}$ |
| $h_{12e}$ | $h_{re}$ | | $3 \times 10^{-4}$ | $\dfrac{h_{ib}h_{ob}}{1+h_{fb}} - h_{rb}$ | $1 - h_{rc}$ | $\dfrac{r_e}{(1-\alpha)r_c}$ |
| $h_{21e}$ | $h_{fe}$ | $A_I$ | 49 | $-\dfrac{h_{fb}}{1+h_{fb}}$ | $-(1+h_{fc})$ | $\dfrac{\alpha}{1-\alpha} = \beta$ |
| $h_{22e}$ | $h_{oe}$ | $\dfrac{1}{r_o}$ | $30 \times 10^{-6}$ mho | $\dfrac{h_{ob}}{1+h_{fb}}$ | $h_{oc}$ | $\dfrac{1}{(1-\alpha)r_c} = \dfrac{1}{r_d}$ |
| $h_{11b}$ | $h_{ib}$ | $r_i$ | $\dfrac{h_{ie}}{1+h_{fe}}$ | 30 ohms | $-\dfrac{h_{ic}}{h_{fc}}$ | $r_e + r_b(1-\alpha)$ |
| $h_{12b}$ | $h_{rb}$ | | $\dfrac{h_{ie}h_{oe}}{1+h_{fe}} - h_{re}$ | $5 \times 10^{-4}$ | $h_{re} - 1 - \dfrac{h_{ic}h_{oc}}{h_{fc}}$ | $\dfrac{r_b}{r_c}$ |
| $h_{21b}$ | $h_{fb}$ | $A_I$ | $-\dfrac{h_{fe}}{1+h_{fe}}$ | $-0.98$ | $-\dfrac{1+h_{fc}}{h_{fc}}$ | $-\alpha$ |
| $h_{22b}$ | $h_{ob}$ | $\dfrac{1}{r_o}$ | $\dfrac{h_{oe}}{1+h_{fe}}$ | $0.5 \times 10^{-6}$ mho | $\dfrac{h_{oc}}{h_{fc}}$ | $\dfrac{1}{r_c}$ |
| $h_{11c}$ | $h_{ic}$ | $r_i$ | $h_{ie}$ | $\dfrac{h_{ib}}{1+h_{fb}}$ | 1500 ohms | $r_b + \dfrac{r_e + R_l}{1-\alpha} = R_l(\beta+1)$ |
| $h_{12c}$ | $h_{rc}$ | | $1 - h_{re} \approx 1$ | 1 | 1 | $1 - \dfrac{r_e}{(1-\alpha)r_c}$ |
| $h_{21c}$ | $h_{fc}$ | $A_I$ | $-(1+h_{fe})$ | $-\dfrac{1}{1+h_{fb}}$ | $-50$ | $-\dfrac{1}{1-\alpha} = -(\beta+1)$ |
| $h_{22c}$ | $h_{oc}$ | $\dfrac{1}{r_o}$ | $h_{oe}$ | $\dfrac{h_{ob}}{1+h_{fb}}$ | $30 \times 10^{-6}$ mho | $\dfrac{1}{(1-\alpha)r_c}$ |
| | $\alpha$ | | $\dfrac{h_{fe}}{1+h_{fe}}$ | $-h_{fb} = -h_{21b}$ | $\dfrac{1+h_{fc}}{h_{fc}}$ | 0.98 |
| | $r_c$ | | $\dfrac{1+h_{fe}}{h_{oe}}$ | $\dfrac{1-h_{rb}}{h_{ob}} = \dfrac{1}{h_{22b}}$ | $-\dfrac{h_{fc}}{h_{oc}}$ | 1.7 MΩ |
| | $r_e$ | | $\dfrac{h_{re}}{h_{oe}}$ | $h_{ib} - \dfrac{h_{rb}}{h_{ob}}(1+h_{fb})$ | $\dfrac{1-h_{rc}}{h_{oc}}$ | 10 Ω |
| | $r_b$ | | $h_{ie} - \dfrac{h_{re}}{h_{oe}}(1+h_{fe})$ | $\dfrac{h_{rb}}{h_{ob}} = \dfrac{h_{12b}}{h_{22b}}$ | $h_{ic} + \dfrac{h_{fc}}{h_{oc}}(1-h_{rc})$ | 1 KΩ |

$r_i$ = input resistance
$r_o$ = output resistance
$A_V$ = voltage amplification
$A_I$ = current amplification

$r_c$ = collector resistance
$r_e$ = emitter resistance
$r_b$ = base resistance
$\alpha$ = short circuit current multiplier

power gain = $G = A_V A_I$
$\beta = \dfrac{\alpha}{1-\alpha}$, $\alpha = \dfrac{\beta}{\beta+1}$
$R_l$ = load resistance

**Table 6.2** *h*-Parameters

| Symbol | Definition | General | Common Emitter | Common Base | Common Collector |
|---|---|---|---|---|---|
| $h_{11}$ | $Z_{in}$ with output shorted | $\left.\dfrac{\Delta V_1}{\Delta I_1}\right\|_{\Delta V_2=0}$ | $\left.\dfrac{\Delta V_{BE}}{\Delta I_B}\right\|_{\Delta V_{CE}=0} = \dfrac{h_{11}}{(1+h_{21})}$ | $\left.\dfrac{\Delta V_{EB}}{\Delta I_E}\right\|_{\Delta V_{CB}=0} = h_{11}$ | $\left.\dfrac{\Delta V_{CB}}{\Delta I_B}\right\|_{\Delta V_{CB}=0} = \dfrac{h_{11}}{(1+h_{21})}$ |
| $h_{12}$ | inverse voltage transfer ratio with input open | $\left.\dfrac{\Delta V_1}{\Delta V_2}\right\|_{\Delta I_1=0}$ | $\left.\dfrac{\Delta V_{BE}}{\Delta V_{CE}}\right\|_{\Delta I_B=0} = \dfrac{(\Delta h_e - h_{12})}{(1+h_{21})}$ | $\left.\dfrac{\Delta V_{EB}}{\Delta V_{CB}}\right\|_{\Delta I_E=0} = h_{12}$ | $\left.\dfrac{\Delta V_{CB}}{\Delta V_{EC}}\right\|_{\Delta I_B=0} = 1$ |
| $h_{21}$ | forward current transfer ratio with output shorted | $\left.\dfrac{\Delta I_2}{\Delta I_1}\right\|_{\Delta V_2=0}$ | $\left.\dfrac{\Delta I_C}{\Delta I_B}\right\|_{\Delta V_{CE}=0} = -\dfrac{h_{21}}{(1+h_{21})}$ | $\left.\dfrac{\Delta I_C}{\Delta I_E}\right\|_{\Delta V_{CB}=0} = h_{21}$ | $\left.\dfrac{\Delta I_E}{\Delta I_B}\right\|_{\Delta V_{CB}=0} = -\dfrac{1}{(1+h_{21})}$ |
| $h_{22}$ | $Y_{out}$ with input open | $\left.\dfrac{\Delta I_2}{\Delta V_2}\right\|_{\Delta I_1=0}$ | $\left.\dfrac{\Delta I_C}{\Delta V_{CE}}\right\|_{\Delta I_B=0} = \dfrac{h_{22}}{(1+h_{21})}$ | $\left.\dfrac{\Delta I_C}{\Delta V_{CB}}\right\|_{\Delta I_E=0} = h_{22}$ | $\left.\dfrac{\Delta I_E}{\Delta V_{EC}}\right\|_{\Delta I_B=0} = \dfrac{h_{22}}{(1+h_{21})}$ |
| $A_V$ | $e_o/e_i$ | $\dfrac{R_l}{h_{11}+\Delta h R_l}$ | $\dfrac{-h_{fe}}{h_{ie}G_l + \Delta h_e}$ | $\dfrac{-h_{fb}}{h_{ib}G_l + \Delta h_b}$ | $\dfrac{h_{fc}}{h_{ic}G_l + \Delta h_c}$ |
| $R_i$ | input resistance | $\dfrac{h_{11}+\Delta h R_l}{1+h_{22}R_l}$ | $h_{ie} - \dfrac{h_{fe}h_{re}}{h_{oe}+G_l}$ | $h_{ib} - \dfrac{h_{fb}h_{rb}}{h_{ob}+G_l}$ | $h_{ic} - \dfrac{h_{fc}h_{rc}}{h_{oc}+G_l}$ |
| $R_o$ | output resistance | $\dfrac{h_{11}+R_g}{\Delta h + h_{22}R_g}$ | $\dfrac{R_g + h_{ie}}{h_{oe}(R_g+h_{ie}) - h_{fe}h_{re}}$ | $\dfrac{h_{ib}+R_g+R_B(1+h_{fb})}{\Delta h_b' + R_g h_{ob}}$ | $\dfrac{R_g + h_{ic}}{R_g h_{oc} + \Delta h_c}$ |
| $A_I$ | $i_o/i_i$ | $\dfrac{h_{21}}{1+h_{22}R_l}$ | $\left[\dfrac{h_{fe}}{1+h_{oe}R_l}\right]\left[\dfrac{R_B}{R_B+R_i}\right]$ | $h_{fb}$ | $\dfrac{h_{fc}}{1+h_{oc}R_E+\Delta h_c}$ |
| $G$ | $\|A_V A_I\|$ | $\dfrac{h_{21}^2 R_l}{(1+h_{22}R_l)(h_{11}+\Delta h R_l)}$ | $\|A_V A_I\|$ | $\left\|\dfrac{-h_{fb}^2}{h_{ib}G_l + \Delta h_b}\right\|$ | $\left\|\dfrac{-h_{fc}^2}{h_{ic}G_l + \Delta h_c}\right\|$ |

$\Delta h_e = h_{ie}h_{oe} - h_{fe}h_{re}$   $\Delta h_b = h_{ib}h_{ob} - h_{fb}h_{rb}$   $\Delta h_c = h_{ic}h_{oc} - h_{fc}h_{rc}$   $Y_2' = \dfrac{1}{R_2} + \dfrac{1}{R_l} = G_l$

$\Delta h_b' = \dfrac{\{h_{ib}(1+h_{ob}R_B) + (1-h_{rb})(1+h_{fb})R_B\}h_{ob} - (h_{rb}+h_{ob}R_B)(h_{fb}-h_{ob}R_B)}{(1+h_{ob}R_B)^2}$

# Common Emitter Equivalent Circuits

- Transistor Circuits

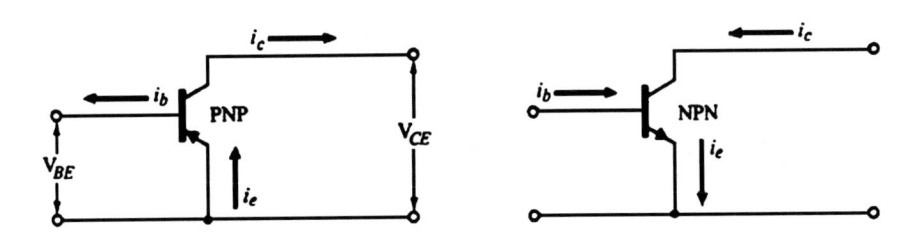

$$i_e = i_b + i_c, \quad i_c = \alpha i_e + I_{co}$$

- T-Equivalent Circuits

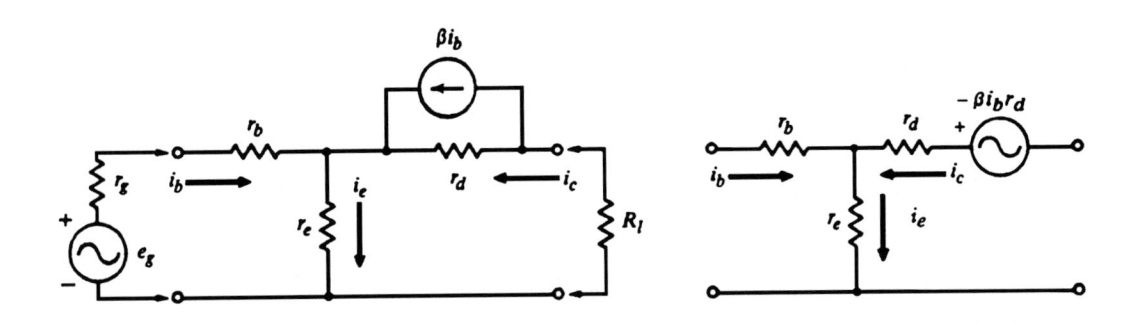

$r_d = r_c(1-\alpha)$ = equivalent emitter-collector transresistance

- Hybrid-Equivalent Circuit

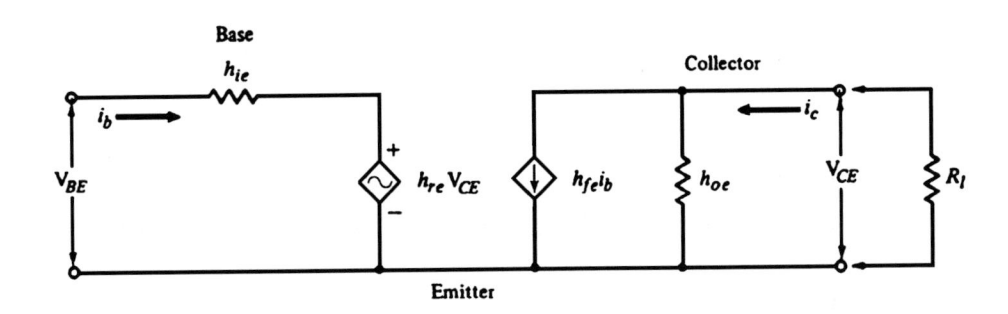

- Simplified Gain Equations (see Tables 6.1 and 6.2)

$$A_V \approx -\frac{h_{fe}R_l}{h_{ie}} = \frac{h_{fe}R_l}{h_{ie}} = -\frac{\beta R_l}{r_b + r_e}$$

$$A_I = \beta = \frac{\alpha}{1-\alpha} = \frac{\beta r_c(1-\alpha)}{r_c(1-\alpha) + R_l}$$

$$G_p = A_V A_I = \frac{\beta^2 R_l}{r_b + r_e}$$

# Common Base Equivalent Circuits

■ Transistor Circuits

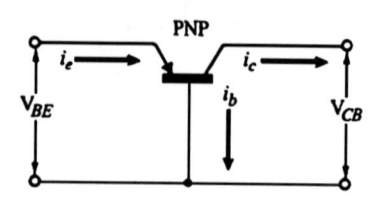

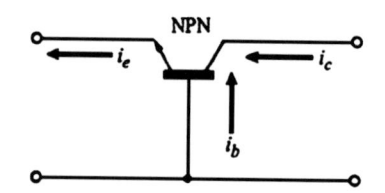

■ T-Equivalent Circuits

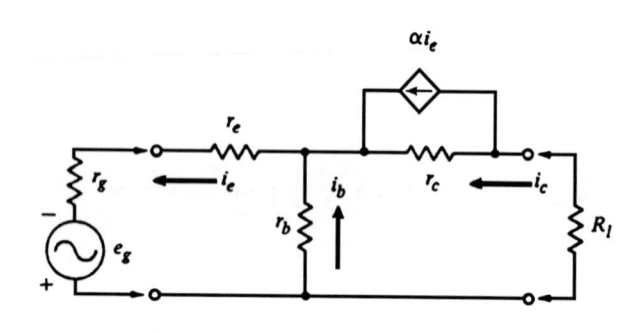

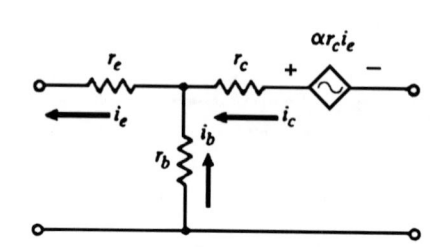

■ Hybrid-Equivalent Circuit

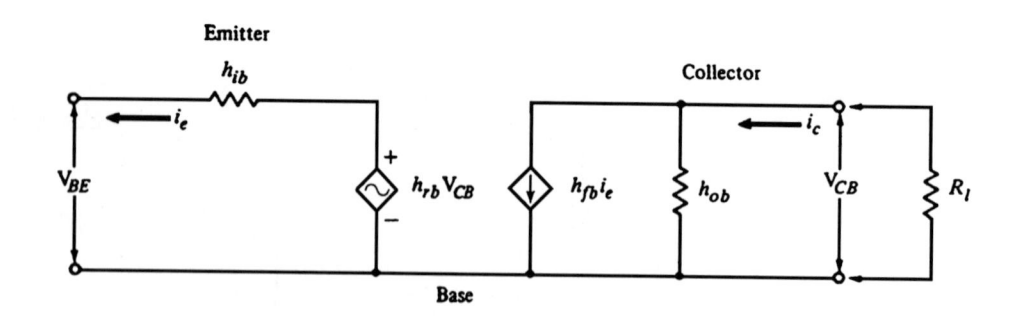

■ Simplified Gain Equations (see Tables 6.1 and 6.2)

$$A_V = \frac{\alpha R_l}{r_e + r_b(1-\alpha)} \approx \frac{\alpha R_l}{r_e}$$

$$A_I = -\alpha$$

$$G_p = A_V A_I = \frac{\alpha^2 R_l}{r_e + r_b(1-\alpha)} \approx \frac{\alpha^2 R_l}{R_e}$$

# Common Collector (Emitter-Follower) Equivalent Circuits

- Transistor Circuits

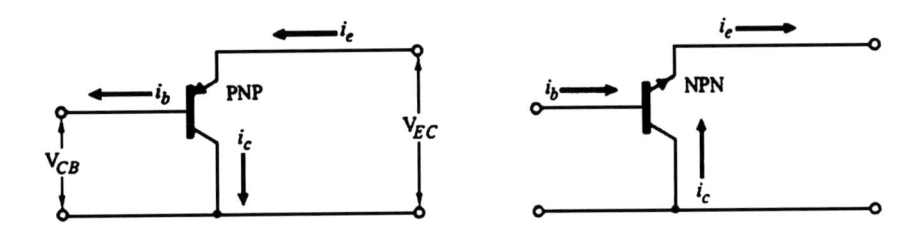

- T-Equivalent Circuit

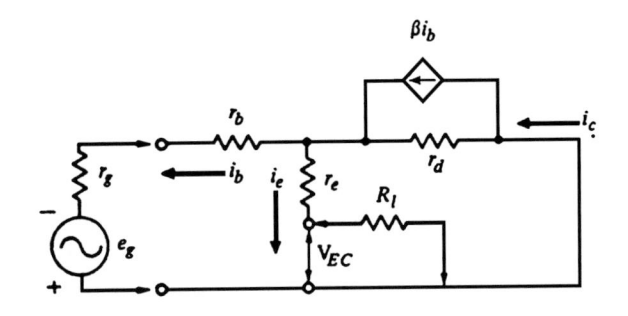

- Hybrid-Equivalent Circuit

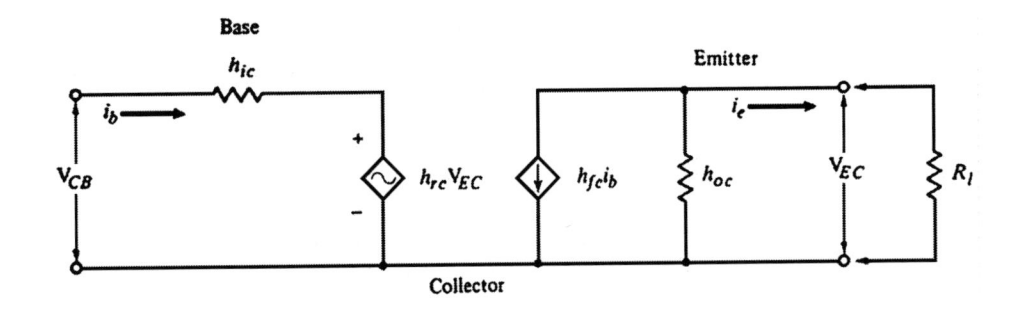

- Simplified Gain Equations (see Tables 6.1 and 6.2)

$$A_V = \frac{-h_{fc}}{h_{ic}G_l + (h_{ic}h_{oc} - h_{rc}h_{fc})} = \frac{R_l}{R_l + r_e + r_b(1-\alpha)} \approx 1$$

$$A_I = -(\beta + 1)$$

$$G_P = A_V A_I = \beta + 1 \frac{1}{(1-\alpha)}$$

## PROBLEM 6.2  One-Stage Transistor Amplifier

The mid-frequency voltage gain $A_V$ of the transistor circuit in Exhibit 6.2 is:
- a. $A_V = -105$
- b. $A_V = -103.5$
- c. $A_V = -98.5$
- d. $A_V = -112$

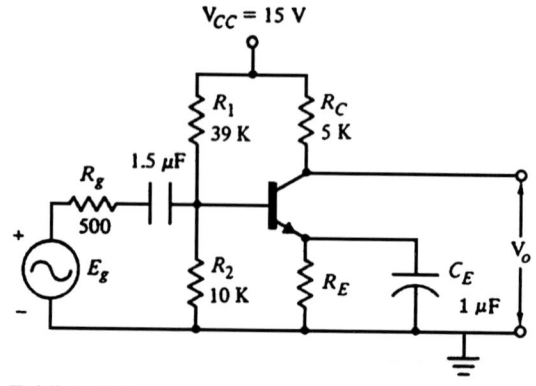

Exhibit 6.2

$$\alpha = 0.98 \qquad h_{ie} = 1500 \; \Omega$$
$$r_e = 10 \; \Omega \qquad h_{re} = 3 \times 10^{-4}$$
$$r_b = 1 \; k\Omega \qquad h_{fe} = 49$$
$$r_c = 1.7 \; M\Omega \qquad h_{oe} = 30 \times 10^{-6} \; \text{siemens}$$

### Solution

Convert the circuit to a hybrid-equivalent as shown in Exhibit 6.2a.

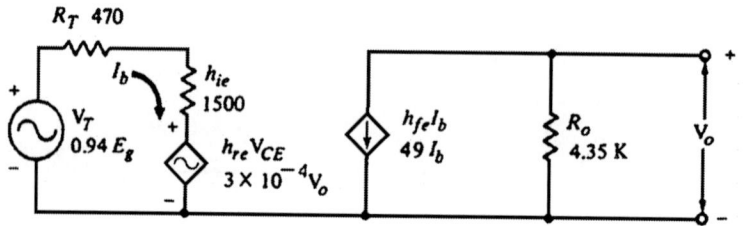

Exhibit 6.2a

$$R_o = \frac{1}{h_{oe}} \| R_C = \frac{(33 \; K)(5 \; K)}{38 \; K} = 4.35 \; k\Omega$$

$$V_T = \frac{R_1 \| R_2 \, E_g}{R_g + R_1 \| R_2} = \frac{7959 \, E_g}{8459} = 0.94 \, E_g$$

$$R_T = \frac{(500)(7959)}{8459} = 470 \; \Omega$$

From Kirchhoff's Laws,

$$V_o = -(4350)(49 I_b)$$
$$0.94 E_g = (470 + 1500) I_b + 3 \times 10^{-4} V_o$$

Solving for $I_b$:

$$0.94E_g = 1970I_b - 3\times 10^{-4} \times 4350 \times 49 I_b$$
$$= I_b(1970-64) = 1906 I_b$$

$$I_b = \frac{0.94}{1906} E_g$$

$$A_V = \frac{V_o}{E_g} = -\frac{(4350)(49)}{E_g}\left(\frac{0.94}{1906}E_g\right) = -105$$

### Alternative Solution

An alternative circuit conversion is shown in Exhibit 6.2b.

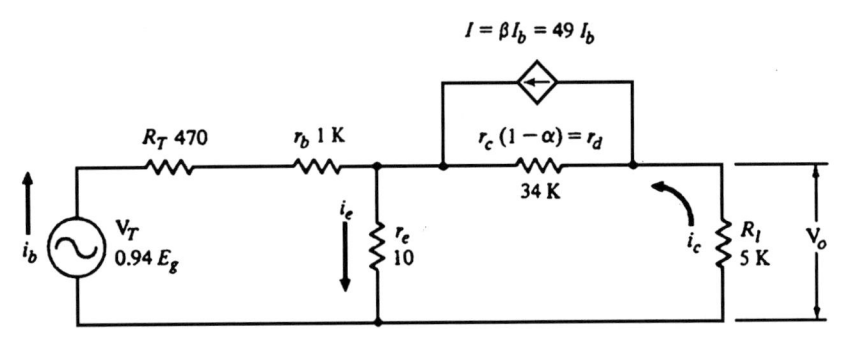

**Exhibit 6.2b**

$$A_I = \frac{I_c}{I_b} = \frac{\beta r_c(1-\alpha)}{r_c(1-\alpha)+R_l} = \frac{49 \times 1.7 \times 10^6 \times 0.02}{1.7 \times 10^6(0.02)+5\times 10^3} = \frac{1.67 \times 10^6}{3.9 \times 10^4} = 42.7$$

$$A_V = \frac{V_o}{E_g} = -0.94\left[\frac{R_l}{R_i + R_T}\right]A_I$$

$$R_i = h_{ie} - \frac{h_{fe}h_{re}}{h_{oe}+G_l} = 1500 - \frac{(40)(3\times 10^{-4})}{30\times 10^{-6}+\frac{1}{5k}} = 1500 - 52 = 1448$$

$$A_V = -0.94\left[\frac{5000}{1448+470}\right]42.8 = -105$$

### Alternative Solution

An alternative circuit conversion is shown in Exhibit 6.2c.

$$R_B = \frac{(39\text{ k})(10\text{ k})}{49\text{ k}} = 7.96\text{ k}$$

$$A_{V_1} = \frac{-h_{fe}}{h_{ie}G_l + [h_{ie}h_{oe}-h_{re}h_{fe}]}$$

$$= \frac{-49}{1500 \times \frac{1}{5k}+[1500\times 30\times 10^{-6}-3\times 10^{-4}\times 49]} = \frac{-49}{0.33} = -148$$

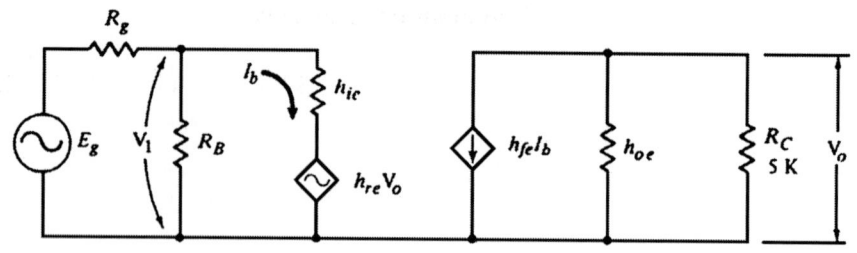

Exhibit 6.2c

$$R_1 = R_i \| R_B = \frac{(1448)(7960)}{1448 + 7960} = 1225$$

$$V_1 = \frac{R_1}{R_g + R_1} E_g = \frac{1225 E_g}{500 + 1225} = 0.71 E_g$$

$$A_V = A_{V_1} \times \frac{V_1}{E_g} = 0.71 A_{V_1} = -0.71 \times 148 = -105$$

The correct answer is a.

*Answer Rationale*

Incorrect solution b is the result of incorrectly using the value of $R_1 \| R_2$ instead of $R_g + R_1 \| R_2$ in the denominator when calculating $R_T$. Incorrect solution c is the result of not using the minus sign for the number $3 \times 10^{-4}$ in the equation when solving for $I_b$. Incorrect solution d is the result of not multiplying by the factor 0.94 in the numerator when solving for $A_V$.

If your answers are correct, go to Problem 6.3.

If your answers are not correct, review the subject of common emitter equivalent circuits in this chapter. For more details, see chapter 4 in reference 1 and chapters 7 and 8 in reference 5.

## PROBLEM 6.3 Common Base Amplifier

Calculate the input impedance for the common base amplifier circuit shown in Exhibit 6.3. Neglect all capacitances.

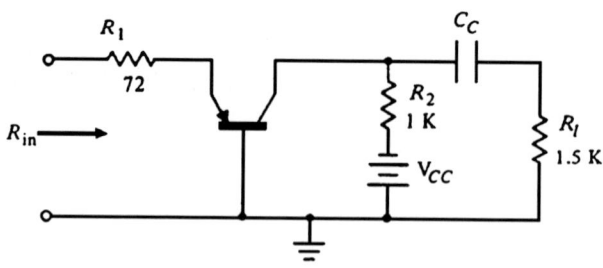

Exhibit 6.3

The transistor *h*-parameters are

$$h_{ib} = 30 \, \Omega$$
$$h_{rb} = 5 \times 10^{-4}$$
$$h_{fb} = -0.98$$
$$h_{ob} = 0.5 \times 10^{-6} \text{ siemens}$$

The input impedance is
- a. $R_{in} = 101.7\ \Omega$
- b. $R_{in} = 102.30\ \Omega$
- c. $R_{in} = 30.29\ \Omega$
- d. $R_{in} = 102.29\ \Omega$

*Solution*

$$R_i = h_{ib} - \frac{h_{fb}h_{rb}}{Y_2' + h_{ob}}, \qquad Y_2' = \frac{1}{R_2} + \frac{1}{R_l} = \frac{1}{1000} + \frac{1}{1500} = \frac{1}{600}\ \text{siemens}$$

$$R_i = 30 - \frac{(-0.98)(5\times 10^{-4})}{\frac{1}{600} + 0.5\times 10^{-6}} = 30 + 0.29 = 30.29$$

$$R_{in} = R_1 + R_i = 72 + 30.29 = 102.29\ \Omega$$

*Alternative Solution*

$$R_i = \frac{\Delta h_b + h_{ib}Y_2'}{h_{ob} + Y_2'}$$

$$\Delta h_b = h_{ib}h_{ob} - h_{fb}h_{rb} = (30)(0.5\times 10^{-6}) - (-0.98)(5\times 10^{-4}) = 5.05\times 10^{-4}$$

$$R_i = \frac{5.05\times 10^{-4} + (30)(1.67\times 10^{-3})}{0.5\times 10^{-6} + 1.67\times 10^{-3}} = 30.29, \qquad R_{in} = 72 + 30.29 = 102.29\ \Omega$$

The correct answer is d.

*Answer Rationale*

Incorrect solution a is the result of incorrectly using the plus sign for the term $h_{fb}h'_{rb}/(y_2 + h_{ob})$ when calculating $R_i$. Incorrect solution b is the result of incorrectly using 5 instead of 0.5 in the denominator when solving for $R_i$. Incorrect solution c is the result of not adding the value of $R_1$ to $R_i$ in the last step when calculating $R_{in}$ (or incorrectly assuming that $R_i = R_{in}$).

If your answers are correct, go to Problem 6.4.

If your answers are not correct, review the subject of common base equivalent circuits in this chapter. For more details, see chapter 4 in reference 1 and chapters 3 and 7 in reference 5.

## PROBLEM 6.4 Two-Stage Transistor Amplifier, Common Emitter

The following information is given for the amplifier circuit shown in Exhibit 6.4.

$h_{ib} = 30\ \Omega, \quad h_{rb} = 5\times 10^{-4}, \quad h_{fb} = -0.98, \quad h_{ob} = 0.5\times 10^{-6}\ \text{siemens}$
$R_{b1} = 4.7\ \text{k}\Omega, \quad R_{l1} = 10\ \text{k}\Omega, \quad R_{b2} = 4.7\ \text{k}\Omega, \quad R_{l2} = 22\ \text{k}\Omega, \quad R_g = 500\ \Omega$

Determine the voltage gain of each stage, overall power gain, inter-stage losses and pre-first stage losses, and total losses.

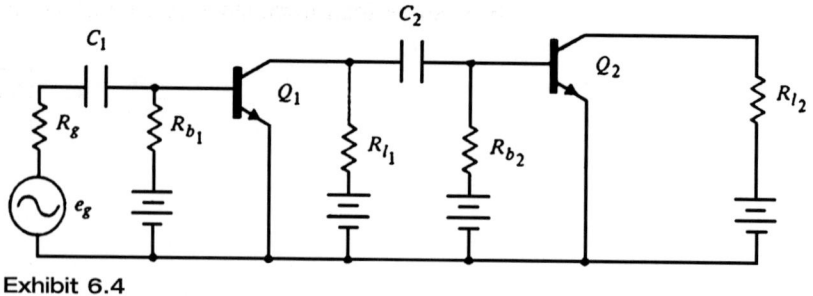

## Exhibit 6.4

### Solution

The transistors are connected in common emitter configuration. Therefore, the $h$-parameters given in common base must be converted as follows:

$$h_{ie} = \frac{h_{ib}}{1+h_{fb}} = \frac{30}{1-0.98} = 1500 \, \Omega$$

$$h_{re} = \frac{h_{ib}h_{ob}}{1+h_{fb}} - h_{rb} = \frac{30 \times 0.5 \times 10^{-6}}{1-0.98} - 5 \times 10^{-4} = 2.5 \times 10^{-4}$$

$$h_{fe} = \frac{-h_{fb}}{1+h_{fb}} = \frac{0.98}{1-0.98} = 49$$

$$h_{oe} = \frac{h_{ob}}{1+h_{fb}} = \frac{0.5 \times 10^{-6}}{1-0.98} = 2.5 \times 10^{-5} \text{ siemens}$$

This problem is best solved by breaking up the circuit and determining the gain of each stage (Exhibit 6.4a).

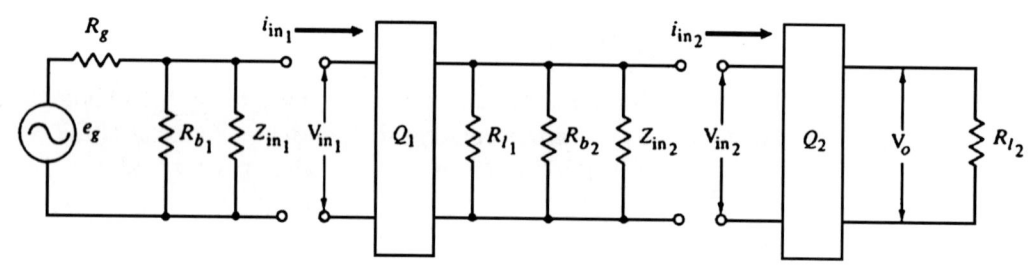

### Exhibit 6.4a

Capacitances are neglected.

The following two equations will be used in voltage gain and impedance calculations:

$$A_V = \frac{-h_{fe}}{h_{ie}G_l + (h_{ie}h_{oe} - h_{re}h_{fe})}$$

$$Z_{in} = h_{ie} - \frac{h_{re}h_{fe}}{h_{oe} + G_l}$$

Second stage voltage gain is

$$A_{V2} = \frac{V_o}{V_{in2}} = \frac{-49}{(1500)(4.545 \times 10^{-5}) + (1500)(2.5 \times 10^{-5}) - (2.5 \times 10^{-4})(49)}$$
$$= -524.48$$

First stage load impedance is the parallel combination of $R_{L1}$, $R_{b2}$, and $Z_{in2}$.

$$Z_{in2} = 1500 - \frac{(2.5 \times 10^{-4})(49)}{(2.5 \times 10^{-5}) + 4.545 \times 10^{-5}} = 1326 \, \Omega$$

Hence

$$G'_{l1} = \frac{1}{R_{l1}} + \frac{1}{R_{b2}} + \frac{1}{Z_{in2}} = \frac{1}{10^4} + \frac{1}{4.7 \times 10^3} + \frac{1}{1.326 \times 10^3} = 1.067 \times 10^{-3} \text{ siemens}$$

First stage voltage gain is

$$A_{V1} = \frac{-49}{(1500)(1.067 \times 10^{-3}) + (1500)(2.5 \times 10^{-5}) - (2.5 \times 10^{-4})(49)}$$
$$= -30.14$$

Total voltage gain = $A_V = (A_{V1})(A_{V2}) = (-30.14)(-524) = 15{,}808$

$$Z_{in1} = 1500 - \frac{(2.5 \times 10^{-4})(49)}{2.5 \times 10^{-5} + 1.067 \times 10^{-3}} = 1500 - 11.22 = 1489 \, \Omega$$

$$P_{in} = \frac{V_i^2}{R_{b1} \| Z_{in1}}, \quad R_{b1} \| Z_{in1} = \frac{(4700)(1489)}{4700 + 1489} = 1130.76 \, \Omega$$

$$P_{in} = \frac{V_i^2}{1130.76}$$

$$P_{out} = \frac{V_o^2}{R_{l2}} = \frac{V_o^2}{22{,}000}$$

Overall power gain is

$$G = \frac{P_{out}}{P_{in}} = \left[\frac{V_o^2}{22{,}000}\right]\left[\frac{1130.76}{V_i^2}\right] = \frac{1130.76}{22{,}000} A_V^2 = \frac{1130.76}{22{,}000}(15{,}808)^2$$
$$= 1.284 \times 10^7$$

In dB,

$$G_{dB} = 10 \log G = 10(7 + 0.109) = 71.09 \text{ dB}$$

Interstage power loss is that power lost in the parallel combination of $R_{l1}$ and $R_{b2}$.

$$R_{l1} \| R_{b2} = \frac{(10\,\text{K})(4.7\,\text{K})}{10\,\text{K} + 4.7\,\text{K}} = 3200 \, \Omega$$

$$P_{IL} = \frac{V_{in2}^2}{3200} = \frac{(30.14 V_{in1})^2}{3200} = 0.28 V_{in1}^2$$

Pre-first stage loss is that power lost in $R_{b1}$:

$$P_{PFSL} = \frac{V_{in1}^2}{R_{b1}} = \frac{V_{in1}^2}{4700} = 2.13 \times 10^{-4} V_{in1}^2$$

Total losses in terms of $e_g$ are calculated as follows: Make a Thevenin's equivalent circuit out of the pre-first stage circuit (Exhibit 6.4b).

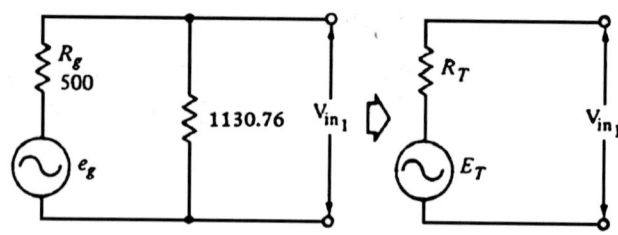

## Exhibit 6.4b

$$R_T = \frac{(1130.76)(500)}{1630.76} = 346.7 \, \Omega$$

$$E_T = \frac{1130.76}{1630.76} e_g = 0.6934 e_g$$

$$P_{TL} = P_{IL} + P_{PFSL} = (0.28 + 2.13 \times 10^{-4})(0.6934)^2 e_g^2$$
$$= 0.1356 \, e_g^2$$

If your answers are correct, go to Problem 6.5.

If your answers are not correct, review the subject of common emitter equivalent circuits and table 6.1 in this chapter. For more details, see chapter 4 in reference 1 and chapter 12 in reference 5.

### PROBLEM 6.5 Two-Stage Voltage Gain, *h*-Parameters

The $h$ = parameters and bias resistor values are given for the two-stage transistor amplifier in Exhibit 6.5. Assume all capacitors are short circuits at signal frequency. The voltage gain, $V_2/V_s$ is:

a. $V_2/V_s = -40.54$      b. $V_2/V_s = -39.73$
c. $V_2/V_s = 39.73$       d. $V_2/V_s = -38.9$

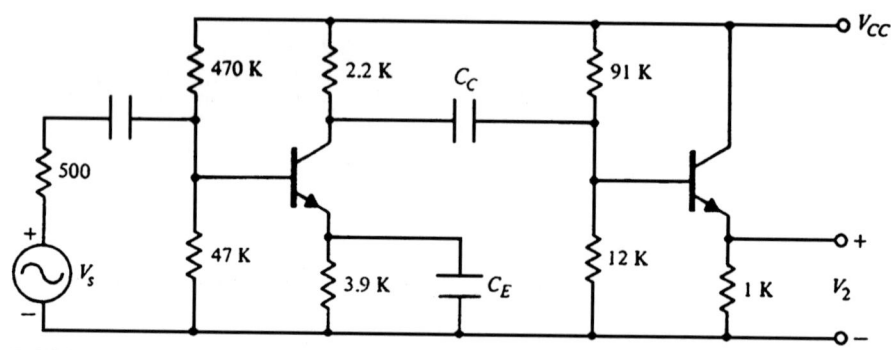

Exhibit 6.5

$$h_{ie} = 1500 \, \Omega$$
$$h_{re} = 3 \times 10^{-4}$$
$$h_{fe} = 49$$
$$h_{oe} = 30 \times 10^{-6} \text{ siemens}$$
$$h_{ic} = 1500 \, \Omega$$
$$h_{rc} = 1$$
$$h_{fc} = -50$$
$$h_{oc} = 30 \times 10^{-6} \text{ siemens}$$

## Solution

Notice that the first stage is CE and the second stage is CC (emitter follower). Draw an equivalent circuit of the first stage (Exhibit 6.5a).

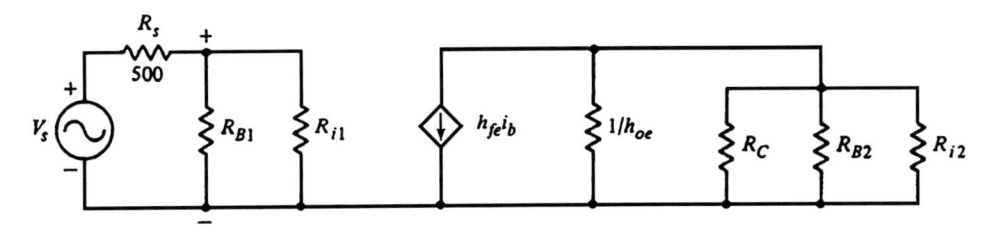

**Exhibit 6.5a**

$$R_{B1} = 470\,k \| 47\,k = 42.73\,k\Omega$$

$$R_{B2} = 91\,k \| 12\,k = 10.6\,k\Omega$$

$$R_{i2} = h_{ic} - \frac{h_{fc}h_{rc}}{h_{oc} + G_{L2}} = 1500 - \frac{(-50)(1)}{30 \times 10^{-6} + 10^{-3}} = 50.04\,k\Omega$$

$$G_{L1} = \frac{1}{R_C} + \frac{1}{R_{B2}} + \frac{1}{R_{i2}} = \frac{1}{2.2\,k} + \frac{1}{10.6\,k} + \frac{1}{50.04\,k} = 0.57 \times 10^{-3}\,\text{siemens}$$

$$R_{i1} = h_{ie} - \frac{h_{fe}h_{re}}{h_{oe} + G_{L1}} = 1500 - \frac{(49)(3 \times 10^{-4})}{30 \times 10^{-6} + 0.57 \times 10^{-3}} = 1475.5\,\Omega$$

$$R_1 = \frac{V_1}{i_1} = R_{B1} \| R_{i1} = \frac{(42.73)(1.4755)}{42.73 + 1.4755} = 1.43\,k\Omega$$

$$V_1 = \frac{R_1}{R_S + R_1} V_S = \frac{1.43}{1.93} V_S = 0.74\,V_S$$

$$R_{L1} = \frac{1}{G_{L1}} = \frac{1}{0.57 \times 10^{-3}} = 1754.4\,\Omega$$

The voltage gain of the first stage (CE) is

$$A_{V1} = \frac{-h_{fe}}{h_{ie}G_{L1} + \Delta_{he}}$$

$$= \frac{-49}{(1500)(0.57 \times 10^{-3}) + [(1500)(30 \times 10^{-6}) - (49)(3 \times 10^{-4})]} = -55.35$$

The voltage gain of the second stage (CC) is

$$A_{V2} = \frac{-h_{fc}}{h_{ic}G_{L2} + \Delta h_c}$$

$$= \frac{-(-50)}{(1500)(10^{-3}) + [(1500)(30 \times 10^{-6}) - (-50)(1)]} = \frac{50}{51.55} = 0.97$$

$$V_2 = V_1 A_{V1} A_{V2} = (0.74\,V_s)(-55.35)(0.97) = -39.73\,V_s$$

$$\frac{V_2}{V_s} = -39.73$$

The correct answer is b.

### Answer Rationale

Incorrect solution a is the result of incorrectly using the value of $h_{fc}$ instead of $h_{fe}$ in the numerator when calculating $A_{V_1}$. Incorrect solution c is the result of not using the minus sign for $h_{fc}$ in the numerator when calculating $A_{V_2}$. Incorrect solution d is the result of incorrectly using the value of $h_{fe}$ instead of $h_{fc}$ in the numerator when calculating $A_{V_2}$.

If your solution is correct, go on to Problem 6.6.

If your solution is not correct, review the subjects of common emitter, common base, and common collector equivalent circuits as well as tables 6.1 and 6.2 in this chapter. For more details, see chapter 4 in reference 1 and chapter 12 in reference 5.

### PROBLEM 6.6 Current Gain, $Z_{OUT}$, h-Parameters

The h-parameters for the single-stage transistor shown in Exhibit 6.6 are

$$h_{ie} = 1400 \, \Omega \quad h_{re} = 4 \times 10^{-4}$$
$$h_{fe} = 50 \quad h_{oe} = 25 \times 10^{-6} \text{ siemens}$$

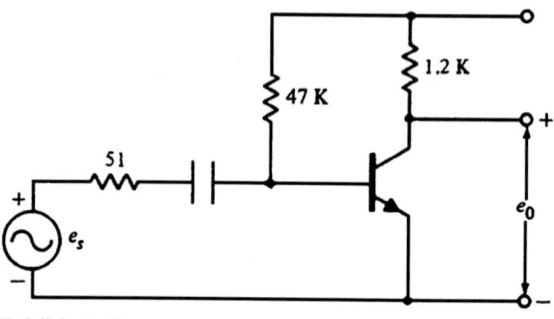

Exhibit 6.6

The current gain and output impedance are:
a. $A_I = 47.14$, $Z_{out} = 89.231 \text{ k}\Omega$
b. $A_I = 32.82$, $Z_{out} = 89.231 \text{ k}\Omega$
c. $A_I = 47.14$, $Z_{out} = 25.784 \text{ k}\Omega$
d. $A_I = 47.16$, $Z_{out} = 89.231 \text{ k}\Omega$

### Solution

$$R_i = h_{ie} - \frac{h_{re} h_{fe} R_L}{1 + h_{oe} R_L} = 1400 - \frac{(4 \times 10^{-4})(50)(1.2 \times 10^3)}{1 + (25 \times 10^{-6})(1.2 \times 10^3)} = 1376.7 \, \Omega$$

$$A_I = \frac{h_{fe}}{1 + h_{oe} R_L} \left[ \frac{R_B}{R_B + R_i} \right] = \frac{50}{1 + (25 \times 10^{-6})(1.2 \times 10^3)} \left[ \frac{47}{47 + 1.3767} \right]$$
$$= 47.16$$

$$Y_{out} = h_{oe} - \frac{h_{fe} h_{re}}{R_s + h_{ie}} = 25 \times 10^{-6} - \frac{(50)(4 \times 10^{-4})}{51 + 1400} = 1.12 \times 10^{-5} \text{ siemens}$$

$$Z_{out} = \frac{1}{Y_{out}} = 89.231 \text{ k}\Omega$$

The correct answer is d.

## Answer Rationale

Incorrect solution a is the result of incorrectly using the value of $h_{oe}$ instead of $h_{re}$ in the numerator when calculating $R_i$. Incorrect solution b is the result of incorrectly using the value of $h_{re}$ instead of $h_{oe}$ in the denominator when calculating $A_I$. Incorrect solution c is the result of incorrectly using the plus sign for the term $h_{fe}h_{re}/(R_s + h_{ie})$ when calculating $Y_{out}$.

If your answers are correct, go on to Problem 6.7.

If your answers are not correct, review the subjects of common emitter, common base, and common collector equivalent circuits as well as tables 6.1 and 6.2 in this chapter. For more details, see chapter 4 in reference 1 and chapters 7 and 8 in reference 5.

## PROBLEM 6.7 Transistor Amp–Upper and Lower Cutoff Frequency

The circuit in Exhibit 6.7 represents one stage of an RC-coupled amplifier:

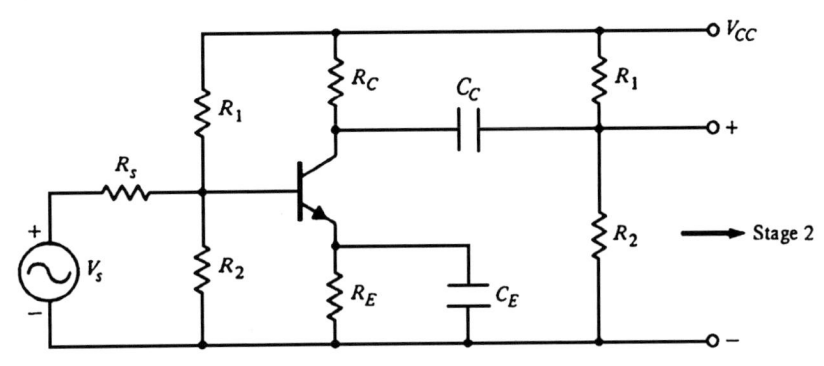

**Exhibit 6.7**

$h_{ie} = 1800\ \Omega$  
$h_{re} = 3 \times 10^{-4}$  
$h_{fe} = 50$  
$h_{oe} = 5 \times 10^{-6}$ siemens  

beta cutoff frequency, $f_B = 2.5$ MHz  
$R_B = R_1 \parallel R_2 = 4.7$ k$\Omega$  
$R_E = 50\ \Omega$  
$C_p = 620$ pF  
$R_s = 2200\ \Omega$

$C_p$ represents the total effective parallel capacitance for transistor and wiring.

1. Calculate $R_C$ to yield an upper cutoff frequency, $f_2$, of 300 kHz.

2. Determine the mid-frequency current gain.

3. Determine the values of $C_C$ and $C_E$ to yield a lower cutoff frequency, $f_1$, of 200 Hz.

## Solution

### 1. Calculation of $R_C$

The equivalent circuit is shown in Exhibit 6.7a. Since $f_\beta$ is well above $f_2$, the output resistance becomes

$$R_o = \frac{1}{2\pi f_2 C_p} = \frac{1}{2\pi \times 300 \times 10^3 \times 620 \times 10^{-12}} = 856\ \Omega$$

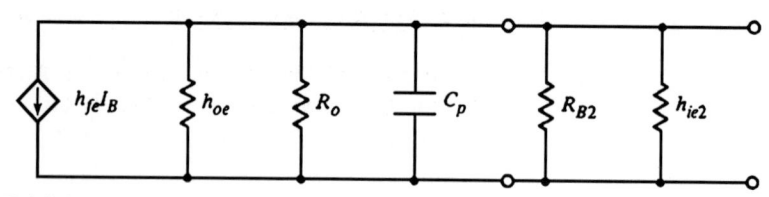

Exhibit 6.7a

$$G_o = \frac{1}{R_o} = \frac{1}{856} = \frac{1}{R_C} + h_{oe} + \frac{1}{R_{B2}} + \frac{1}{h_{ie2}}$$

$$= \frac{1}{R_C} + 5 \times 10^{-6} + \frac{10^{-3}}{4.7} + \frac{1}{1800} = \frac{1}{R_C} + 7.73 \times 10^{-4}$$

$$\frac{1}{R_C} = \frac{1}{856} - 7.73 \times 10^{-4} = 3.95 \times 10^{-4}$$

$$R_C = 2530\ \Omega$$

2. **Mid-Frequency Current Gain**

$$A_{I\,mid} = -\frac{h_{fe} R_o}{h_{ie}} = -\frac{50 \times 856}{1800} = -23.78$$

3. **Values of $C_C$ and $C_E$ to Yield Lower Cutoff Frequency of 200 Hz**
The equivalent circuit is shown in Exhibit 6.7b.

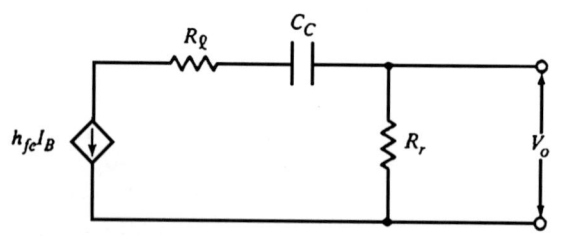

Exhibit 6.7b

$$R_\ell = \frac{R_C / h_{oe}}{R_C + 1/h_{oe}} = \frac{2530/5 \times 10^{-6}}{2530 + \frac{1}{5 \times 10^{-5}}} = 2498.4\ \Omega$$

$$R_r = \frac{R_B h_{ie}}{R_B + h_{ie}} = \frac{4.7 \times 10^3 \times 1800}{4.7 \times 10^3 + 1800} = 1301.5\ \Omega$$

$$C_C = \frac{1}{2\pi f_1 (R_\ell + R_r)} = \frac{1}{2\pi \times 200(2498.4 + 1301.5)} = 0.21\ \mu F$$

$$C_E = \frac{1 + h_{fe}}{2\pi f_1 R_s} = \frac{1 + 50}{2\pi \times 200 \times 2200} = 18.4\ \mu F$$

If your answers are correct, go on to Problem 6.8.

If your answers are not correct, review the subjects of common emitter, common base, and common collector equivalent circuits as well as tables 6.1 and 6.2 in this chapter. For more details, see chapter 4 in reference 1 and chapter 11 in reference 5.

## PROBLEM 6.8 Darlington/Transient Problem

Determine the voltage across the capacitor 0.25 second after the switch is open in the darlington amplifier circuit shown in Exhibit 6.8.

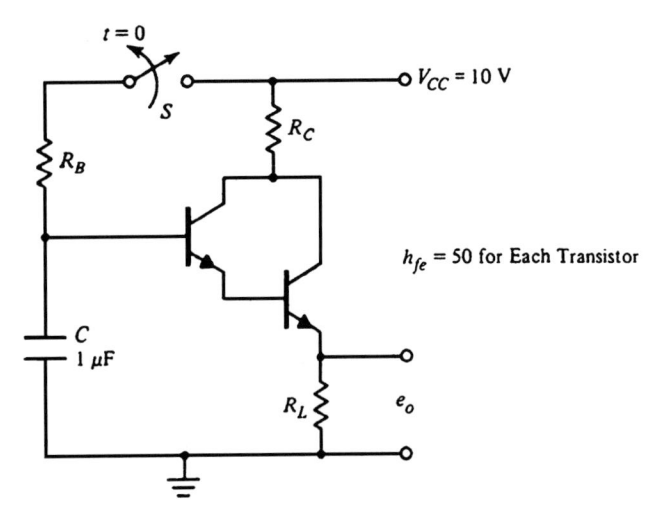

**Exhibit 6.8**

Assuming the switch has been closed for a long time, the voltage across the capacitor 0.25 second after the switch is open is:
  a. $V_C = 3.7$ volts     b. $V_C = 0$ volts
  c. $V_C = 9.999$ volts   d. $V_C = 27.2$ volts

### Solution

Assume capacitor is initially charged to $V_{CC} = 10$ volts. Current gain of a darlington is $\beta^2 = 50^2 = 2500$. Exhibit 6.8a shows an equivalent circuit with $R_L$ reflected into the input.

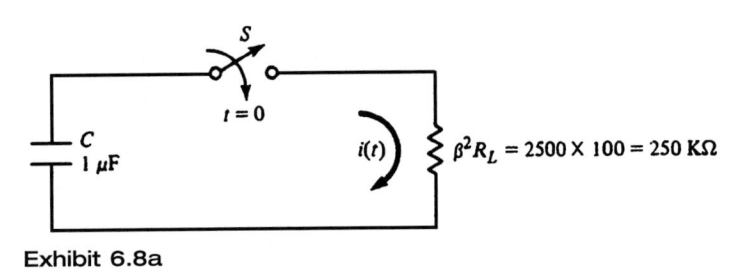

**Exhibit 6.8a** ($\beta^2 R_L = 2500 \times 100 = 250$ K$\Omega$)

$$\tau = RC = 250 \times 10^3 \times 10^{-6} = 0.25 \text{ sec.}$$
$$V_C(t) = 10 e^{-t/\tau} = 10 e^{-4t}$$

For $t = 0.25$ seconds, $V_C = 10 e^{-1} = 3.7$ volts.
The correct answer is a.

### Answer Rationale

Incorrect solution b is the result of not multiplying by $10^3$ in calculating $\tau$. Incorrect solution c is the result of not multiplying by $10^{-6}$ in calculating $\tau$. Incorrect solution d is the result of incorrectly using the plus sign in raising $e$ to the power of 1 when calculating $V_C$.

If your answer is correct, go on to the next section.

If your answer is not correct, review the subjects of circuit element equations, transients, and circuit examples in chapter 1. For more details, see chapter 4 in reference 1 and chapter 12 in reference 5.

## BIASING AND STABILITY

All transistors in linear circuit applications require some form of biasing in order to establish collector-base-emitter voltage and current relationships at the *operating point* (quiescent point, *Q*-point, no-signal point) of the circuit. The actual circuit configuration and bias circuit values are selected on the basis of dynamic current conditions (desired output voltage swing, expected input signal level, class of operation, desired gain, supply voltages, transistor type, input and output impedances, etc.). The basic bias network must maintain the desired base current in the presence of temperature and frequency changes (referred to as bias stability).

The following problems illustrate biasing techniques and stability factor calculations.

### PROBLEM 6.9  Transistor Curves and Load Line

Exhibit 6.9 shows a common emitter transistor circuit. The transistor characteristic curves are shown in Exhibit 6.9a. It is desired to operate the transistor at $V_{CE} = 10$ V dc when no signal is applied.

1. Draw the dc load line.
2. Calculate $R_B$.
3. Draw the ac load line.
4. Calculate $h_{fe}$ and $h_{oe}$.

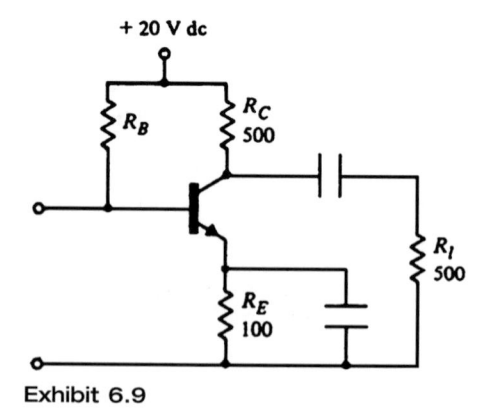

Exhibit 6.9

*Solution*

**1. The dc Load Line**

$$I_C\big|_{V_{CE}=0} = \frac{20}{(100+500)} = 33.3 \text{ mA}$$

$$V_{CE}\big|_{I_C=0} = 20 \text{ V}$$

Thus, the dc load line intersects the ordinate at 33.3 mA and the abscissa at 20 V.

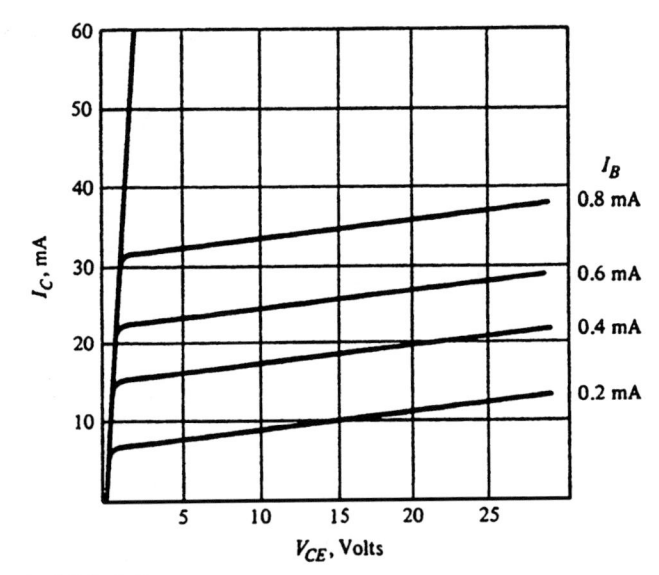

Exhibit 6.9a

### 2. Calculate $R_B$

The $Q$-point is indicated on the dc load line at $V_{CE} = 10$ V and $I_C = 16.7$ mA. At this point, $I_B = 0.4$ mA. The three transistor voltages may be determined as follows:

$$V_C = 20 - (500)(16.7 \times 10^{-3}) = 20 - 8.3 = 11.7 \text{ V}$$

$$V_E = (100)(16.7 \times 10^{-3}) = 1.7 \text{ V}$$

$V_B$ is one diode drop (assume 0.6 V) above $V_E$.

$$\therefore V_B = 1.7 + 0.6 = 2.3 \text{ V}$$

Thus,

$$R_B = \frac{(20 - 2.3)}{0.4 \times 10^{-3}} = 44.25 \text{ }\Omega$$

### 3. The ac Load Line

$$R_{ac} = 500 \| 500 = 250 \text{ }\Omega$$

The ac load line has the slope, $-1/R_a = -1/250 = -4$ mA/V and passes through $Q$ (Exhibit 6.9b).

### 4. Calculate $h_{fe}$ and $h_{oe}$

At $V_{CE} = 10$ volts,

$$h_{fe} = \frac{\Delta I_C}{\Delta I_B}\bigg|_{\Delta V_{CE}=0} = \frac{24.4 - 9}{(0.6 - 0.2)} = \frac{15.4}{0.4} = 38.5$$

$$h_{oe} = \frac{\Delta I_C}{\Delta V_{CE}}\bigg|_{\Delta I_B=0} = \frac{(17.7 - 15.7)}{15 - 5} \times 10^{-3} = \frac{2 \times 10^{-3}}{10} = 0.2 \times 10^{-3} \text{ siemens}$$

If your answers are correct, go on to Problem 6.10.

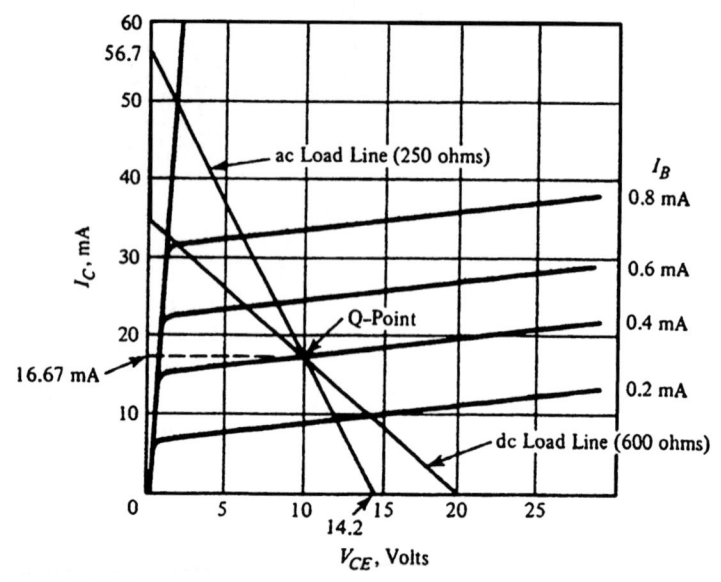

## Exhibit 6.9b

If your answers are not correct, review tables 6.1 and 6.2 in this chapter. For more details, see chapter 4 in reference 1 and chapter 4 in reference 5.

### PROBLEM 6.10  Transistor Stability

The silicon NPN common emitter circuit configuration shown in Exhibit 6.10 has the following given information:

$$Q\text{-point: } V_{CE} = 5 \text{ volts}, I_c = 1 \text{ mA}$$
$$R_1 \| R_2 = R_B = 4000 \ \Omega$$

$C_E$ and $C_C$ are negligible at the operating frequency but block dc currents.

$$V_{BE} = 0.7 \text{ volt}$$

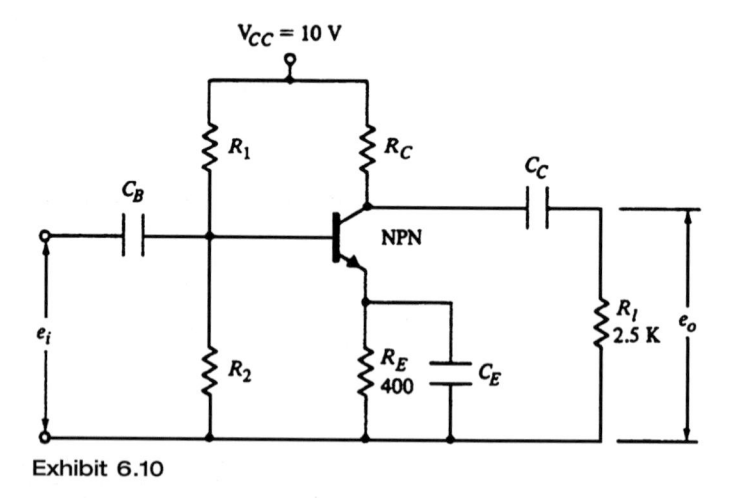

Exhibit 6.10

1. Calculate $R_C$, $R_1$, and $R_2$.
2. Calculate the stability factor $S$.
3. Determine the new $Q$-point if $I_{CO}$ increases by 50 $\mu$A.

*Solution*

1. **Calculate $R_C$, $R_1$, and $R_2$**

$$I_C(R_E + R_C) = V_{CC} - V_{CE}$$

$$I_E = \frac{I_C}{\alpha} = \frac{10^{-3}}{0.98} = 1.02 \times 10^{-3}$$

$$V_E = 400 I_E = 0.408 \text{ volt}$$

$$R_C = \frac{V_{CC} - V_C}{I_C} = \frac{10 - (5 + 0.408)}{10^{-3}} = 4592 \; \Omega$$

$$V_B = V_E + V_{BE} = 0.408 + 0.7 = 1.108 \text{ volts}$$

$$\frac{10}{R_1 + R_2} = \frac{1.108}{R_2}$$

$$10 R_2 = 1.108(R_1 + R_2)$$

$$9.03 R_2 = R_1 + R_2$$

$$R_1 = 8.03 R_2$$

$$\frac{R_1 R_2}{R_1 + R_2} = 4000 = \frac{8.03 R_2^2}{9.03 R_2} = 0.89 R_2$$

$$R_2 = \frac{4000}{0.89} = 4489 \; \Omega$$

$$R_1 = 8.03 R_2 = 36{,}120 \; \Omega$$

2. **Calculate Stability Factor $S$**

Transistor stability is a function of collector leakage current ($I_{CO}$) change with temperature, and its effect on $I_C$. Stability factor, $S$, may be defined by the following equations:

$$S = \frac{\Delta I_C}{\Delta I_{CO}} = \frac{1 + R_E/R_B}{1 - \alpha + R_E/R_B} = \frac{R_B + R_E}{R_B(1 - \alpha) + R_E} = \frac{\beta + 1}{1 + \frac{\beta R_E}{R_E + R_B}}$$

It is desirable to have as low a value of $S$ as possible. For this problem,

$$S = \frac{R_B + R_E}{R_B(1-\alpha) + R_E} = \frac{4000 + 400}{4000(1 - 0.98) + 400} = 9.14$$

3. **Determine New Q-Point If $I_{CO}$ Increases by 50 $\mu$A**

$$\Delta I_C = S(\Delta I_{CO}) = (9.17)(50 \times 10^{-6}) = 4.58 \times 10^{-4} \text{ amp}$$

$$\text{new } I_C = 10^{-3} + 0.458 \times 10^{-3} = 1.458 \text{ mA}$$

$$I_C(R_E + R_C) = V_{CC} - V_{CE}$$

$$V_{CE} = V_{CC} - I_C(R_E + R_C) = 10 - 1.458 \times 10^{-3}(400 + 4592)$$
$$= 2.72 \text{ volts}$$

Thus, it is seen that the *Q*-point has shifted considerably.

If your answers are correct, go on to Problem 6.11.

If your answers are not correct, review the preceding solution. For more details, see chapter 4 in reference 1 and chapter 4 in reference 5.

## PROBLEM 6.11  Transistor Specs, Stability

The single-stage transistor amplifier in Exhibit 6.11 has the following *h*-parameters:

$$h_{ie} = 2500 \; \Omega$$
$$h_{fe} = 120$$
$$h_{oe} = 10^{-5} \; \text{siemens}$$
$$h_{FE} = 145$$

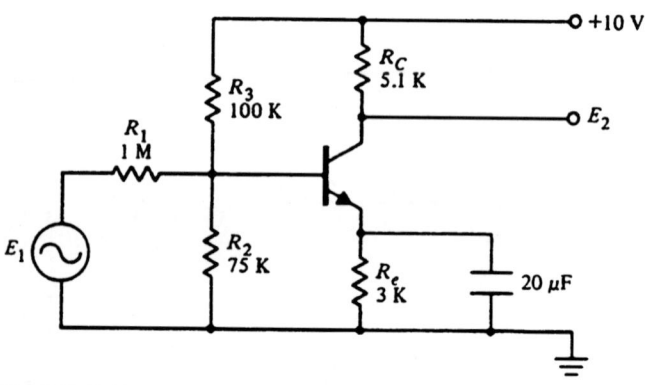

Exhibit 6.11

The voltage gain $E_2/E_1$, current gain, and stability factor are:

a. $A_I = 130.4, A_V = 6.52, S = 13.92$

b. $A_I = 107.9, A_V = 0.55, S = 13.67$

c. $A_I = 107.9, A_V = 0.55, S = 15.39$

d. $A_I = 107.9, A_V = 0.55, S = 13.92$

*Solution*

Draw an equivalent circuit (Exhibit 6.11a).

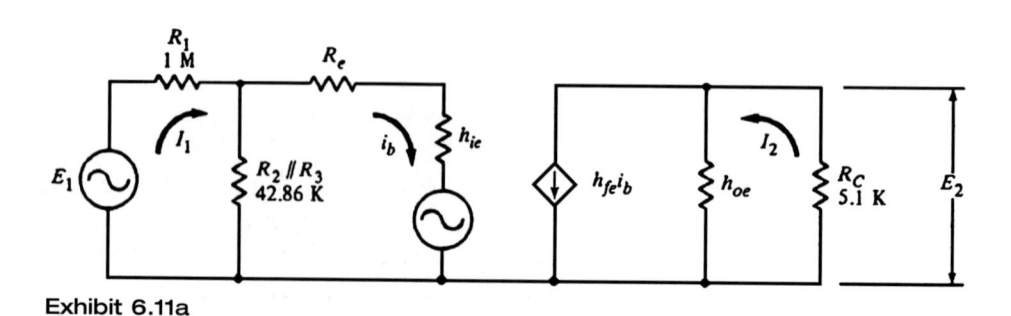

Exhibit 6.11a

Assume $h_{re} = 0$ since it is not given, and $R_e$ is bypassed.

$$i_b = I_1 \frac{42.86}{42.86 + h_{ie}} = 0.945\, I_1$$

$$I_2 = -120 i_b \left( \frac{100}{5.1 + 100} \right) = 120(0.945 I_1)(0.9515) = 107.9\, I_1$$

$$A_I = \frac{I_2}{I_1} = 107.9$$

$$E_2 = I_2 (5.1\text{ K}) = (107.9\, I_1)(5.1\text{ K}) = 550{,}290\, I_1$$

$$E_1 = 10^6\, I_1$$

$$A_V = \frac{E_2}{E_1} = \frac{550{,}290}{10^6} = 0.55$$

$$S = \frac{\beta + 1}{1 + \dfrac{\beta R_E}{R_E + R_B}} = \frac{145 + 1}{1 + \dfrac{(145)(3)}{3 + 42.86}} = 13.92$$

The correct answer is d.

*Answer Rationale*

Incorrect solution a is the result of incorrectly using the value of $h_{FE} = 145$ instead of $h_{fe} = 120$ in the equation when calculating $I_2$. Incorrect solution b is the result of incorrectly using the value of $h_{fe} = 120$ instead of $h_{FE} = 145$ in the numerator when calculating $S$. Incorrect solution c is the result of not adding 1 in the denominator when calculating $S$.

If your answers are correct, go on to the next section.

If your answers are not correct, review the subject of biasing and stability as well as problem 6.10 in this chapter. For more details, see chapter 4 in reference 1 and chapters 4 and 8 in reference 5.

# FIELD EFFECT TRANSISTORS

The field effect transistor (FET) is different from the bipolar junction transistor in the following important ways:

1. FET operation depends upon the flow of majority carriers only. Therefore, it is a unipolar device.

2. It has a much higher input resistance, typically many mega ohms.

3. It is less noisy than a bipolar transistor.

4. It is simpler to fabricate and occupies less area when used in LSI technology.

5. Thermal runaway is not a problem.

The main disadvantage is the relatively small gain-bandwidth product.

Figure 6.5 shows the typical circuit symbol for an *n*-channel FET.

Figures 6.6 and 6.7 show low-frequency and high-frequency (including node capacitors), small-signal FET models.

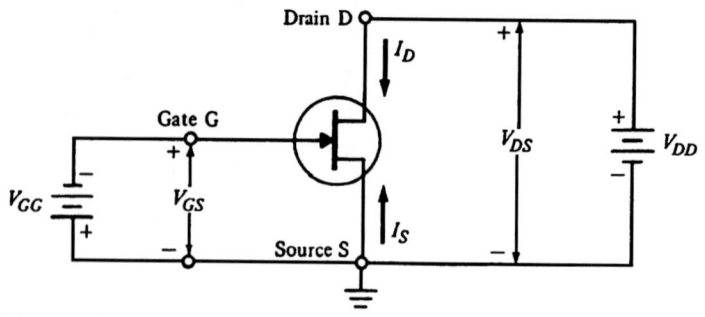

**Figure 6.5** An n-channel FET

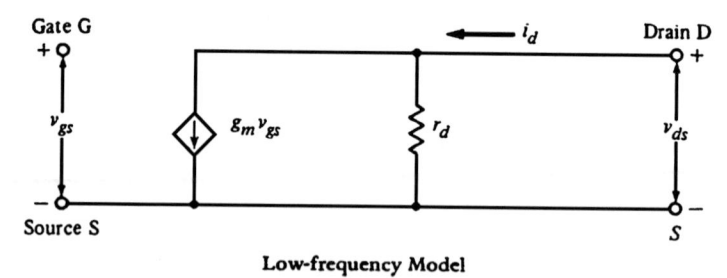

**Figure 6.6** Low-frequency, small-signal FET

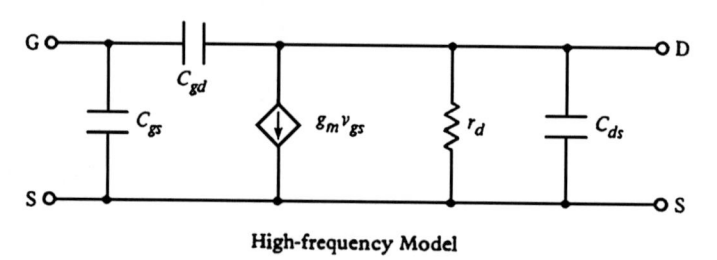

**Figure 6.7** High-frequency, small-signal FET

Because of the internal capacitances of a FET, feedback exists between output and input circuits, and voltage amplification drops off rapidly with frequency. Typical parameters for junction FETs and MOSFETs are given in the table below:

| Parameter | JFET | MOSFET |
| --- | --- | --- |
| $g_m$ | 0.1–10 mA/volt | 0.1–50 mA/volt |
| $r_d$ | 0.1–1 m$\Omega$ | 1–50 k$\Omega$ |
| $C_{ds}$ | 0.1–1 pF | 0.1–1 pF |
| $C_{gs}, C_{gd}$ | 1–10 pF | 1–10 pF |
| $r_{gs}$ | >$10^8$ $\Omega$ | >$10^{10}$ $\Omega$ |
| $r_{gd}$ | >$10^8$ $\Omega$ | >$10^{14}$ $\Omega$ |

The FET may be connected in common source, CS, or common drain, CD, configuration, as shown in Figure 6.8. No biasing is shown.

Compared to a bipolar transistor, CS is equivalent to CE, and CD is equivalent to CC. The circuit in Figure 6.9 shows a single power supply biased FET circuit in the CS configuration.

Field Effect Transistors   **187**

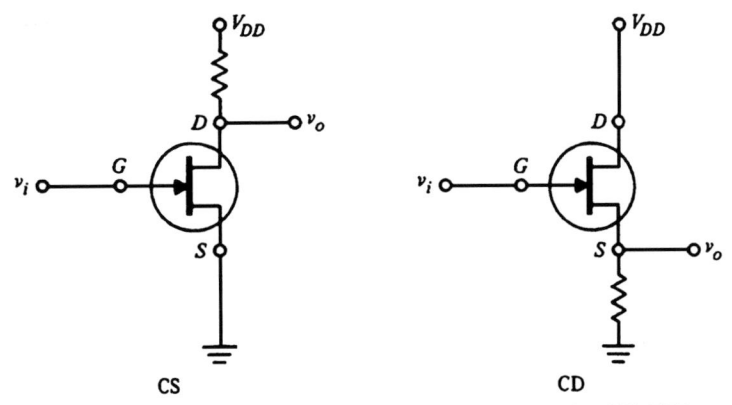

**Figure 6.8**  Common source (CS) and common drain (CD) FET configurations

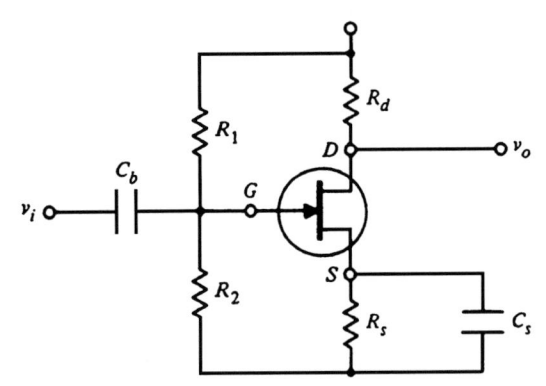

**Figure 6.9**  Single source, supply biased FET in the CS configuration

## PROBLEM 6.12  Field Effect Transistor Amplifier

The circuit of a two-stage FET amplifier is shown in Exhibit 6.12. Each FET has the following characteristics:

$$g_m = 3 \times 10^{-3} \text{ siemens}, \quad r_d = 6.8 \text{ k}\Omega, \quad R_d = 10 \text{ k}\Omega, \quad R_g = 47 \text{ k}\Omega$$
$$C_g = 0.01 \text{ }\mu\text{F}, \quad C_{sb} = 40 \text{ pF}$$

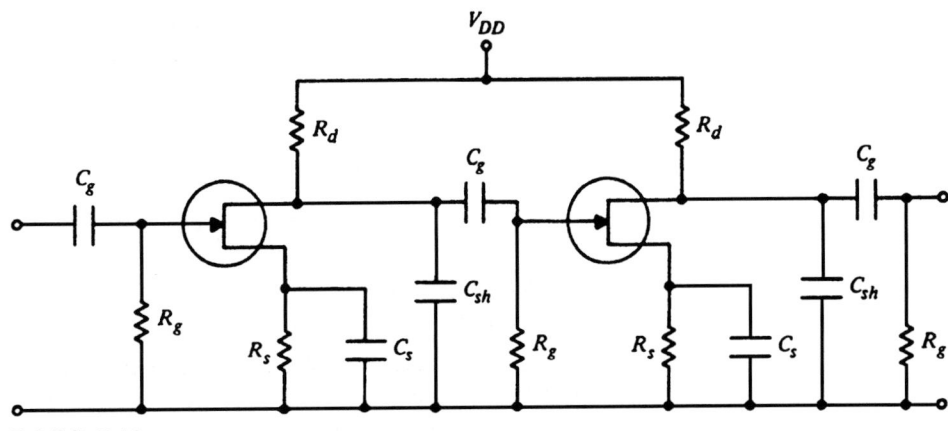

**Exhibit 6.12**

Determine the overall mid-band voltage gain, lower 3 dB frequency, and upper 3 dB frequency.

*Solution*

A junction FET has properties analogous to a vacuum tube. Its voltage gain is given by the formula:

$$A_V = -g_m R$$

In this case, $R = r_d \| R_d \| R_g$

$$\frac{1}{R} = \frac{1}{6.8 \times 10^3} + \frac{1}{10^4} + \frac{1}{47 \times 10^3} = 2.683 \times 10^{-4}$$

$$R = 3727 \ \Omega$$

$$A_V = -3 \times 10^{-3} \times 3727 = 11.18 \text{ per stage}$$

Overall gain is: $A_{VT} = A_V^2 = 11.18^2 = 125$

In db, $A_{VT} = 20 \ \log 125 = 42 \text{ dB}$

Lower 3 dB frequency, $f_1$, for each stage is

$$f_1 = \frac{1}{2\pi R_1 C_g},$$

where

$$R_1 = R_g + r_d \| R_d$$

$$= 47 \times 10^3 + \frac{(6.8 \times 10^3)(10^4)}{16.8 \times 10^3}$$

$$= 51.05 \ k\Omega$$

$$f_1 = \frac{1}{2\pi \times 51.05 \times 10^3 \times 10^{-8}} = 311.76 \text{ Hz}$$

Overall 3 dB frequency is

$$f_{1n} = \frac{f_1}{\sqrt{2^{1/n} - 1}} = \frac{311.76}{\sqrt{2^{1/2} - 1}} = 484.41 \text{ Hz}$$

Upper 3 dB frequency, $f_2$, for each stage is

$$f_2 = \frac{1}{2\pi R C_{sh}} = \frac{1}{2\pi \times 3727 \times 40 \times 10^{-12}} = 1.07 \text{ MHz}$$

Overall 3 dB frequency is

$$f_{2n} = f_2 \sqrt{2^{1/2} - 1} = 1.07 \times 10^6 \sqrt{2^{1/2} - 1} = 687 \text{ kHz}$$

If your answers are correct, go to the next section.

## OPERATIONAL AMPLIFIERS

A typical operational amplifier circuit application is shown in Figure 6.10.
Since the amplifier is ideal, $i- = 0 = i+$ and $v- = v+$ is forced by the amplifier.

$$V_+ = \frac{Z_4}{Z_3 + Z_4} V_B \quad \text{(voltage devider method)}$$

$$V_- = V_A \frac{Z_2}{Z_1 + Z_2} + V_o \frac{Z_1}{Z_1 + Z_2} \quad \text{(superposition method)}$$

$$\frac{Z_4}{Z_3 + Z_4} V_B = \frac{Z_2 V_A + Z_1 V_o}{Z_1 + Z_2}, \quad V_o = \frac{Z_1 + Z_2}{Z_1} \left[ V_B \frac{Z_4}{Z_3 + Z_4} - V_A \frac{Z_2}{Z_1 + Z_2} \right]$$

$$\therefore V_o = V_B \left[ \frac{Z_1 + Z_2}{Z_1} \right] \left[ \frac{Z_4}{Z_3 + Z_4} \right] - V_A \frac{Z_2}{Z_1}$$

### Special Cases

Following is a collection of operational amplifier circuit applications.

1. Differential amplifier

$$\frac{Z_2}{Z_1} = \frac{Z_4}{Z_3}, \quad V_o = (V_B - V_A) \frac{Z_2}{Z_1}$$

2. Inverting amplifier

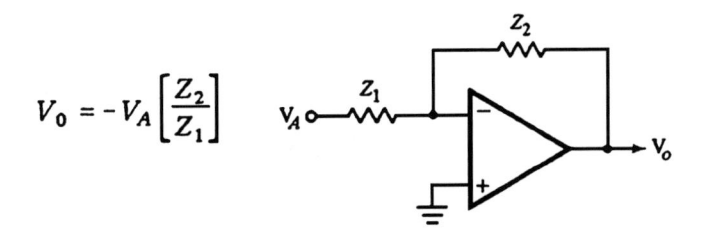

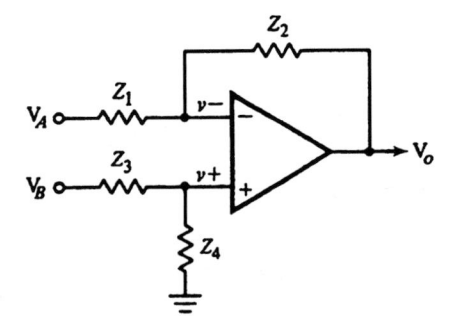

$$V_o = -V_A \left[ \frac{Z_2}{Z_1} \right]$$

| Parameter | Ideal | μA741 |
|---|---|---|
| $A_v$ | ∞ | $10^5$ V/V |
| $R_i$ | ∞ | $10^5$ Ω |
| $i_-, i_+$ | 0 | $10^{-7}$ A |

**Figure 6.10** Typical operational amplifier circuit

3. Non-inverting amplifier

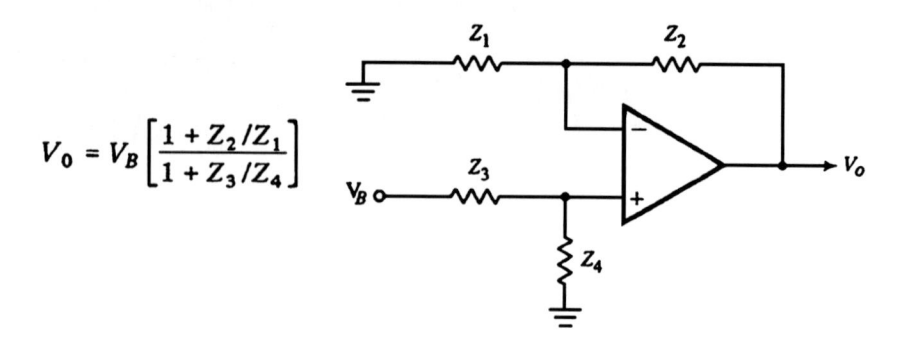

$$V_0 = V_B \left[ \frac{1 + Z_2/Z_1}{1 + Z_3/Z_4} \right]$$

4. Voltage follower

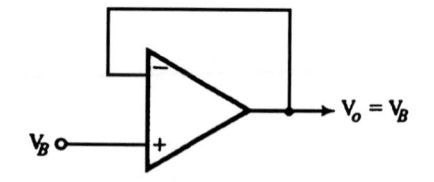

$V_o = V_B$

5. Low pass inverting amplifier

$$\frac{V_o}{V_i}(s) = \frac{-R_f/R_i}{1 + sR_f C_f}$$

1st order lag at $\omega = \dfrac{1}{R_f C_f}$

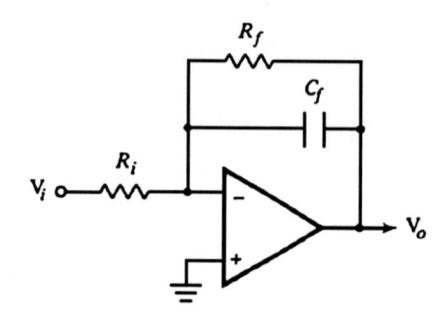

6. Integrator

$$\frac{V_o}{V_i}(s) = \frac{-1}{sRC}; \quad V_o(t) = -\frac{1}{RC}\int V_i \, dt$$

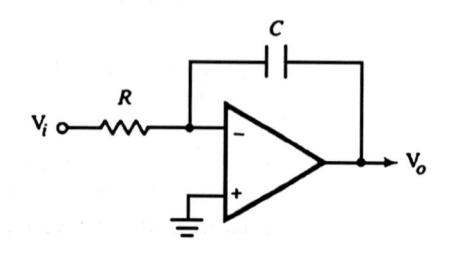

7. Differentiator

$$\frac{V_o}{V_i}(s) = -sRC; \quad V_o(t) = -RC\frac{dV_i(t)}{dt}$$

(noise sensitive due to input $C$)

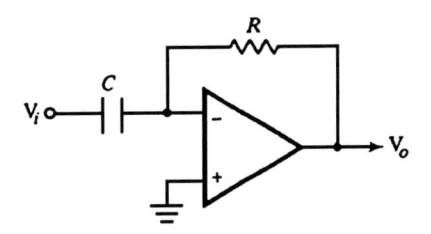

For additional op amp circuits, see chapters 14 and 15 in reference 5.

## PROBLEM 6.13  Operational Amplifier

Determine the output voltage for the circuit shown in Exhibit 6.13 if the input is 0.2 sin $\omega t$ volts.

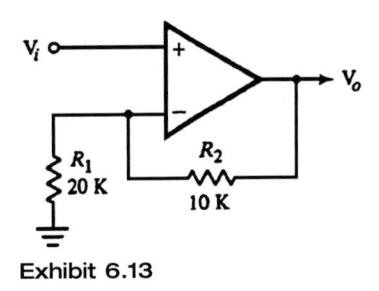

**Exhibit 6.13**

### Solution

This circuit is a non-inverting amplifier. Output voltage is given by the formula:

$$V_o = V_i\left[1 + \frac{R_2}{R_1}\right]$$

$$\therefore V_o = 0.2\sin\omega t(1+0.5) = 0.3\sin\omega t \text{ volts}$$

If your answer is correct, go on to Problem 6.14.

If your answer is not correct, review the subject of operational amplifiers in this chapter. For more details, see chapter 4 in reference 1 and chapters 14 and 15 in reference 5.

## PROBLEM 6.14  ac/dc Converter

For the ac/dc converter circuit shown in Exhibit 6.14, determine the values for $R_1$, $R_2$, and $R_3$ that will provide a dc voltage equal to the RMS value of an input sine wave ac voltage. Assume $V_i = V_m \sin \omega t$, where $V_m \leq 18$ volts, and the ideal op amps limit at ±15 volts.

**Exhibit 6.14**

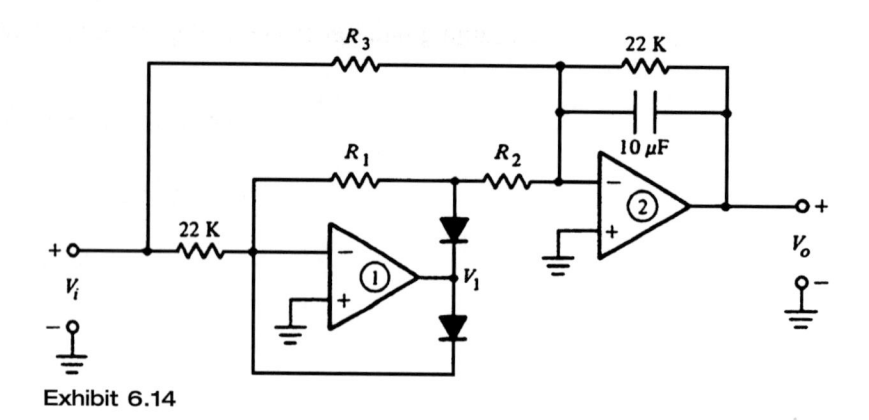

*Solution*

Op amp 1 is an inverting rectifier. Op amp 2 is an inverting summer. For a sinusoid, RMS value = 0.707 $V_m$. Select $R_1$ such that $V_1 = -15$ V when $V_i = 18$ V.

$$V_1 = V_i \frac{R_1}{22\text{ k}} = -18 \frac{R_1}{22\text{ k}} = -15$$

$$R_1 = \frac{15}{18} \times 22\text{ k} = 18.3\text{ k}$$

Half-wave rectified sine wave:

$$V_1 = -V_m K_1 \left[ \frac{1}{\pi} + \frac{1}{2} \sin \omega t - \frac{2}{\pi} \left( \frac{\cos 2\omega t}{3} \right) + \cdots \right]$$

$$K_1 = \frac{R_1}{22\text{ k}} = 0.833$$

Summer:

$$V_o = -\frac{22\text{ K}}{R_3} (V_m \sin \omega t) - \frac{22\text{ K}}{R_2} V_1$$

$$V_o = -\frac{22\text{ K}}{R_3} (V_m \sin \omega t) + \frac{22\text{ K}}{R_2} (0.833 V_m) \left[ \frac{1}{\pi} + \frac{1}{2} \sin \omega t - \frac{2}{\pi} \left( \frac{\cos 2\omega t}{3} \right) + \cdots \right]$$

$$= \frac{22\text{ K}}{R_2} (0.833 V_m) \frac{1}{\pi} + \left[ \frac{22\text{ K}}{2 R_2} (0.833 V_m) - \frac{22\text{ K}}{R_3} V_m \right] \sin \omega t + \cdots$$

Choose $R_2$ and $R_3$ such that the sin term cancels out:

$$\frac{22\text{K}}{2 R_2} (0.833 V_m) = \frac{22\text{ K}}{R_3} V_m$$

$$R_3 = 2.4 R_2$$

$$V_o = \frac{22\text{ K}}{R_2} (0.833 V_m) \frac{1}{\pi} + \text{harmonic terms}$$

Assuming harmonics are negligible because of the filter:

$$V_o = 0.707 V_m = \frac{22\,\text{k}}{R_2}(0.833 V_m)\frac{1}{\pi}$$

$$R_2 = \frac{22\,\text{K}\,(0.833)}{0.707\,\pi} = 8.25\,\text{k}$$

$$R_3 = 2.4 R_2 = 19.8\,\text{k}$$

If your answers are correct, go on to the next section.

If your answers are not correct, check your calculations. For more details, see chapter 4 in reference 1 and chapters 14 and 15 in reference 5.

## AMPLIFIER CLASS

Amplifier stages can be classified according to *class* of operation in response to a sinusoidal input, as indicated in Figure 6.11.

In general, class A is the most linear (contains the least distortion) and is the least efficient. Class C is the other extreme.

If the input to an amplifier is a pure sine wave of single frequency, but the output contains higher harmonics (which can be expressed in Fourier series form), then harmonic distortion is present.

If

$$V_o(t) = K\,[V_1\cos(\omega t + \theta_1) + V_2\cos(2\omega t + \theta_2) + V_3\cos(3\omega t + \theta_3) \\ + \cdots + V_n\cos(n\omega t + \theta_n)]$$

is produced from an input of $V_1 \cos \omega t$, then harmonic distortion is

$$\%\ \text{2nd H.D.} = \frac{V_2}{V_1} \times 100$$

$$\%\ \text{total H.D.} = \frac{\sqrt{V_2^2 + V_3^2 + \cdots + V_n^2}}{V_1} \times 100$$

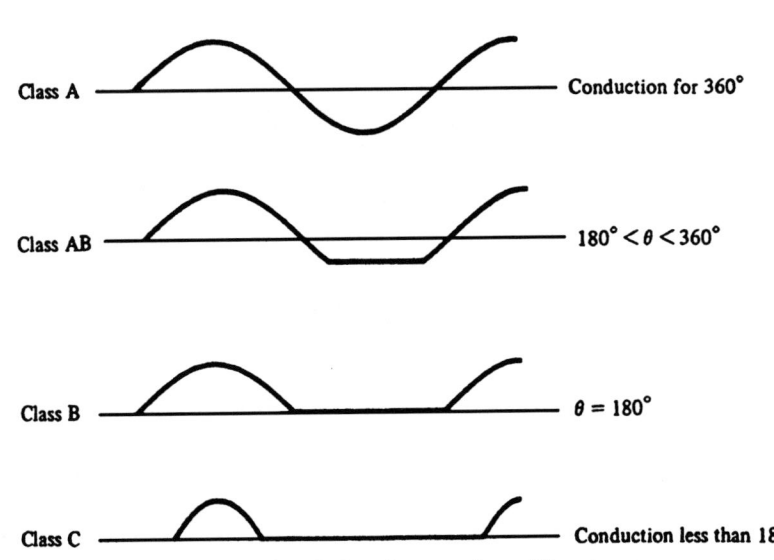

**Figure 6.11** Sinusoidal inputs for classes of amplifier stages

- Class A — Conduction for 360°
- Class AB — 180° < $\theta$ < 360°
- Class B — $\theta$ = 180°
- Class C — Conduction less than 180°

## PROBLEM 6.15  Amplifier Class

The common emitter connected transistor amplifier circuit shown in Exhibit 6.15 has the following characteristics:

$$h_{FE} = 50$$
$$V_{BE} = 0.6 \text{ volt}$$
$$I_B = 20 \text{ } \mu A$$

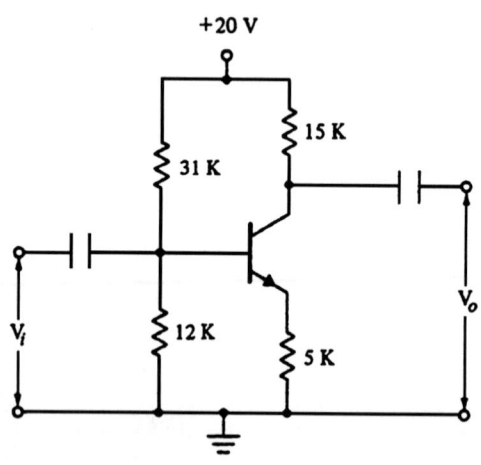

Exhibit 6.15

The $Q$-point, amplifier class, voltage gain, and $V_o$ are:
a. Class B, $A_v = -2.9$, $V_o = -2.9 V_i$
b. Class C, $A_v = -2.9$, $V_o = -2.9 V_i$
c. Class B, $A_v = -0.97$, $V_o = -0.97 V_i$
d. Class B, $A_v = -2.9$, $V_o = -2.9 V_i$

*Solution*

$$I_C = h_{FE} I_B = (50)(20 \times 10^{-6}) = 1 \text{ mA}$$
$$V_C = V_{CC} - I_C R_C = 20 - 15 = 5 \text{ volts}$$
$$V_E = I_E R_E \approx I_C R_E = (10^{-3})(5 \times 10^3) = 5 \text{ volts}$$

Since $V_C = V_E = 5$ volts, $V_{CE} = 0$ and the transistor is saturated.

Thus, the $Q$-point is at $V_{CE} = 0$ and $I_C = 1$ mA.

The circuit is operating as a class B amplifier, since the output changes only when the input goes negative, as shown in Exhibit 6.15a.

$$A_V = -\frac{h_{FE} R_l}{hie} = \frac{-h_{FE} R_l}{R_b + \dfrac{R_e}{1-\alpha}} = \frac{V_o}{V_i}$$

$$h_{FE} = 50$$
$$R_b = \frac{(31 \text{ k})(12 \text{ k})}{43 \text{ k}} = 8.65 \text{ k}\Omega$$
$$R_l = 15 \text{ k}\Omega$$
$$R_e = 5 \text{ k}\Omega$$

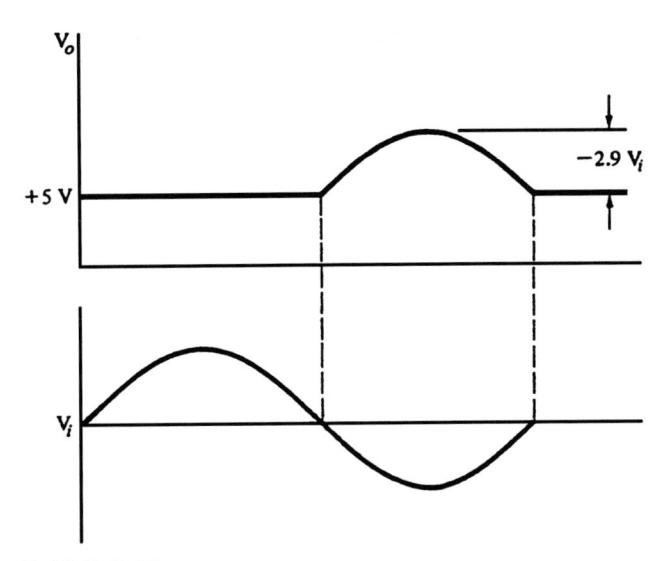

**Exhibit 6.15a**

$$\alpha = \frac{\beta}{\beta+1} = \frac{50}{51} = 0.98$$

$$A_V = \frac{(-50)(15 \times 10^3)}{8650 + \frac{5000}{0.02}} = -2.9$$

$$V_o = (A_V)(V_i) = -2.9\ V_i$$

The correct answer is a.

*Answer Rationale*

Incorrect solution b is the result of incorrectly assuming that it is a class C amplifier. Incorrect solution c is the result of incorrectly using the value of 5 k instead of 15 k in the numerator when calculating $A_V$. Incorrect solution d is the result of not using the minus sign for the value 2.9 multiplied by $V_i$ in the expression for $V_o$.

If your answers are correct, go on to the next section.

If your answers are not correct, review problem 6.5 in this chapter. For more details, see chapter 4 in reference 1 and chapter 16 in reference 5.

# POWER SUPPLY CIRCUITS

Design of power supplies and power supply circuits is an old and diverse specialty. The following problems illustrate only the simplest type of ac/dc converters.

### PROBLEM 6.16  Zener Regulator

Design a zener diode voltage regulator circuit having the following specifications:

$$V_{out} = 7.5 \text{ VDC}, \quad R_L = 250\ \Omega, \quad V_{in} = 15 \text{ VDC}$$

Define the zener diode characteristics and series resistor.

## Solution

Exhibit 6.16 shows the circuit.

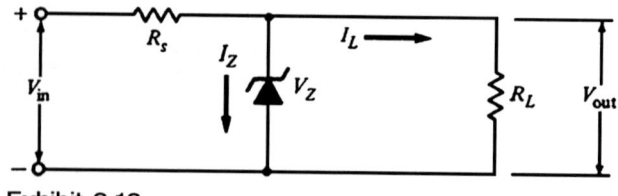

**Exhibit 6.16**

$$V_Z = V_{out} = 7.5 \text{ volts}$$

$$I_L = \frac{V_Z}{R_L} = \frac{7.5}{250} = 30 \text{ mA}$$

$$P_L = V_Z I_L = 7.5 \times 30 = 225 \text{ mW}$$

For a safety margin, the zener diode should be rated at about three times the maximum load. Thus,

$$P_Z = 3P_L = 675 \text{ mW}$$

so specify the next higher practical value, namely, 1 watt.

For safe zener current rating, $I_Z = 3I_L = 90$ mA. Select $I_Z = 100$ mA.

$$R_s = \frac{(V_{in\,max} - V_Z)^2}{P_Z} = \frac{(15 - 7.5)^2}{0.675} = 83.33 \, \Omega$$

To assure safe current limiting, select $R_s = 91 \, \Omega$.

$$P_{R_s} = \frac{(15 - 7.5)^2}{R_s} = \frac{7.5^2}{91} = 0.62 \text{ watt}$$

Select a 1 watt resistor for $R_s$.

To summarize, the zener diode is 7.5 volts, 100 mA, 1 watt. The series resistor is 91 ohms, 1 watt.

If your answers are correct, go on to Problem 6.18.

If your answers are not correct, review an appropriate text on zener diode regulation. For more details, see chapter 4 in reference 1 and chapters 1 and 2 in reference 5.

## PROBLEM 6.17  Ripple Factor

An audio amplifier is to be powered from a dc supply. The power supply uses a series choke in a full-wave rectifier circuit as shown in Exhibit 6.17.

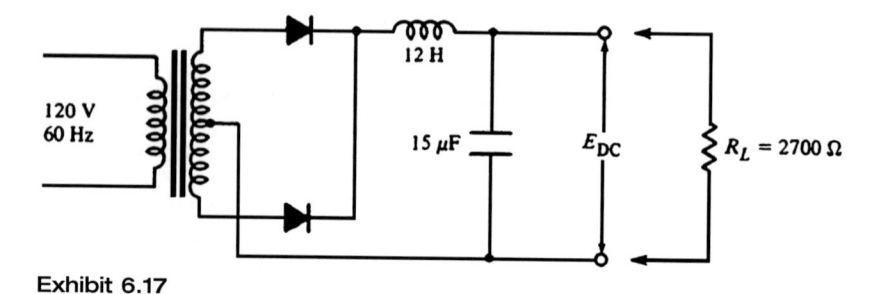

**Exhibit 6.17**

The transformer is 120:480 volt center-tapped, and the drop across each diode is 10 volts. The output voltage and ripple factor are:
  a. $E_{DC}$ = 209.7 volts, Ripple Factor = 0.474%
  b. $E_{DC}$ = 146.4 volts, Ripple Factor = 0.474%
  c. $E_{DC}$ = 209.7 volts, Ripple Factor = 0.469%
  d. $E_{DC}$ = 209.7 volts, Ripple Factor = 47.4%

## Solution

The output of a rectifier circuit without a series choke is equal to the peak of the sine wave (ignoring ripple). However, with a choke in the circuit as shown, the output voltage becomes the average of a sine wave peak. If $X_C \ll R_L$, the capacitor effectively bypasses the ac current around the load. In this case:

$$X_C = \frac{1}{2\omega C} = \frac{1}{4\pi \times 60 \times 15 \times 10^{-6}} = 88.4 \, \Omega$$

($2\omega$ is used because the ripple frequency of a full-wave rectifier is twice the line frequency.)

Thus, it is seen that $X_C$ is, indeed, much less than $R_L$.

$$E_{peak} = \sqrt{2} \times 240 = 339.4$$

$$E_{DC} = \frac{2}{\pi}[E_{peak} - V_{diode}] = \frac{2}{\pi}[339.4 - 10] = 209.7 \text{ volts}$$

$$\text{Ripple Factor} = \frac{0.48}{(2\omega)^2 LC - 1} = \frac{0.48}{5.68 \times 10^5 \times 12 \times 15 \times 10^{-6} - 1}$$
$$= 0.474\%$$

The correct answer is a.

## Answer Rationale

Incorrect solution b is the result of not multiplying by $\sqrt{2}$ in calculating $E_{peak}$. Incorrect solution c is the result of not subtracting 1 in the denominator when calculating the ripple factor. Incorrect solution d is the result of incorrectly showing the percentage as 47.4 instead of 0.474.

If your answers are correct, go on to the next chapter.

If your answers are not correct, review an appropriate text on rectifiers. For more details, see chapter 4 in reference 1 and chapters 2 and 19 in reference 5.

# CHAPTER 7

# Communications

**OUTLINE**

LOW-FREQUENCY TRANSMISSION  199

RF TRANSMISSION  203
Open-Wire Line Parameters ■ Coax Line Parameters ■ Zero Dissipation Line Constants ■ Zero Dissipation Line Voltages and Currents ■ Standing Waves ■ Zero-Dissipation-Line Input Impedance ■ Reflection Losses at Any Point ■ Power ■ The Eighth-Wave Line ■ The Quarter-Wave Line ■ The Half-Wave Line ■ Impedance Matching ■ Smith Chart ■ Impedance Transfer

ATTENUATION  220

ANTENNAS  224

Signals produced by various information sources are not always suitable for direct transmission over a given channel. These signals are typically modified to facilitate transmission. This conversion process is known as modulation. In this process, the transmitted signal is used to modify certain parameters of a high-frequency carrier signal.

A carrier is a sinusoidal signal of high frequency. One of its parameters, such as amplitude, frequency, or phase, is varied in proportion to the transmitted signal. As such, we can have amplitude modulation (AM), frequency modulation (FM), or phase modulation (PM).

This chapter deals with communications problems that fall into either of two frequency bands: low frequency or radio frequency. The low-frequency band is generally that range between 500 Hz and 1 MHz; this includes telephone transmission problems. The RF range is considered to be anything above 1 MHz; this includes radio and television broadcasting, antennas, and waveguides.

Each band has its own transmission line characteristics, permitting certain unique assumptions to be made.

## LOW-FREQUENCY TRANSMISSION

In this section, the transmission line is treated as a circuit having distributed parameters defined as follows:

$R$ = series resistance, ohms per unit length of line (includes both wires)
$L$ = series inductance, henries per unit length of line

$C$ = capacitance between conductors, farads per unit length of line
$G$ = shunt leakage conductance between conductors, siemens per unit length of line
$\omega L$ = series reactance, ohms per unit length of line
$Z = R + j\omega L$ = series impedance, ohms per unit length of line
$\omega C$ = shunt susceptance, siemens per unit length of line.

$$Z_o = \sqrt{\frac{Z}{Y}} = \sqrt{\frac{j\omega L}{j\omega C}} = \sqrt{\frac{L}{C}} = \sqrt{Z_{oc} Z_{sc}} = \text{characteristic impedance} = \sqrt{R_g R_L}$$

$Y = G + j\omega C$ = shunt admittance, siemens per unit length of line
$s$ = distance to point of observation, measured from receiving (or sending) end of the line
$I$ = current at any point in the line
$E$ = voltage between the conductors at any point
$l$ = length of the line
$\gamma$ = propagation constant, per unit length of line
$\quad = \sqrt{ZY} = \alpha + j\beta$
$\alpha$ = attenuation constant, dB or neper per unit length of line
$\quad$ (1 neper = 8.686 dB)
$\beta$ = phase constant, angle per unit length of line
$E = E_s e^{-\gamma s} = E_s e^{-\alpha s} e^{-j\beta s}$
$I = I_s e^{-\gamma s} = I_s e^{-\alpha s} e^{-j\beta s}$
$E^{-\alpha s}$ = attenuation
$E^{-j\beta s}$ = phase factor
$\lambda$ = wavelength = $2\pi/\beta = v/f$
$v = \omega/\beta$ = velocity of propagation along the line, miles per second if $\beta$ is in radians per mile, or meters per second if $\beta$ is in radians per meter
$\omega = 2\pi f$ in radians
$f$ = frequency in Hz

Velocity $v$ is reduced by $1/\sqrt{\varepsilon}$ in a medium having a dielectric constant of $\varepsilon$.

### PROBLEM 7.1  Communication Line Characteristics

An open-wire communication line is 5 miles long and is terminated in its characteristic impedance, $Z_o$. The line parameters per mile are

$$R = 75 \, \Omega \quad C = 0.1 \, \mu F$$
$$L = 1 \, mH \quad G = 0.05 \times 10^{-6} \text{ siemens}$$

Determine the following at 1 kHz:

1. Characteristic impedance $Z_o$
2. Propagation constant $\gamma$
3. Attenuation $\alpha$
4. Phase shift $\beta$
5. Velocity $v$
6. wavelength $\lambda$

*Solution*

1. $Z = R + j\omega L = 75 + j2\pi \times 10^3 \times 10^{-3} = 75 + j6.28 = 75.26\angle 4.79°\ \Omega$

    $Y = G = j\omega C = 0.05 \times 10^{-6} + j2\pi \times 10^3 \times 10^{-7} = 0.05 \times 10^{-6} + j6.28 \times 10^{-4}$
    $= 6.28 \times 10^{-4}\angle 90°$ siemens

    $Z_o = \sqrt{\dfrac{Z}{Y}} = \sqrt{\dfrac{78.26\angle 4.79°}{6.28 \times 10^{-4}\angle 90°}} = 346.18\angle -42.61°\ \Omega$

2. $\gamma = \sqrt{ZY} = \sqrt{(75.26\angle 4.79°)(6.28 \times 10^{-4}\angle 90°)} = 0.2174\angle 47.4° = 0.147 + j0.16$

    $\gamma = \alpha + j\beta$

3. $\alpha = 0.147\ \dfrac{\text{neper}}{\text{mile}} \times \dfrac{8.686\ \text{dB}}{\text{neper}} = 1.28\ \text{dB/mile}$

4. $\beta = \dfrac{0.16\ \text{radian}}{\text{mile}}$

5. $v = \dfrac{\omega}{\beta} = \dfrac{2\pi \times 10^3}{0.16} = 39{,}270$ miles/second

6. $\lambda = \dfrac{2\pi}{\beta} = \dfrac{6.28}{0.16} = 37.27$ miles

    Also, $\lambda = \dfrac{v}{f} = \dfrac{39{,}270}{10^3} = 39.27$ miles

If your answers are correct, continue to the next problem.

If your answers are not correct, review the subject of low-frequency transmission in this chapter. For more details, see chapter 8 in reference 1 and chapter 25 in reference 3.

### PROBLEM 7.2  LF Transmission Line Power

A two-volt generator supplies power to a 150-mile open-wire line terminated in its characteristic impedance and having the following characteristics:

$$R = 10\ \Omega/\text{mile}$$
$$L = 4\ \text{mH/mile}$$
$$G = 1\mu\ \text{siemens/mile}$$
$$C = 0.01\mu\text{F/mile}$$

The receiving end power at a frequency of 800 Hz is:
  a.  $P_R = 0.575$ mW     c.  $P_R = 0.135$ mW
  b.  $P_R = 0.575$ W      d.  $P_R = 0.263$ mW

*Solution*

$Z = R + j\omega L = 10 + j2\pi \times 800 \times 4 \times 10^{-3} = 10 + j20.11 = 22.46\angle 63.56\ \Omega/\text{mile}$
$Y = G + j\omega C = 10^{-6} + j2\pi \times 800 \times 10^{-8} = (1 + j50.3)10^{-6}$
$= 50.3 \times 10^{-6}\angle 90°$ siemens/mile

$$Z_o = \sqrt{\frac{Z}{Y}} = \sqrt{\frac{22.46\angle 63.56°}{50.3\times 10^{-6}\angle 90°}} = 668.4\angle -13.22°\ \Omega$$

$$I_S = \frac{E_S}{Z_o} = \frac{2\angle 0°}{668.4\angle -13.22} = 3\angle 13.22°\ \text{mA}$$

$$\gamma = \sqrt{ZY} = 0.0336\angle 76.8°/\text{mile}$$
$$\alpha = 0.0336\ \cos 76.8° = 0.0077\ \text{neper/mile}$$
$$\beta = 0.0336\ \sin 76.8° = 0.0327\ \text{rad/mile}$$

$$\frac{I_R}{I_S} = e^{-\gamma} = e^{-\alpha l}e^{-j\beta l} = e^{-1.16}e^{-j4.91}$$

$$e^{-j4.91} = \angle -4.91\ \text{rad} = \angle -281°$$
$$I_R = I_s e^{-1.16}\angle -281° = (3\angle 13.22°)(0.31\angle -281°)$$
$$= 0.94\angle -268°\ \text{mA}$$

$$E_R = I_R Z_o = (0.94\times 10^{-3}\angle -268°)(668.4\angle 13°)$$
$$= 0.628\angle -281°\ \text{volt}$$

$$P_R = E_R I_R \cos\theta = 0.628 \times 0.94 \times 10^{-3}\cos 13.22°$$
$$= 0.575\ \text{mW}$$

The correct answer is a.

*Answer Rationale*

Incorrect solution b is the result of not multiplying by $10^{-3}$ in the last step when calculating $P_R$. Incorrect solution c is the result of incorrectly using the sine of the angle 13.22 instead of the cosine in the last step when calculating $P_R$. Incorrect solution d is the result of incorrectly using the angle of $Z = 63.56$ instead of 13.22 in the last step when calculating $P_R$.

If your answer is correct, go on to Problem 7.3.

If your answer is not correct, review the subject of low-frequency transmission and Problem 7.1 in this chapter. For more details, see chapter 8 in reference 1 and chapter 25 in reference 3.

### PROBLEM 7.3  Transmission Line Maximum Power

Exhibit 7.3 shows the circuit for a low-frequency transmission line. Determine the value of the terminating resistor, $R_L$, in order to develop maximum power in the load. Generator frequency is 1500 Hz.

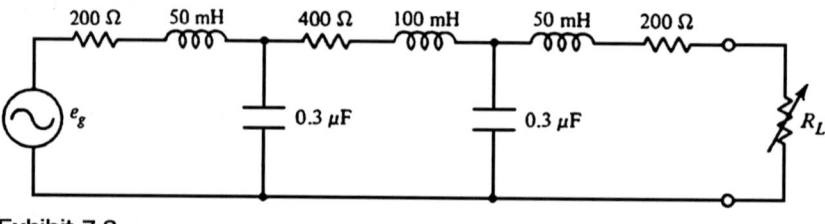

Exhibit 7.3

The value of the terminating resistor is:
 a. $R_L = 425.51$ ohms   b. $R_L = 1066.6$ ohms
 c. $R_L = 601.8$ ohms    d. $R_L = 13.45$ ohms

*Solution*

$$Z = 400 + j\omega L = 400 + j2\pi \times 1500 \times 100 \times 10^{-3}$$
$$= 400 + j942.5 = 1023.87\angle 67° \; \Omega$$
$$Y = G + j\omega C = j2\pi \times 1500 \times 0.6 \times 10^{-6}$$
$$= 5.655 \times 10^{-3} \angle 90° \text{ siemens}$$
$$Z_o = \sqrt{\frac{Z}{Y}} = \sqrt{\frac{1023.85\angle 67°}{5.66 \times 10^{-3} \angle 90°}} = 425.51\angle -11.5° \; \Omega$$

Maximum power is achieved when $R_L = |Z_o| = 425.51$ ohms
The correct answer is *a*.

*Answer Rationale*

Incorrect solution b is the result of not multiplying by $2\pi$ in calculating $Y$. Incorrect solution c is the result of incorrectly using the capacitance value of 0.3 instead of 0.6 in calculating $Y$. Incorrect solution d is the result of not multiplying by $10^{-3}$ in the denominator in the last step when calculating $Z_o$.

If your answer is correct, go on to the next section.

If your answer is not correct, review Problem 7.1 and maximum power transfer theory corollary in chapter 1. For more details, see chapter 8 in reference 1 and chapter 25 in reference 3.

# RF TRANSMISSION

In the preceding section, very few simplifying assumptions could be made since, in the low-frequency range, resistances and reactances are of similar order of magnitude. In the RF range, skin effect and other phenomena unique to high frequencies allow certain simplifying assumptions to be made. These apply both to open-wire and coaxial transmission lines and include the following:

1. Skin effect is significant, so that current may be assumed to flow on conductor surfaces, internal inductance then being zero.

2. $\omega L \gg \omega R$ when computing $Z$, because resistance increase because of skin effect with $\sqrt{f}$, while reactance increases directly with $f$.

3. Shunt $G$ may be considered to be zero.

If $R$ is small, the line is said to have *small dissipation*. Small-dissipation lines are not discussed here; however, this is a useful concept when lines are used as circuit elements or where resonance is involved. Reference to a text on the subject will supply information as required.

If $R$ is zero, the line has *zero dissipation (dissipationless line)*. Under these conditions, transmission of power is highly efficient. The dissipationless line is assumed in the following paragraphs and examples.

Two types of transmission lines are used in this section: open-wire and coaxial. Their parameters are discussed in the next two paragraphs.

When discussing the transmission line, the model of a T-section, shown in Figure 7.1, is used

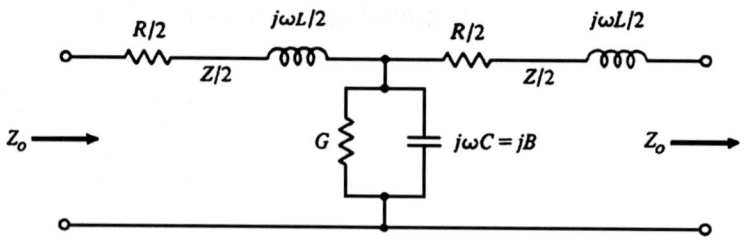

Figure 7.1  T-section circuit model

## Open-Wire Line Parameters

$$L = \frac{\mu_0}{2\pi} \ln \frac{d}{a} \times 2 \text{ wires} = 4 \times 10^{-7} \ln \frac{d}{a} \text{ H/m}$$

where
$\mu_0 = 4\pi \times 10^{-7}$ = permeability of air

Generally, magnetic permeability $\mu$ is

$$\mu = \mu_r \mu_v$$

where
$\mu_r$ is relative permeability of a particular material
$\mu_v$ is magnetic permeability of space = $4\pi \times 10^{-7}$ MKS
$\mu_0$ is magnetic permeability of air = $4\pi \times 10^{-7}$ MKS

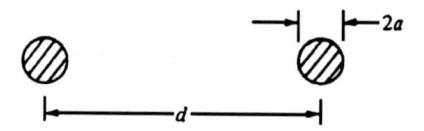

Internal flux and inductance of wires is assumed zero.

$$C = \frac{\pi \varepsilon}{\ln \frac{d}{a}} \text{ farad/meter} = \frac{27.7}{\ln \frac{d}{a}} \text{ pF/m}$$

where
$\varepsilon = \varepsilon_r \varepsilon_v$
$\varepsilon_r$ is dielectric constant or relative permittivity = 1 for air
$\varepsilon_v$ is permittivity of space = $10^{-9}/36\pi = 8.85 \times 10^{-22}$ MKS

For round copper conductors,

$$\frac{R_{ac}}{R_{dc}} = \frac{a\sqrt{\pi f \mu \sigma}}{2} = 7.53 a \sqrt{f}$$

where
$\mu = 4\pi \times 10^{-7}$ (Cu is same as space)
$\sigma = 5.75 \times 10^7$ siemens/m (conductivity of Cu at 20°C)

$$R_{ac} = \frac{8.33 \times 10^{-8} \sqrt{f}}{a} \, \Omega/\text{m of line for spacing} > 20\,a$$

## Coax Line Parameters

$$L = \frac{\mu_v}{2\pi} \ln \frac{b}{a} = 2 \times 10^{-7} \ln \frac{b}{a} \text{ henry/meter}$$

Capacitance is not affected by frequency (except as frequency may alter the relative permittivity of the dielectric):

$$C = \frac{2\pi\varepsilon}{\ln \frac{b}{a}} = \frac{55.5\varepsilon_r}{\ln \frac{b}{a}} \text{ pF/m}$$

$$R_{ac} = 4.16 \times 10^{-8} \sqrt{f} \left[ \frac{1}{a} + \frac{1}{b} \right]$$

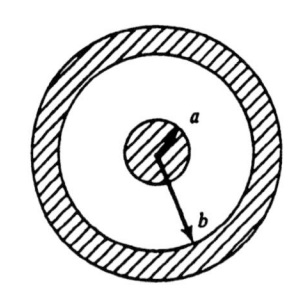

where
 $a$ is outer radius of inner conductor, in meters
 $b$ is inner radius of outer conductor, in meters

Shunt losses of air dielectric lines are zero, but where solid dielectric materials are used, conductance losses sometimes exist, especially at very high frequencies. The quality of the dielectric may be measured in terms of power factor. The shunt susceptance is

$$Y = G + j\omega C$$

and the power factor is

$$PF = \frac{G}{\sqrt{G^2 + \omega^2 C^2}}$$

If $G \ll \omega C$,

$$PF = \frac{G}{\omega C} \quad \text{and} \quad G = \omega C \times PF$$

The quality of the dielectric may be expressed in terms of *dissipation factor*, which is the ratio of energy dissipated to energy stored in the dielectric per cycle. For good dielectrics with small PF angles ($G \ll \omega C$), the dissipation factor and $PF$ are equal in magnitude.

## Zero Dissipation Line Constants

The line parameters for a zero dissipation line are

$$Z = j\omega L$$
$$Y = j\omega C$$

so that the characteristic impedance $Z_0$ may be written

$$Z_0 = \sqrt{\frac{Z}{Y}} = \sqrt{\frac{j\omega L}{j\omega C}} = \sqrt{\frac{L}{C}} \text{ ohms}$$

This value is wholly resistive and may be given the symbol $R_0$. Propagation constant is

$$\gamma = \sqrt{ZY} = \sqrt{-\omega^2 LC} = j\omega\sqrt{LC} = \alpha + j\beta$$

where

$\alpha = 0$ (attenuation constant)

$\beta = \omega\sqrt{LC}$ rad/m (phase constant)

$$= \frac{2\pi f}{v} = \frac{2\pi}{\lambda}$$

Velocity of propagation is

$$v = \frac{\omega}{\beta} = \frac{1}{\sqrt{LC}} \text{ m/s}$$

For an *open-wire* line the velocity of propagation is

$$v = 3 \times 10^8 \text{ m/sec}$$

and

$$R_0 = 120 \ln\frac{d}{a} \, \Omega, \quad \text{for } \frac{d}{a} > 10$$

For a *coax* line,

$$v = \frac{3 \times 10^8}{\sqrt{\varepsilon_r}} \text{ m/s}$$

and

$$R_0 \frac{60}{\sqrt{\varepsilon_r}} \ln\frac{b}{a} \, \Omega$$

where $\varepsilon_r$ is the dielectric constant between conductors.

## Zero Dissipation Line Voltages and Currents

In a dissipationless line, attenuation is zero, $Z_0 = R_0$, and voltage and current may be found at any point $s$ units distant from the receiving end of a transmission line from the following equations:

$$E = E_R \cos\frac{2\pi s}{\lambda} + jI_R R_0 \sin\frac{2\pi s}{\lambda}$$

$$I = I_R \cos\frac{2\pi s}{\lambda} + j\frac{E_R}{R_0}\sin\frac{2\pi s}{\lambda}$$

$$Z = E/I$$

where
> $E_R$ is the receiving end voltage
> $I_E$ is the receiving end current
> $\lambda$ is the wavelength in meters

If the line is open-circuited, $I_R = 0$, and

$$E_{oc} = E_R \cos \frac{2\pi s}{\lambda}$$

$$I_{oc} = \frac{jE_R}{R_0} \sin \frac{2\pi s}{\lambda}$$

The current and voltage are in quadrature everywhere, and no power is transmitted along the line. If the line is short-circuited, $E_R = 0$, and,

$$E_{sc} = jI_R R_0 \sin \frac{2\pi s}{\lambda}$$

$$I_{sc} = I_R \cos \frac{2\pi s}{\lambda}$$

Again, the current and voltage are in quadrature, but the current and voltage waves are shifted $\lambda/4$ from the positions for the open-circuit case. If the line is terminated in a resistance $R_R$ greater than $R_0$, the *reflection coefficient K* will be positive and the voltage and current conditions on the line will be intermediate to the open circuit and $R_0$-terminated conditions. $K$ is defined by the formula

$$K = \frac{Z_R - Z_0}{Z_R + Z_0} \angle \phi = \frac{\text{reflected voltage at load}}{\text{incident voltage at load}}$$

For example, if $R_R = 3R_0$, the value of $K$ is 0.5 and the *incident wave* has an amplitude twice that of the *reflected wave*.

The term "incident wave" refers to a voltage or current wave progressing from the source toward the load. The term "reflected wave" is the voltage or current wave moving from the load toward the source. The magnitude of the reflected wave is dependent on the value of $K$. The actual voltage at any point on the transmission line is the *vector sum* of the incident and reflected wave voltage at that point. The resultant total voltage wave appears to stand still on the line, oscillating in magnitude with time but having fixed positions of maxima and minima. Such a wave is known as a *standing wave*.

If the line is terminated in $Z_R = R_0$, $K$ and the reflected wave becomes zero, the voltage on the line is

$$E = E_R \varepsilon^{j\beta s}$$

where
> $\beta = 2\pi/\lambda$, the phase constant in radians/meter
> $s$ = distance from the receiving end in meters

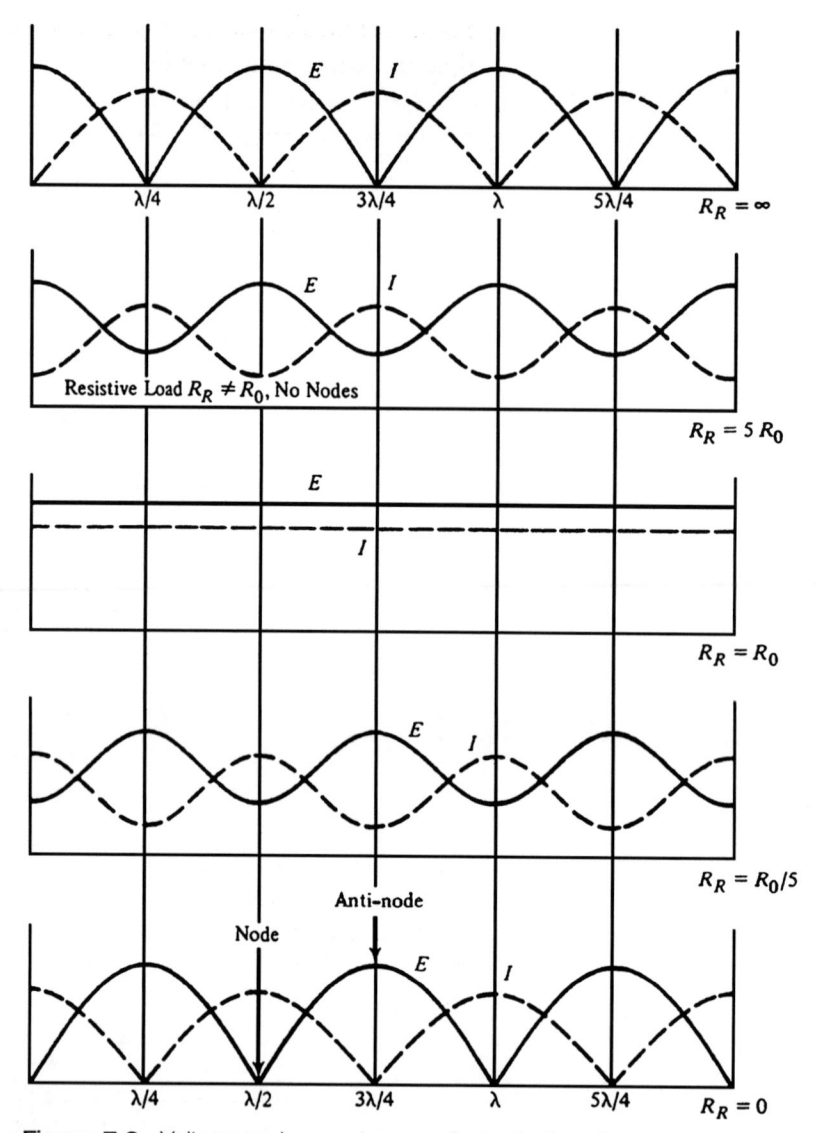

**Figure 7.2** Voltage and currents on a dissipationless line

The previous equation represents a constant voltage magnitude with continuously varying phase angle along the line. Similarly,

$$I = I_R \varepsilon^{j\beta s}$$

Voltage and current waveforms for different load terminators are shown in Figure 7.2.

## PROBLEM 7.4  Reflection Coefficient

The load end reflection coefficient on a transmission line is given by the following formula:

$$K = \left| \frac{Z_L - Z_o}{Z_L + Z_o} \right| \angle \phi°$$

The ratio of reflected voltage to incident voltage $s$ meters from the load for a zero dissipation transmission line is:

a. $K_s = \left|\dfrac{Z_L - Z_o}{Z_L + Z_o}\right| e^{2\alpha s} \angle \phi - 2\beta s$

b. $K_s = \left|\dfrac{Z_L - Z_o}{Z_L + Z_o}\right| e^{-2\alpha s} \angle \phi$

c. $K_s = \left|\dfrac{Z_L - Z_o}{Z_L + Z_o}\right| e^{-2\alpha s} \angle \phi - \beta s$

d. $K_s = \left|\dfrac{Z_L - Z_o}{Z_L + Z_o}\right| e^{-2\alpha s} \angle \phi - 2\beta s$

*Solution*

$$K = \frac{\text{reflected voltage at load}}{\text{incident voltage at load}} = \frac{E''_L}{E'_L}$$

At $s$ meters from the load end,

$$E''_s = E''_L e^{-(\alpha + j\beta)s}$$
$$E'_s = E'_L e^{(\alpha + j\beta)s}$$

where
 $E'$ is incident voltage
 $E''$ is reflected voltage
 $\alpha$ is attenuation constant in nepers/meter
 $\beta$ is phase constant in radians

$$K_s = \frac{E''_s}{E'_s} = \frac{E''_L e^{-(\alpha+j\beta)s}}{E'_L e^{(\alpha+j\beta)s}} = \left|\frac{Z_L - Z_o}{Z_L + Z_o}\right| e^{-2\alpha s} \angle \phi - 2\beta s$$

The correct answer is d.

*Answer Rationale*

Incorrect solution a is the result of incorrectly using the plus sign for the power in the term $e^{-2\alpha s}$. Incorrect solution b is the result of not subtracting $2\beta s$ in the angle of the final answer. Incorrect solution c is the result of not multiplying the term $\beta s$ by 2 in the angle of the final answer.

If your answer is correct, go on to the next section.

If your answer is not correct, review the subject of zero dissipation line voltages and currents in this chapter. For more details, see chapter 8 in reference 1 and chapter 25 in reference 3.

## Standing Waves

In standing-wave situations, *nodes* are points of zero voltage or current, as occurs in the case of an open-circuit or a short-circuit load. In cases where there is a finite load where $R_R \neq R_0$, there is a standing wave with maximum and minimum points, but not nodes. A line terminated in $R_0$ has no standing waves and thus no nodes; it is called a *smooth* line. The ratio of maximum to minimum magnitudes of voltage or current is called the *standing wave ratio S*, where

$$S = \left|\frac{E_{max}}{E_{min}}\right| = \left|\frac{I_{max}}{I_{min}}\right| = \frac{1+|K|}{1-|K|}$$

or

$$|K| = \frac{S-1}{S+1} = \frac{|E_{max}|-|E_{min}|}{|E_{max}|+|E_{min}|}$$

where

$$E_{max} = \frac{E_R(Z_R+Z_0)}{2Z_R}(1+|K|)$$

$$E_{min} = \frac{E_R(Z_R+Z_0)}{2Z_R}(1-|K|)$$

$S$ is always expressed as a number $\geq 1$.

For the special case of the resistive load,

$$S = \frac{R_R}{R_0} \quad \text{for } R_R > R_0$$

$$S = \frac{R_0}{R_R} \quad \text{for } R_R < R_0$$

## Zero-Dissipation-Line Input Impedance

The input impedance of a dissipationless line is

$$Z_S = \frac{E_S}{I_S} = R_0\left[\frac{Z_R + jR_0\tan\beta s}{R_0 + jZ_R\tan\beta s}\right]$$

Another convenient expression for $Z_S$ is

$$Z_s = R_0\left[\frac{1+|K|\angle\phi-2\beta s}{1-|K|\angle\phi-2\beta s}\right]$$

where $\phi$ is the angle of $K$.

At values of $s = (\phi/2\beta) + (n\lambda/4)$, the numerator and denominator terms are in phase ($n = 0, 1, 2, \ldots$). At these points the input of the line is purely resistive, with min and max values occurring every quarter wavelength.

Maximum input impedance (resistive) at $s = (\phi/2\beta) + (n\lambda/2)$ (phasors coincident) is

$$R_{max} = R_0\left[\frac{1+|K|}{1-|K|}\right] = SR_0$$

Minimum input impedance (resistive) at $s = (\phi/2\beta) + (2n - 1) \lambda/4$ (phasors coincident) is

$$R_{min} = R_0 \left[\frac{1-|K|}{1+|K|}\right] = \frac{R_0}{S}$$

$$= R_0 \left[\frac{Z_R + jR_0 \tan\frac{2\pi s_2}{\lambda}}{R_0 + jZ_R \tan\frac{2\pi s_2}{\lambda}}\right]$$

where $s_2$ is distance of the first voltage minimum from the load.

Solving for load impedance yields

$$Z_R = R_0 \left[\frac{1 - jS\tan\frac{2\pi s_2}{\lambda}}{S - j\tan\frac{2\pi s_2}{\lambda}}\right]$$

The point of voltage minimum is measured rather than voltage maximum because it is usually possible to determine the exact point of the minimum with greater accuracy.

For a short-circuited line, the input impedance is

$$Z_{sc} = jR_0 \tan\beta s$$

For an open-circuited line, the input impedance is

$$Z_{oc} = -jR_0 \cot\beta s$$

The characteristic impedance of the line is

$$Z_0 = \sqrt{Z_{oc} Z_{sc}}$$

Both cases are purely reactive with no power dissipation, alternating between inductance and capacitance each quarter wavelength.

## Reflection Losses at Any Point

If a line is not matched to its load, the energy delivered by the line to the load is less than if the impedances were matched. The reflection due to the mismatch results in the following relations:

$$|E_{max}| = |E_i| + |E_r|$$
$$|E_{min}| = |E_i| - |E_r|$$

$$S = \frac{E_{max}}{E_{min}} = \frac{|E_i| + |E_r|}{|E_i| - |E_r|}$$

## Power

The expression for power passing along the line and delivered to the load is

$$P = \frac{|E_{max}| \cdot |E_{min}|}{R_0} = |I_{max}| \cdot |I_{min}| R_0 = \frac{|E_i|^2 - |E_r|^2}{R_0}$$

where

$$E_{max} = \frac{I_R |Z_R + Z_0|}{2}(1 + |K|)$$

$$I_{max} = \frac{E_{max}}{R_0}$$

$$R_0 = \frac{E_{max}}{I_{max}} = \frac{E_{min}}{I_{min}}$$

$$E_{min} = \frac{I_R |Z_R + R_0|}{2}(1 - |K|)$$

$$I_{min} = \frac{E_{min}}{R_0}$$

$$R_{max} = \frac{E_{max}}{I_{min}} = SR_0$$

$$R_{min} = \frac{E_{min}}{I_{max}} = \frac{R_0}{R}$$

The greatest amount of power is transmitted if $|E_{max}| = |E_{min}|$, or the line is smooth with no standing wave (*i.e.*, $S = 1$) and with $R_0$ termination.

The ratio of power $P$ delivered to the load to the power $P_i$ transmitted by the incident wave is

$$\frac{P}{P_i} = \frac{P_i - P_r}{P_i} = \frac{|E_i|^2 - |E_r|^2}{E_i^2} = 1 - |K^2| = \frac{4S}{(S+1)^2}$$

All of the power is absorbed by the load when $S = 1$.

## The Eighth-Wave Line

The input impedance of a line of length $s = \lambda/8$ is

$$Z_S = R_0 \left[ \frac{Z_R + jR_0}{R_0 + jZ_R} \right]$$

If the line is terminated in a pure resistance $R_R$, then the numerator and denominator have identical magnitudes and,

$$|Z_S| = R_0$$

Thus, an eighth-wave line may be used to transform any resistance to an impedance with a magnitude equal to $R_0$ of the line.

## The Quarter-Wave Line

For $s = \lambda/4$,

$$Z_S = \frac{R_0^2}{Z_R}$$

A quarter-wave section may be used as a transformer to match a load of $Z_R$ to a source of $Z_S$.

A quarter-wave section may be considered to be an impedance inverter, since it can transform a low impedance into a high impedance and vice versa. A $\lambda/4$ short-circuited line looks like an open circuit at the source, and an open-circuited $\lambda/4$ line looks like a short circuit at the source.

## The Half-Wave Line

The half-wave line has an input impedance of

$$Z_S = R_0 \left[ \frac{Z_R + jR_0 \tan \pi}{R_0 + jZ_R \tan \pi} \right] = Z_R$$

A half-wavelength line may be thought of as a one-to-one transformer. It is useful in connecting a load to a source in cases where the load and source cannot be made adjacent.

## Impedance Matching

For greatest efficiency and delivered power, an RF transmission line should be operated as a smooth line with an $R_0$ termination. If it is not possible to terminate in $R_0$, then an impedance transforming section added between line and load can be used (discussed under the topic of impedance transformer in this chapter). Another technique is to use an open or closed stub line of suitable length as a reactance shunted across the line at a designated distance from the load, as shown in Figure 7.3.

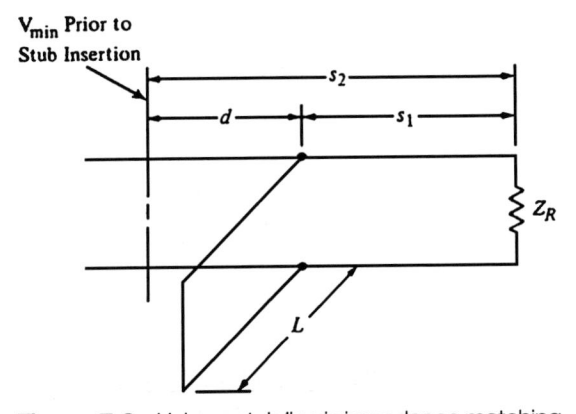

**Figure 7.3** Using a stub line in impedance matching

The stub should be located at a distance $d$ measured in either direction from a voltage minimum. Ordinarily, the stub is placed on the load side of the minimum nearest the load. A short-circuited stub is ordinarily preferred to an open-circuited stub because of greater ease in construction and because of the inability to assure high enough insulation resistance for an open circuit. A shorted stub also has a lower loss of energy due to radiation. Dimensions for length and location of the short-circuited stub are given by the following formulas:

$$S = \frac{V_{max}}{V_{min}}$$

$$d = s_2 - s_1 = \left[\arccos\left(\frac{S-1}{S+1}\right)\right]\left(\frac{\lambda}{4\pi}\right)$$

$$L = \frac{\lambda}{2\pi} \arctan\left(\frac{\sqrt{S}}{S-1}\right)$$

when stub is placed toward the load from $V_{min}$. When stub is placed toward the source from $V_{min}$,

$$L' = \frac{\lambda}{s} - L$$

## PROBLEM 7.5  Stub Match

Standing wave measurements made on an RF transmission line with unknown load impedance yield $S = 3.0$. The distance from the load to the nearest minimum = 25 cm = $s_2$. The distance between the standing wave minima = 40 cm. $Z_0 = 200\ \Omega$.

Neglecting line dissipation, determine:

1. Frequency
2. Load parameters
3. Ratio of power transmitted to power delivered
4. Design and location of stub to match load to line

*Solution*

1. Wavelength, $\lambda = 2 \times 40$ cm $= 80$ cm or 0.8 meter.

$$f = \frac{v}{\lambda} = \frac{3 \times 10^8}{0.8} = 375\ \text{MHz}$$

2. 
$$Z_R = Z_0 \left[\frac{1 - jS\tan\left(\frac{360 s_2}{\lambda}\right)^\circ}{S - j\tan\left(\frac{360 s_2}{\lambda}\right)^\circ}\right]$$

$$\tan\frac{360^\circ \times 25}{80} = \tan 112.5^\circ = 2.41$$

$$Z_R = 200\left(\frac{1 + j3 \times 2.41}{3 + j2.41}\right) = 379.22\angle 43.35^\circ\ \Omega = 276 + j260\ \Omega = R_R + jX_R$$

3. $$\frac{\text{power delivered}}{\text{power transmitted}} = \frac{4S}{(S+1)^2} = \frac{4\times 3}{(4)^2} = \frac{12}{16} = 0.75 \text{ or } 75\%$$

4. $$d = \arccos\left(\frac{S-1}{S+1}\right)\left(\frac{\lambda}{4\pi}\right)$$

$$d = \arccos\left(\frac{3-1}{3+1}\right)\left(\frac{\lambda}{4\pi}\right) = \arccos(0.5)\left(\frac{80}{4\pi}\right)$$

$$d = \left(60°\frac{\pi}{180°}\text{rad}\right)(6.37 \text{ cm}) = 1.05 \times 6.37 = 6.67 \text{ cm}$$

$$L = \frac{\lambda}{2\pi}\arctan\left(\frac{\sqrt{S}}{S-1}\right) = \frac{80}{2\pi}\arctan 0.87 = 12.73\left(40.89°\frac{\pi}{180°}\right)$$

$$L = (12.73 \text{ cm})(0.717 \text{ rad}) = 9.09 \text{ cm}$$

$$L' = \frac{\lambda}{2} - L = 40 - 9.09 = 30.9 \text{ cm}$$

If your answers are correct, continue to the next section.

If your answers are not correct, review the subjects of zero dissipation line voltages and currents, zero dissipation line input impedance, reflection losses at any point, power, and impedance matching in this chapter. For more details, see chapter 8 in reference 1 and chapter 25 in reference 3.

## Smith Chart

A Smith chart is used in many instances to determine the parameters of a lossless transmission line. The Smith chart has the following properties:

1. All possible values of impedance are contained inside the outer circle of unit radius.

2. $\beta s$ increments are indicated around the outer edge of the chart in terms of wavelengths.

3. A straight edge pivoted at the center and marked in terms of $S$ serves as a distance coordinate to any point on the chart and has the effect of adding constant-$S$ circles to the chart without actually complicating the figure with additional lines.

4. The impedance of a transmission line may be read at any point on the appropriate $S$-circle.

5. The point at the center of the chart represents the impedance of the line terminated in its characteristic impedance, where $Z/R_0 = 1$ for all distances.

6. The point at the extreme left of the resistance $r_a$ axis represents a short circuit (zero impedance), and the point at the extreme right represents an open circuit (infinite impedance).

7. The outer circle represents $S = \infty$.

8. The chart may be used for admittance as well as impedance, the $r_a$ and $x_a$ axes becoming $g_a$ and $b_a$ axes, with the convention that capacitive susceptance

is positive (above) and inductive susceptance is negative (below); the leftmost point is then an open circuit (zero conductance), and the rightmost point is a short circuit (infinite conductance).

9. $V_{min}$ occurs on the real axis. When using impedances, $V_{min}$ occurs on the left half; when using conductances, $V_{min}$ occurs on the right half.

Figure 7.4 is a copy of a Smith chart. The following problems may be solved with its aid.

### PROBLEM 7.6 Normalized Sending Impedance

*Given*

$$Z_R/R_0 = 2.5 + j1.1$$
$$\text{line length} = 30°$$

Find the normalized $Z_s$. (*Note:* Smith chart problems require a compass and a straight edge.)

*Solution*

- Locate load point $A$ at $2.5 + j1.1$ on Smith chart.
- Draw constant-$\beta s$ line from origin through $A$ to outer circle reading $0.224\lambda$ (toward generator). See Exhibit 7.6.
- Calculate line length in terms of $\lambda$,

$$\frac{30°}{360°} = 0.083\lambda$$

and move this distance toward generator along outer circle to the point $(0.224 + 0.083)\lambda = 0.307\lambda$.

- Draw another constant-$\beta s$ line from origin to $0.307\lambda$ on outer circle. See Exhibit 7.6.
- Draw a portion of a constant-$S$ circle (center at origin) from point $A$ to point $B$, where it intersects with the constant $\beta s$ line drawn in step 4.
- Read normalized input impedance at point $B$ as $Z_S/R_0 = 1.59 - j1.37$ (capacitive reactance).

If your answer is correct, go to the next problem.
If your answer is not correct, review the subject of Smith chart in this chapter. For more details, see chapter 8 in reference 1 and chapter 23 in reference 2.

### PROBLEM 7.7 Normalized $Z_R$, $Z_S$, and $K$

*Given*

$$S = 2.4$$
$$V_{min} \text{ occurs at } 0.2\lambda \text{ from load}$$
$$\text{line length} = 0.38\lambda$$

Find the normalized $Z_R$ and $Z_S$ and reflection coefficient, $K$.

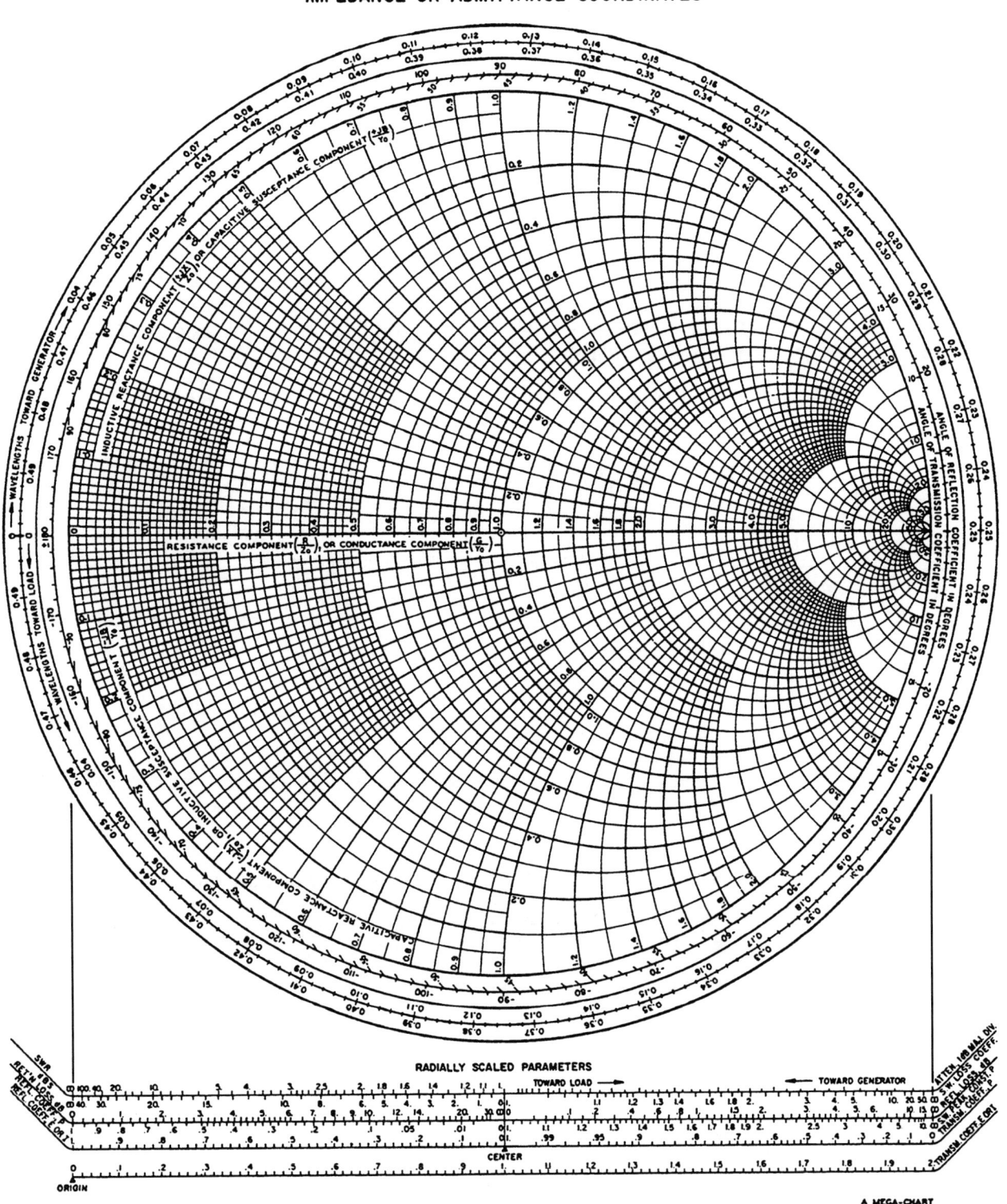

**Figure 7.4** Smith Chart (Reprinted by permission of Kay Elemetrics Corp.)

**218** Chapter 7 Communications

**Exhibit 7.6**

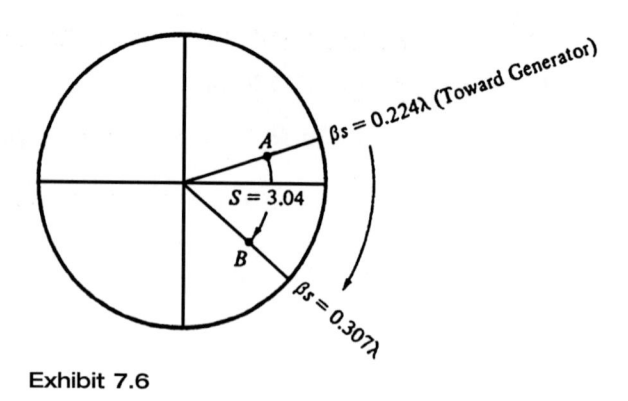

*Solution*

$V_{min}$ occurs where $S$ intersects the left half of the resistance axis (least resistance for least voltage).

- Locate $V_{min}$ as point $A$ on Smith chart as shown in Exhibit 7.7.

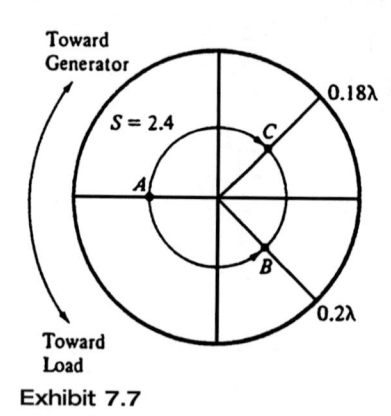

**Exhibit 7.7**

- Move toward load (counter clockwise) to point $B$ where constant $\beta s$ line = $0.2\lambda$ intersects with $S = 2.4$ circle.

- Read $Z_R/R_0 = 1.65 - j0.97$ at point $B$.

- Move toward generator (clockwise) from point $B$ to point $C$ a distance of $0.38\lambda$ along $S = 2.4$ circle and read $Z_S/R_0 = 1.3 + j1.0$.

- $|K| = \dfrac{S-1}{S+1} = \dfrac{1.4}{3.4} = 0.412$

- Read angle of $K$ from $\beta s$ line at load on the third from outermost scale of chart, giving $-36°$. $\therefore K = 0.413 \angle -36°$

If your answers are correct, go to the next problem.

If your answers are not correct, review the subjects of standing waves, zero dissipation line input impedance, and Smith chart in this chapter. For more details, see chapter 9 in reference 1 and chapter 23 in reference 2.

## PROBLEM 7.8 Stub Match, Smith Chart

*Given*

$$Y_R/G_0 = 2.8 + j1.7 \text{ (capacitive load using admittance form)}.$$

Find the short-circuit stub position and length.

*Solution*

- Locate normalized load admittance at point $A$ on chart $(2.8 + j1.7)$ as shown in Exhibit 7.8.

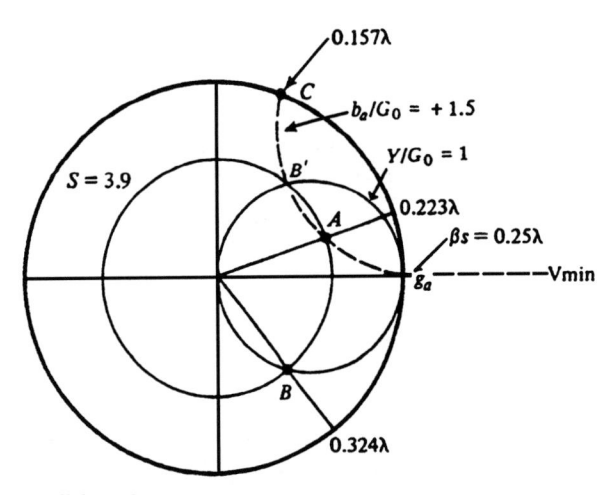

**Exhibit 7.8**

- Drawing a constant-$S$ circle through $A$ shows $S = 3.9$ before use of the stub.

- Locate the $Y/G_0 = 1$ circle on the chart. This is the locus of all points for which the real part of the line conductance is unity, the desired condition at the point of stub connection. Thus, the intersection of the $S$ circle with the $Y/G_0 = 1$ circle at point $B$ is the proper location of the stub toward generator (cw).

- Since the load is located on the $A$ line ($\beta s = 0.223\lambda$) and the stub is located on the $B$ line ($\beta s = 0.324\lambda$), the stub is located $(0.324 - 0.223)\lambda = 0.101\lambda$ toward the generator from the load.

- The $b_a$ value at $B$ represents the line susceptance at the stub connection and is $-1.5$ inductive. The stub must cancel out this imaginary component with a capacitive susceptance of $+1.5$. Find the electrical length of the short circuited stub by locating the intersection of the $b_a/G_0 = +1.5$ circle on the $\beta s$ scale ($= 0.157\lambda$) at point $C$. This intersection occurs $(0.157 + 0.25)\lambda = 0.407\lambda$ from a short circuit at the right end of the real axis or infinite admittance point (measuring toward the load). Thus, a short-circuited stub line $0.407\lambda$ in length would have the required capacitive susceptance. An open-circuit stub of $0.157\lambda$ would also provide proper match, although open-circuit matching is not as desirable, for reasons given previously.

If your answers are correct, continue to the next section.

If your answers are not correct, review the above solution. For more details, see chapter 8 in reference 1 and chapter 23 in reference 2.

## Impedance Transfer

There are situations when it is desirable to match a source impedance $Z_S$ to a load impedance $Z_R$ by inserting a quarter-wave (or any odd multiple of a quarter-wave) line in series with the transmission line. Such an application would be to couple a transmission line to the resistive load of an antenna. The quarter-wave matching section must be designed to have the characteristic impedance $R'_0$ so that the antenna resistance $R_A$ is transformed to a value equal to the characteristic impedance $R_0$ of the transmission line. The line is then terminated in its characteristic impedance and is operated under conditions of no reflection. In this case,

$$R'_0 = \sqrt{Z_A R_0}$$

This is the value required to achieve critical coupling and maximum power transfer from the transmission line to the load. In general, $R'_0$ of the matching section should equal the geometric mean of the source and load impedance and is equal to

$$R'_0 = \left| \sqrt{Z_S Z_R} \right|$$

# ATTENUATION

In many phases of electrical engineering, particularly communications, the use of logarithmic units for power ratios is convenient. If $P_1$ is the power at one place and $P_2$ is the power at a second place, they can be related by the formula

$$\text{bels} = \log \frac{P_1}{P_2}$$

or

$$\text{decibels} = \text{dB} = 10 \log \frac{P_1}{P_2}$$

Since $P = V^2/R$,

$$\text{dB} = 10 \log \frac{V_1^2}{V_2^2} \times \frac{R_2}{R_1}$$

or

$$\text{dB} = 20 \log \frac{V_1}{V_2} + 10 \log \frac{R_1}{R_2}$$

If $R_1 = R_2$,

$$\text{dB} = 20 \log \frac{V_1}{V_2}$$

In filter circuits it is sometimes more convenient to use natural logs rather than base-10 (common) logs. In this case the neper, rather than decibel is used. The neper is defined as

$$N \text{ nepers} = \ln\left|\frac{V_1}{V_2}\right| = \ln\left|\frac{I_1}{I_2}\right|$$

or

$$\left|\frac{V_1}{V_2}\right| = \left|\frac{I_1}{I_2}\right| = e^N$$

Similarly for power ratios

$$\frac{P_1}{P_2} = e^{2N} = 10^{\#\text{dB}/10}$$

Taking the log of both sides yields

$$1 \text{ neper} = 8.686 \text{ dB}$$

A common piece of test equipment in a communication lab is the attenuator. This is a device for attenuating the power entering a given load resistance by a known number of dB without changing the total power absorbed by the load resistance and attenuator combined. By use of an attenuator, the power *to the load* is changed, but the power *from the amplifier* is not changed. A simple attenuator circuit is shown in Figure 7.5. Formulas related to this circuit are

$$R_l = R_1 + \frac{R_2 R_l}{R_2 + R_l}$$

$$R_1 = \frac{R_l^2}{R_2 + R_l}$$

From these formulas it is possible to determine the values for $R_1$ and $R_2$ that give the desired power to the load while keeping $P_{in}$ constant and equal to

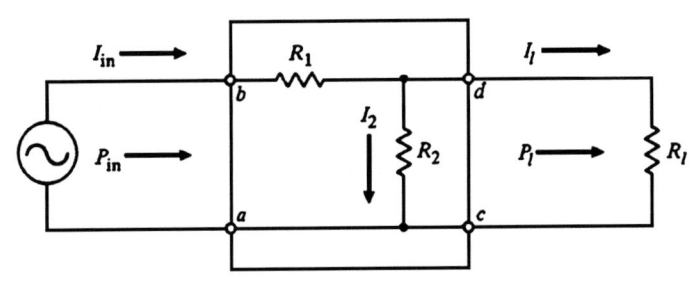

$$P_{in} = I_{in}^2 R_{in} = I_{in}^2 R_l$$

**Figure 7.5** A simple attenuator circuit

If

$$\frac{P_{in}}{P_l} = \frac{I_{in}^2 R_l}{I_l^2 R_l} = \left(\frac{I_{in}}{I_l}\right)^2$$

then

$$dB = 20\log\frac{I_{in}}{I_l}$$

Since $R_1$ and $R_2$ are in parallel, we can write the voltage between $c$ and $d$ as

$$I_2 R_2 = I_1 R_l, \quad \text{or} \quad I_2 = \left(\frac{I_1 R_l}{R_2}\right)$$

Thus,

$$I_{in} = I_1 + I_2 = I_1\left(1 + \frac{R_l}{R_2}\right)$$

and

$$dB = 20\log\left(1 + \frac{R_l}{R_2}\right)$$

The equation for $R_2$ then becomes

$$R_2 = \frac{R_l}{[10^{dB/20} - 1]}$$

## PROBLEM 7.9  Attenuator, L-Section

Design a fixed attenuator to give a loss of 10 dB between the source and a 500-ohm load.

The resistance values for $R_1$ and $R_2$ are:

a. $R_1 = 341.89\,\Omega$ and $R_2 = 231.24\,\Omega$
b. $R_1 = 450\,\Omega$ and $R_2 = 55.56\,\Omega$
c. $R_1 = 379.88\,\Omega$ and $R_2 = 158.11\,\Omega$
d. $R_1 = 73.13\,\Omega$ and $R_2 = 231.24\,\Omega$

*Solution*

$$R_2 = 500\frac{1}{10^{1/2} - 1} = \frac{500}{2.16} = 231.24\,\Omega$$

$$R_1 = \frac{R_l^2}{R_2 + R_l} = \frac{500^2}{231.24 + 500} = 341.89\,\Omega$$

The correct answer is a.

## Answer Rationale

Incorrect solution b is the result of not raising 10 to the power of ½ in the denominator when calculating $R_2$. Incorrect solution c is the result of not subtracting 1 in the denominator when calculating $R_2$. Incorrect solution d is the result of incorrectly multiplying by $R_2^2$ instead of $R_1^2$ in the numerator when calculating $R_1$.

If your answers are correct, go on to the next problem.

If your answers are not correct, review the subject of attenuation in this chapter. For more details, see chapter 8 in reference 1 and chapter 23 in reference 2.

## PROBLEM 7.10   Attenuator, Π Section

Design a Π-section attenuator for a signal generator feeding a resistive load of 72 ohms. Voltage attenuation shall be 12 dB.

The resistance values for $R_1$ and $R_2$ are:

a.  $R_1 = 47.33\,\Omega$ and $R_2 = 120.3\,\Omega$

b.  $R_1 = 86.62\,\Omega$ and $R_2 = 60.15\,\Omega$

c.  $R_1 = 102.74\,\Omega$ and $R_2 = 60.15\,\Omega$

d.  $R_1 = 134.3\,\Omega$ and $R_2 = 60.15\,\Omega$

### Solution

This problem makes use of a somewhat more efficient attenuator than that of the preceding problem. The Π-section circuit is of the form shown in Exhibit 7.10.

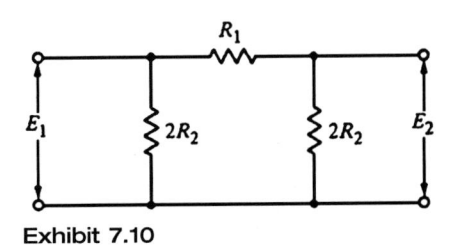

**Exhibit 7.10**

$$dB = 20\log\frac{E_1}{E_2} = 20\log K$$

where $K$ is the reflection coefficient.

$$12 = 20\log K$$
$$K = 3.981$$

$$2R_2 = R_l\left[\frac{K+1}{K-1}\right] = 72\left[\frac{3.981+1}{3.981-1}\right] = 72\left[\frac{4.981}{2.981}\right] = 120.3\,\Omega$$

$$R_2 = 60.15\,\Omega$$

$$R_1 = \sqrt{\frac{R_l R_2}{1+(R_l/4R_2)}}$$

From this,

$$R_1 = \frac{R_l^2}{R_2 - (R_l^2/4R_2)}$$

$$R_1 = \frac{72^2}{60.15 - 72^2/240.6} = 134.3 \, \Omega$$

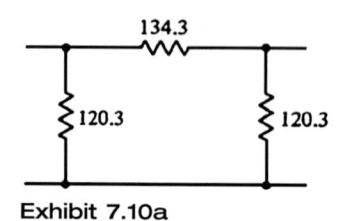

Thus, the Π-section is as shown in Exhibit 7.10a. As a point of information, for a T-filter,

$$R_l = \sqrt{R_1 R_2 \left[1 + \frac{R_1}{4R_2}\right]}$$

The correct answer is d.

*Exhibit 7.10a* (134.3; 120.3; 120.3)

### Answer Rationale

Incorrect solution a is the result of not dividing by 2 in the last step when calculating $R_2$.

Incorrect solution b is the result of not squaring the value 72 in the denominator when calculating $R_1$. Incorrect solution c is the result of incorrectly using the value of $R_l = 72$ instead of $R_2$ in the denominator when calculating $R_1$.

If your answers are correct, go on to the next section.

If your answers are not correct, review the above solution. For more details, see chapter 8 in reference 1 and chapter 23 in reference 2.

## ANTENNAS

A text on networks, lines, and fields, is a good reference for background theory and formulas on antennas.

The following is useful information for working the problems in this section.

- Impedance of free space = 377 Ω.

- *Isotropic* means identical in all directions.

- Radiation resistance of a half-wave ($\lambda/2$) dipole = 73.26 Ω and is best fed with 72-Ω cable.

- Gain of a $\lambda/2$ dipole=unity (reference) when the dipole is oriented to produce its maximum gain in the same direction as the actual antenna.

- Surface area of a cylinder = $2\pi rb$.

- Surface area of a sphere = $4\pi r^2$.

- Gain of a microwave antenna is given by the formula

$$G = \frac{4\pi A}{\lambda^2}$$

where

$\lambda$ = wavelength = $\frac{v}{f} = \frac{3 \times 10^8}{f}$ meters

$A$ = effective area = $\eta A_{actual}$

$\eta$ = antenna efficiency

- Power gain of an actual antenna is the ratio of Poynting vector produced by actual antenna in a particular direction, to value of Poynting vector generated in all directions by an isotropic source of equal power.

- Poynting vector is $P_i = \dfrac{W}{4\pi r^2} = 1 \times H$ watts/m$^2 = P_{\text{incident}}$

- Path Loss = –Path Gain = $-10 \log\left[\dfrac{P_{\text{incident}}}{P_{\text{radiated}}}\right] = -10 \log G = 10 \log \dfrac{4\pi r^2}{W} P_r$

## PROBLEM 7.11   Vertical Antenna

A newly installed vertical antenna at an AM broadcast station transmitter facility has an effective half-power radiation cross section as shown in Exhibit 7.11.

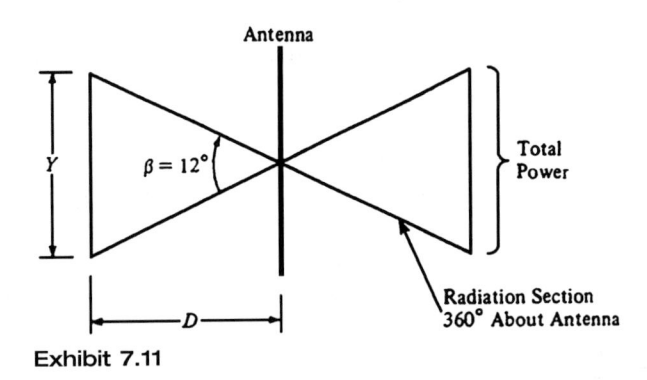

**Exhibit 7.11**

At 80 meters from the antenna, a field strength meter indicates the maximum rms value of field strength is 30 volts/meter. The total value of time-average power that the antenna radiates under these conditions is:

a. $P_{\text{Total}} = 10{,}059$ watts   c. $P_{\text{Total}} = 336.33$ watts
b. $P_{\text{Total}} = 475.64$ watts   d. $P_{\text{Total}} = 20{,}337$ watts

*Solution*

At 80 meters, $E_{\text{rms max}} = 30$ v/m

At half-power points, $E_{\text{rms}} = \left[\dfrac{30}{\sqrt{2}}\right]$ v/m $= E$

$$P = \dfrac{E^2}{377} = \left[\dfrac{30}{\sqrt{2}}\right]^2 \cdot \dfrac{1}{377} = 1.19 \text{ watts/m}^2$$

where 377 Ω is the impedance of three space

$$\tan\dfrac{\beta}{2} = 0.5\dfrac{Y}{80}, \quad Y = 160 \tan 6° = 16.82 \text{ m}$$

Total area of the cylinder of revolution = $2\pi D Y = 2\pi 80 \times 16.82 = 8453$ m$^2$

$$P_{\text{Total}} = 8453 \times 1.19 = 10{,}059 \text{ watts}$$

The correct answer is a.

*Answer Rationale*

Incorrect solution b is the result of not squaring the term $30/\sqrt{2}$ in calculating $P$. Incorrect solution c is the result of not taking the square root of 2 in the denominator when calculating $P$. Incorrect solution d is the result of incorrectly using the angle of 12 instead of 6 when calculating $Y$.

If your answer is correct, go on to Problem 7.12.

If your answer is not correct, review related material in a text on networks, lines, and fields. For more details, see chapter 8 in reference 1 and chapter 6 in reference 6.

## PROBLEM 7.12  Microwave Antenna

Two microwave stations operating at 5 GHz are 50 kilometers apart. Each has an antenna whose gain is 50 dB greater than isotropic. If 6 watts is applied to the input of the transmitting antenna, what is the signal level at the output terminal of the receiving antenna under free-space conditions? What is the path loss?

The signal level and path loss values are:

a. $P_{RCVR} = 9.12 \times 10^{-3}$ watts   and   Path loss = 140.4 dB
b. $P_{RCVR} = 5.47 \times 10^{-4}$ watts   and   Path loss = 80.4 dB
c. $P_{RCVR} = 5.47 \times 10^{-4}$ watts   and   Path loss = 6.83 dB
d. $P_{RCVR} = 5.47 \times 10^{-4}$ watts   and   Path loss = 140.4 dB

*Solution*

$$G = 50 \text{ dB} = 10^5, \quad f = 5 \times 10^9 \text{ Hz}, \quad P_T = 6 \text{ watts}$$

$$\lambda = \frac{v}{f} = \frac{3 \times 10^8}{5 \times 10^9} = 6 \times 10^{-2} \text{ m}$$

$$A = \frac{\lambda^2 G}{4\pi} = \frac{(6 \times 10^{-2})^2 10^5}{4\pi} = 28.65 \text{ m}^2 \quad \text{effective antenna area}$$

$$\text{watts/m}^2 \text{ @ 50 kM} = \frac{P_T G}{\text{surface area of sphere}} = \frac{6 \times 10^5}{4\pi (50 \times 10^3)^2}$$
$$= 1.91 \times 10^{-5} \text{ watts/m}^2$$

$$P_{RCVR} = A[\text{w/m}^2 \text{ @ 50 kM}] = 28.65 \times 1.91 \times 10^{-5} = 5.47 \times 10^{-4} \text{ watt}$$

$$\text{Path Loss} = -10\log G = -10\log \frac{4\pi}{\lambda^2}(4\pi r^2) = -10\log \frac{4\pi r^2}{\lambda}$$

$$= -20\log \frac{4\pi \times 50 \times 10^3}{6 \times 10^{-2}} = -20\log 1.05 \times 10^7$$

$$= -140\log 1.05 = 140.4 \text{ dB}$$

The correct answer is d.

*Answer Rationale*

Incorrect solution a is the result of not squaring the term $6 \times 10^{-2}$ in calculating $A$. Incorrect solution b is the result of not multiplying by $10^3$ in the numerator when calculating the path loss. Incorrect solution c is the result of incorrectly using the natural logarithm instead of log in the last step when calculating the path loss.

If your answers are correct, go on to Problem 7.13.

If your answers are not correct, review the introductory material for this section. For more details, see chapter 8 in reference 1 and chapter 6 in reference 6.

## PROBLEM 7.13  Two Half-Wave Antennas

Exhibit 7.13 shows a phased array consisting of two half-wave antennas located a half-wave apart.

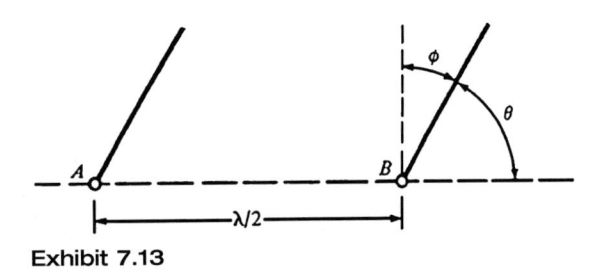

Exhibit 7.13

Antenna currents $I_A$ and $I_B$ are equal, and $I_A$ leads $I_B$ by 90°. Maximum field strength of each individually excited antenna is 250 mV/m at a distance of 40 kM. Radiation resistance of each antenna is 73.1 Ω (the same as a half-wave dipole). The angle $\theta$, at which the resultant field strength is maximum, and field strength of the array for $\theta = 90°$ at a distance of 25 kM are:

   a.  $\theta = 60°$  and  $\varepsilon_i = 452.5$ mV/m
   b.  $\theta = 30°$  and  $\varepsilon_i = 452.5$ mV/m
   c.  $\theta = 60°$  and  $\varepsilon_i = 282.8$ mV/m
   d.  $\theta = 60°$  and  $\varepsilon_i = 69.04$ mV/m

*Solution*

The rms value of the resultant electric field is given by the formula

$$\varepsilon_r = \varepsilon_{rms} \cos\left(\pi n \sin\phi + \frac{\delta}{2}\right)$$

where

$\delta$ is the phase angle between $I_A$ and $I_B$, and is positive when $I_A$ leads $I_B$
$n$ is the distance between $A$ and $B$ in wavelengths, and is usually fractional
angle $\phi$ is measured from the normal to the line of the antennas

Thus,

$$\varepsilon_r = \varepsilon_{rms} \cos\left(\frac{\pi}{2}\sin\phi + \frac{\pi}{4}\right)$$

For maximum $\varepsilon_r$

$$\cos\left(\frac{\pi}{2}\sin\phi + \frac{\pi}{4}\right) = 1$$

or

$$\left(\frac{\pi}{2}\sin\phi + \frac{\pi}{4}\right) = 0$$

$$\sin\phi = 0.5, \phi = 30°$$
$$\therefore \theta = 60°$$
At $\theta = 90°$,

$$\cos\left(\frac{\pi}{2}\sin 0° + \frac{\pi}{4}\right) = \cos\frac{\pi}{4} = 0.707$$

The field strength of the array at 25 kM distance is

$$\varepsilon_i = 0.707 \times 250 \, \frac{mV}{m}\left[\frac{40}{25}\right]^2 = 452.5 \text{ mV/m}$$

The correct answer is a.

Incorrect solution b is the result of incorrectly using the angle of 30 instead of 60 (which is 90 − 30) in calculating $\theta$. Incorrect solution c is the result of not squaring the term 40/25 in the last step when calculating $\varepsilon_i$. Incorrect solution d is the result of incorrectly using the factor 25/40 instead of 40/25 in the last step when calculating $\varepsilon_i$.

If your answers are correct, go on to Problem 7.14.

If your answers are not correct, review the introductory material for this section. For more details, see chapter 8 in reference 1 and chapter 6 in reference 6.

## PROBLEM 7.14  Waveguide Power Rating

A standard brass (Cu-Zn) waveguide has the following outer dimensions and wall thickness (inches): 6.66 × 3.41 × 0.08. Assuming the breakdown strength of air is at least 15,000 volts/cm, what is the theoretical maximum power rating for the lowest operating frequency?

### Solution

The handbook *Reference Data for Radio Engineers* lists the characteristics of this waveguide. The last column shows the maximum power rating to be 11.9 megawatts for the lowest operating frequency of 1.12 GHz.

*Note:* This type of problem found in the PE Exam does not require calculations. Rather, it requires that the examinee have sufficient background in the field to know there is a handbook available containing the proper information. With a handbook, you can solve the problem in less than five minutes.

If your answer is correct, go on to the next chapter.
If your answer is not correct, review the above solution.

# CHAPTER 8

# Logic

**OUTLINE**

DEFINITION OF TERMS, POSTULATES, PROPERTIES  230
Connectives ■ Truth Table of Binary Connectives ■ Boolean Functions: Terms and Synonyms ■ Huntington's Postulates ■ Duality ■ DeMorgan's Theorem ■ Associative Laws ■ Venn Diagrams Using Sets *A* and *B* ■ Prime Implicant ■ Maxterm (Canonic Sum, Standard Sum) ■ Combinational ■ Sequential ■ Gates ■ Flip-Flops

PROBLEMS  236

This chapter presents general principles of digital logic. Everyone preparing to take the PE exam should be familiar with the following subjects:

- Number systems
    - Bases other than 10
    - Conversion between bases
    - Negative numbers
- Truth functions
    - Logical reasoning
    - Binary connectives
    - Physical realizations (AND, NOT, inclusive and exclusive OR)
- Boolean algebra
    - Huntington's postulates and duality
    - Truth calculus vs. Boolean algebra ($+$, $\times$, $\cdot$)
    - DeMorgan's theorem
    - Venn diagrams
    - Boolean simplification
- Switching devices
    - Gates
    - Flip-flops

- Minimization of Boolean functions
  - Terms (literal, product, sum, normal, conjunction, disjunction)
  - Minterms and maxterms, prime implicants
  - Karnaugh map (graphical method of minimization)
  - Quine-McClusky method (tabular method of minimization)
- Codes and special realizations
  - Error detection and correction
  - Parity check
  - Adder (half, full)
  - BCD
  - Gray (reflected)
  - Excess-3
  - Hamming
  - Distance
- Sequential circuits
  - Shift registers
  - Counters
  - Clocked circuits, sequencers
  - Timing considerations
  - State diagrams
  - Mealy-Moore translations and circuits
  - Cycles, races, hazards

## DEFINITION OF TERMS, POSTULATES, PROPERTIES

The following terms are used in the field of logic design.

### Connectives

| Term | Symbols |
| --- | --- |
| AND | $\wedge, \cdot, \cap$ |
| (Inclusive) OR | $\vee, +, \cup$ |
| Exclusive OR | $\oplus, \forall$ |
| NOT | $-, \sim, ', *$ |
| NOT AND (NAND) | $\uparrow$ Sheffer's |
| NOT OR (NOR) | $\downarrow$ Strokes |
| If $A$ is true, then $B$ is true | $\supset$ |

## Truth Table of Binary Connectives

| Connective | | ∧ | | A | | B | ⊕ | ∨ | ↓ | ≡ | $\bar{B}$ | | $\bar{A}$ | ⊃ | ↑ | |
|---|---|---|---|---|---|---|---|---|---|---|---|---|---|---|---|---|
| A | B | 0 | 1 | 2 | 3 | 4 | 5 | 6 | 7 | 8 | 9 | 10 | 11 | 12 | 13 | 14 | 15 |
| 0 | 0 | 0 | 0 | 0 | 0 | 0 | 0 | 0 | 0 | 1 | 1 | 1 | 1 | 1 | 1 | 1 | 1 |
| 0 | 1 | 0 | 0 | 0 | 0 | 1 | 1 | 1 | 1 | 0 | 0 | 0 | 0 | 1 | 1 | 1 | 1 |
| 1 | 0 | 0 | 0 | 1 | 1 | 0 | 0 | 1 | 1 | 0 | 0 | 1 | 1 | 0 | 0 | 1 | 1 |
| 1 | 1 | 0 | 1 | 0 | 1 | 0 | 1 | 0 | 1 | 0 | 1 | 0 | 1 | 0 | 1 | 0 | 1 |

## Boolean Functions: Terms and Synonyms

| Term | Definition | Synonym |
|---|---|---|
| Literal | A variable or its complement $(A, \bar{A}, B, \bar{B})$ | |
| Product term | A series of literals related by AND $(A\bar{B}D)$ | Conjunction |
| Sum term | A series of literals related by OR $(\bar{A}+B+\bar{D})$ | Disjunction |
| Normal term | A product or sum term in which no variable appears more than once | |

## Huntington's Postulates

I. There exists a set of $K$ objects or elements, subject to an equivalence relation, denoted "=", which satisfies the principle of substitution. Substitution means: if $a = b$, then $a$ may be substituted for $b$ in any expression involving $b$ without affecting the validity of the expression.

IIa. A rule of combination "+" is defined such that $a + b$ is in $K$ whenever $a$ or $b$ are in $K$.

IIb. A rule of combination "·" is defined such that $a \cdot b$ (abbreviated $ab$) is in $K$ whenever both $a$ and $b$ are in $K$.

IIIa. There exists an element 0 in $K$ such that, for every $a$ in $K$, $a + 0 = a$.

IIIb. There exists an element 1 in $K$ such that, for every $a$ in $K$, $a \cdot 1 = a$.

IVa. $a + b = b + a$
IVb. $a \cdot b = b \cdot a$ } commutative laws.

Va. $a + (b \cdot c) = (a + b) \cdot (a + c)$
Vb. $a \cdot (b + c) = (a \cdot b) + (a \cdot c)$ } distributive laws.

VI. For every element $a$ in $K$ there exists an element $\bar{a}$ such that

and
$$a \cdot \bar{a} = 0$$
$$a + \bar{a} = 1$$

VII. There are at least two elements $X$ and $Y$ in $K$ such that $X \neq Y$.

## Duality

Every theorem that can be proved for Boolean algebra has a dual that is also true, as shown below:

$$a + (b \cdot c) = (a+b) \cdot (a+c)$$
$$a \cdot (b+c) = (a \cdot b) + (a \cdot c)$$

## DeMorgan's Theorem

For every pair of elements $a$ and $b$ in $K$,

$$\overline{(A \cdot B)} = \overline{A} + \overline{B}$$
$$\overline{(A + B)} = \overline{A} \cdot \overline{B}$$

## Associative Laws

For any three elements, $a$, $b$, and $c$ in $K$,

$$a + (b+c) = (a+b) + c$$

and

$$a \cdot (b \cdot c) = (a \cdot b) \cdot c$$

## Venn Diagrams Using Sets A and B

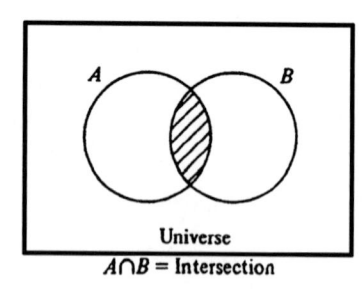

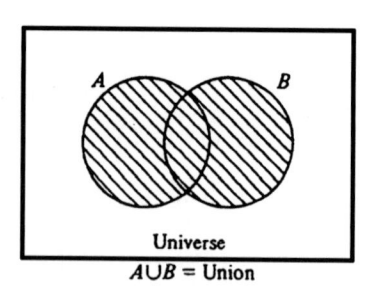

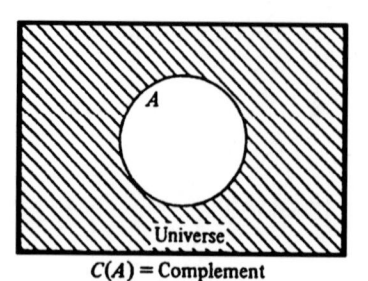

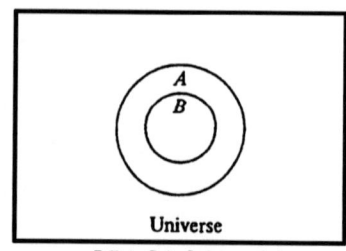

$A \cap B$ = Intersection

$A \cup B$ = Union

$C(A)$ = Complement

$B \subset A$; $B$ is Subset of $A$

If a set $A$ contains all objects of the universe, it is called a universal set, $A_U$. If a set $A$ contains no objects of the universe, it is called a null set $A_Z$.

## Prime Implicant

A *prime implicant* is any sphere of a function that is not totally contained in some larger sphere of the function.

## Maxterm (Canonic Sum, Standard Sum)

If a sum contains as many literals as there are variables in a function, then it is called a maxterm.

## Combinational

If at any particular time the present value of the outputs is determined solely by the present value of the inputs, such a system is combinational.

## Sequential

If the present value of the outputs is dependent not only on the present value of the inputs, but also on the past history of the system, then such a system is sequential.

## Gates

Combinational logic is made up of groups of AND and OR gates, with their many variations. These have been implemented over the years as diode logic, resistor-transistor logic (RTL), diode-transistor logic (DTL), direct-coupled transistor logic (DCTL), transistor-transistor logic (TTL), emitter-coupled logic (ECL), and others. There is negative logic ($V_- = 1$, $V_+ = 0$) and positive logic ($V_- = 0$, $V_+ = 1$); the voltage swing away from ground can be either positive or negative. Transistors used can be either PNP or NPN. No matter how gates are implemented hardware-wise, their inputs and outputs can assume only one of two states at any one time. Circuits for various gate hardware mechanizations are shown below.

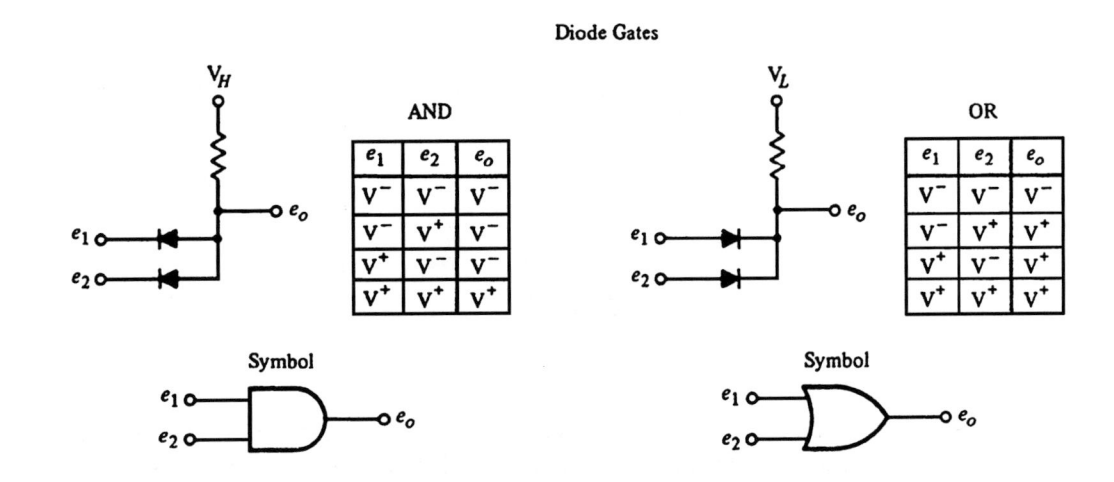

**Diode Gates**

AND

| $e_1$ | $e_2$ | $e_o$ |
|---|---|---|
| $V^-$ | $V^-$ | $V^-$ |
| $V^-$ | $V^+$ | $V^-$ |
| $V^+$ | $V^-$ | $V^-$ |
| $V^+$ | $V^+$ | $V^+$ |

OR

| $e_1$ | $e_2$ | $e_o$ |
|---|---|---|
| $V^-$ | $V^-$ | $V^-$ |
| $V^-$ | $V^+$ | $V^+$ |
| $V^+$ | $V^-$ | $V^+$ |
| $V^+$ | $V^+$ | $V^+$ |

## 234 Chapter 8 Logic

### DTL Gate

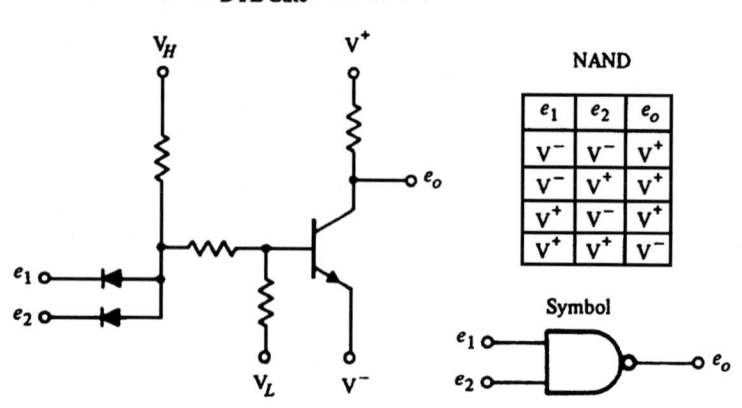

**NAND**

| $e_1$ | $e_2$ | $e_o$ |
|---|---|---|
| $V^-$ | $V^-$ | $V^+$ |
| $V^-$ | $V^+$ | $V^+$ |
| $V^+$ | $V^-$ | $V^+$ |
| $V^+$ | $V^+$ | $V^-$ |

Symbol

### Typical RTL Gate

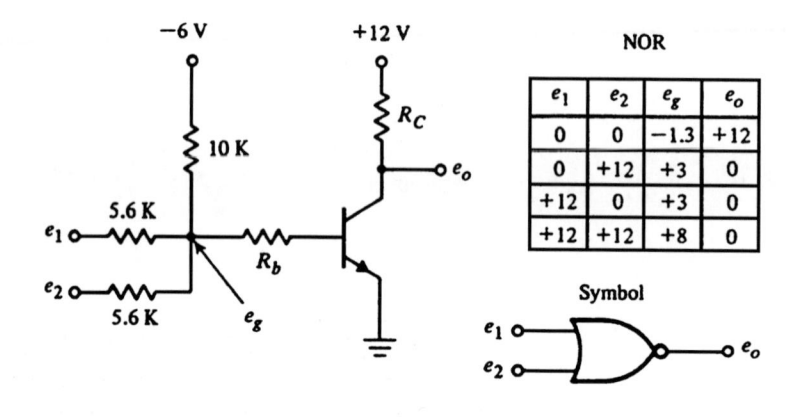

**NOR**

| $e_1$ | $e_2$ | $e_g$ | $e_o$ |
|---|---|---|---|
| 0 | 0 | −1.3 | +12 |
| 0 | +12 | +3 | 0 |
| +12 | 0 | +3 | 0 |
| +12 | +12 | +8 | 0 |

Symbol

### TTL Gates (7400 Series)

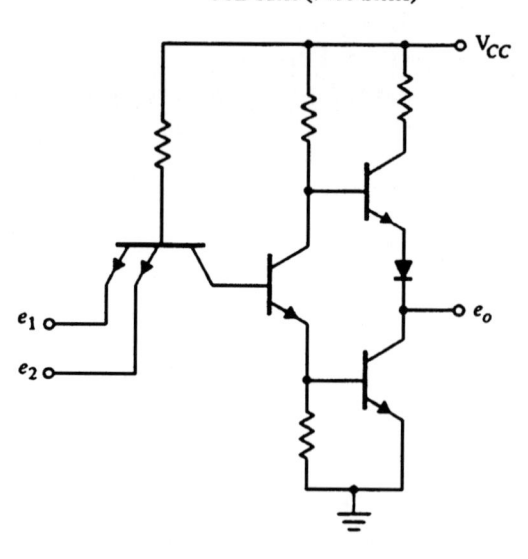

**NAND**

| $e_1$ | $e_2$ | $e_o$ |
|---|---|---|
| 0 | 0 | 1 |
| 0 | 1 | 1 |
| 1 | 0 | 1 |
| 1 | 1 | 0 |

NOR

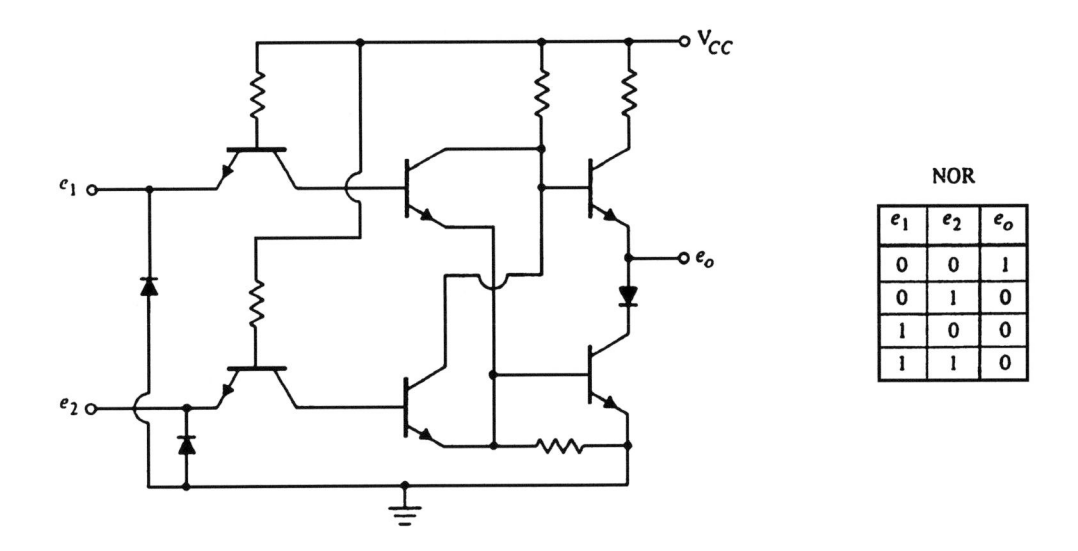

| $e_1$ | $e_2$ | $e_o$ |
|---|---|---|
| 0 | 0 | 1 |
| 0 | 1 | 0 |
| 1 | 0 | 0 |
| 1 | 1 | 0 |

## Flip-Flops

The ability to store information is an important characteristic of a digital system. The most common hardware storage device is the flip-flop, or bistable multivibrator. The following are various types of flip-flops that are implemented in hardware:

- **D.** A flip-flop whose output is a function of the input that appeared just prior to the clock pulse.

- **J-K.** A flip-flop having two inputs, $J$ and $K$. At the application of a clock pulse, 1 on the $J$ input will set the output to 1; 1 on the $K$ input will reset the output to 0; 1 on both inputs will cause the output to change state (toggle); 0 on both inputs results in no change in the output state.

- **R-S.** A flip-flop having two inputs, $R$ and $S$. Operation is the same as the $J$-$K$ flip-flop except that 1 on both inputs is illegal.

- **R-S-T.** A flip-flop having three inputs, $R$, $S$, and $T$. The $R$ and $S$ inputs produce outputs as described for the $R$–$S$ flip-flop; the $T$ input causes the flip-flop to toggle.

- **T.** A flip-flop having only one input. A pulse appearing on the input causes it to toggle.

Truth tables for basic flip-flops are shown in the following diagrams:

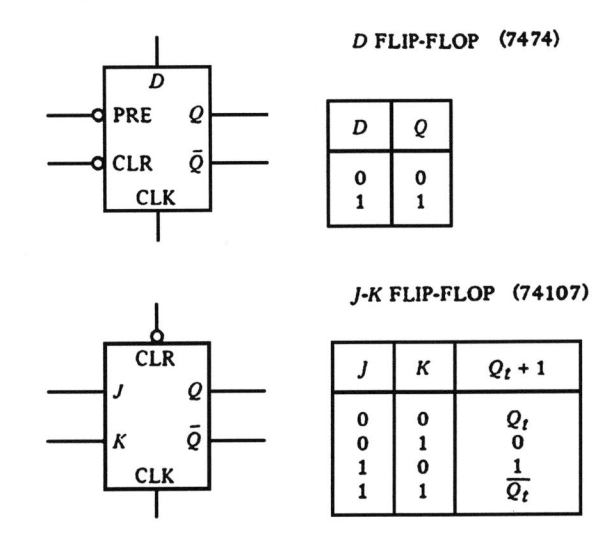

D FLIP-FLOP (7474)

| D | Q |
|---|---|
| 0 | 0 |
| 1 | 1 |

J-K FLIP-FLOP (74107)

| J | K | $Q_t + 1$ |
|---|---|---|
| 0 | 0 | $Q_t$ |
| 0 | 1 | 0 |
| 1 | 0 | 1 |
| 1 | 1 | $\overline{Q_t}$ |

**R-S FLIP-FLOP (LOW ACTIVE INPUTS)**

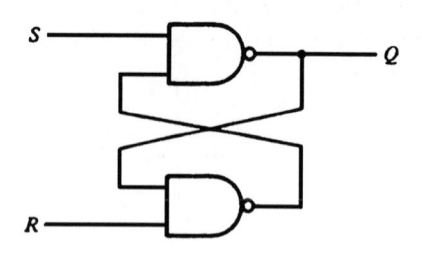

| S | R | $Q_{t+1}$ |
|---|---|---|
| 0 | 0 | illegal |
| 0 | 1 | 1 |
| 1 | 0 | 0 |
| 1 | 1 | $Q_t$ |

**R-S FLIP-FLOP (HIGH ACTIVE INPUTS)**

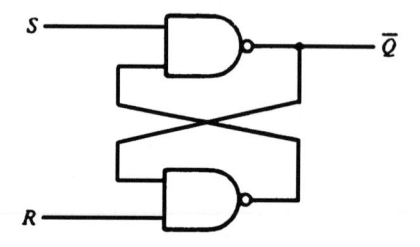

| S | R | $\overline{Q_{t+1}}$ |
|---|---|---|
| 0 | 0 | $\overline{Q_t}$ |
| 0 | 1 | 1 |
| 1 | 0 | 0 |
| 1 | 1 | illegal |

# PROBLEMS

## PROBLEM 8.1  DeMorgan's Theorem

Using DeMorgan's theorem, show that the gate pairs in Exhibit 8.1 are identical.

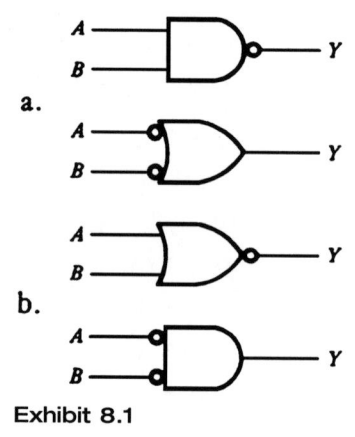

Exhibit 8.1

### Solution

The first gate is a two-input NAND (e.g., 7400); the second gate is the same part used to perform an OR function (Exhibit 8.1a). Since they are identical parts, the outputs must be the same if the inputs are the same.

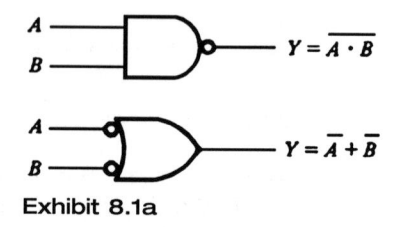

$Y = \overline{A \cdot B}$

$Y = \overline{A} + \overline{B}$

Exhibit 8.1a

By DeMorgan's theorem, $\overline{A \cdot B} = \overline{A} + \overline{B}$.

The second gate pair (Exhibit 8.1b) is a two-input NOR (e.g., 7402) used in two different ways.

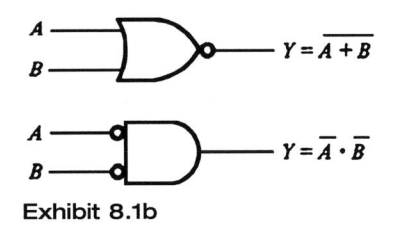

$$Y = \overline{A + B}$$

$$Y = \overline{A} \cdot \overline{B}$$

**Exhibit 8.1b**

By DeMorgan's theorem, $\overline{A + B} = \overline{A} \cdot \overline{B}$.

If your answers are correct, go to the next problem.

If your answers are not correct, review the subject of DeMorgan's theorem in this chapter. For more details, see chapter 7 in reference 1 and chapter 4 in reference 7.

## PROBLEM 8.2  Logic Simplification, Karnaugh Map

Simplify the following equation and show the final logic diagram:

$$Y = \overline{ABC} + \overline{AB}\overline{C} + \overline{A}B\overline{C} + A\overline{B}\overline{C} + A\overline{B}C + AB\overline{C}$$

The simplified equation is:
    a.  $Y = \overline{B} + \overline{C}$
    b.  $Y = \overline{A}\overline{B} + A\overline{B} + \overline{C}$
    c.  $Y = \overline{B} + \overline{A} + A$
    d.  $Y = B + \overline{B}$

*Solution*

Use a Karnaugh map to assist in simplification (Exhibit 8.2).

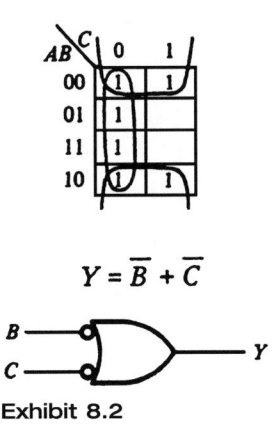

$$Y = \overline{B} + \overline{C}$$

**Exhibit 8.2**

The correct answer is a.

*Answer Rationale*

Incorrect solution b is the result of incorrectly creating three independent groups of ones in the Karnaugh map. Incorrect solution c is the result of incorrectly creating three independent groups of ones in the Karnaugh map, which are not the same as the ones created in choice b. Incorrect solution d is the result of incorrectly grouping the ones in the two groups that were created.

If your answer is correct, go to the next problem.

If your answer is not correct, review the above solution. For more details, see chapter 7 in reference 1 and chapter 4 in reference 7.

## PROBLEM 8.3  Timing Diagram, Truth Table

For a two-input NAND gate, if the $A$ and $B$ inputs are as shown in the timing diagram in Exhibit 8.3, what is the waveform for output $Y$?

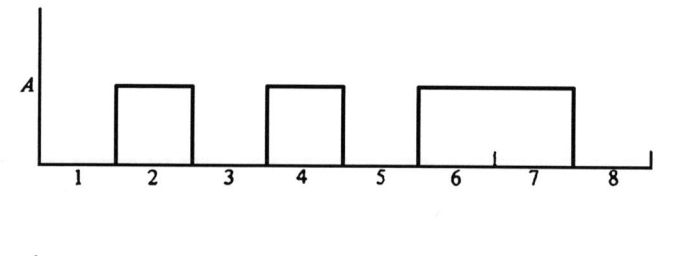

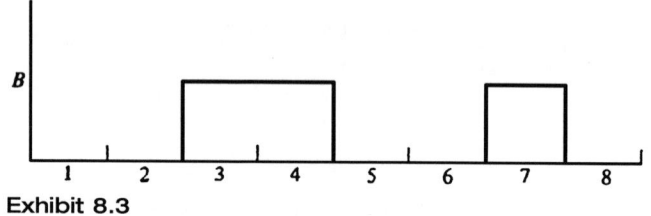

Exhibit 8.3

The truth table is

a.

|   | A | B | Y |
|---|---|---|---|
| 1 | 0 | 0 | 1 |
| 2 | 1 | 0 | 1 |
| 3 | 0 | 1 | 1 |
| 4 | 1 | 1 | 0 |
| 5 | 0 | 0 | 1 |
| 6 | 1 | 0 | 1 |
| 7 | 1 | 1 | 0 |
| 8 | 0 | 0 | 1 |

b.

|   | A | B | Y |
|---|---|---|---|
| 1 | 0 | 0 | 1 |
| 2 | 1 | 0 | 1 |
| 3 | 0 | 1 | 1 |
| 4 | 1 | 1 | 1 |
| 5 | 0 | 0 | 1 |
| 6 | 1 | 0 | 1 |
| 7 | 1 | 1 | 0 |
| 8 | 0 | 0 | 1 |

c.

|   | A | B | Y |
|---|---|---|---|
| 1 | 0 | 0 | 1 |
| 2 | 1 | 0 | 1 |
| 3 | 0 | 1 | 1 |
| 4 | 1 | 1 | 0 |
| 5 | 0 | 0 | 1 |
| 6 | 1 | 0 | 1 |
| 7 | 1 | 1 | 1 |
| 8 | 0 | 0 | 1 |

d.

|   | A | B | Y |
|---|---|---|---|
| 1 | 0 | 0 | 1 |
| 2 | 1 | 0 | 1 |
| 3 | 0 | 1 | 1 |
| 4 | 1 | 1 | 0 |
| 5 | 0 | 0 | 1 |
| 6 | 1 | 0 | 1 |
| 7 | 1 | 1 | 0 |
| 8 | 0 | 0 | 0 |

*Solution*

For each interval, Y is as shown in Exhibit 8.3a.

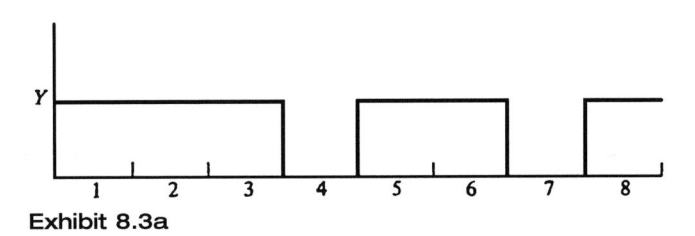

**Exhibit 8.3a**

The correct answer is a.

*Answer Rationale*

Incorrect solution b is the result of incorrectly predicting that the output at time interval 4 would be 1 instead of 0. Incorrect solution c is the result of incorrectly predicting that the output at time interval 7 would be 1 instead of 0. Incorrect solution d is the result of incorrectly predicting that the output at time interval 8 would be 0 instead of 1.

If your answers are correct, go to the next problem.

If your answers are not correct, review the above solution. For more details, see chapter 7 in reference 1 and chapter 5 in reference 7.

## PROBLEM 8.4 Logic Sequencer Input Logic

Exhibit 8.4 shows the block diagram of a sequencer that implements the control loop of Exhibit 8.4a.

**240** Chapter 8 Logic

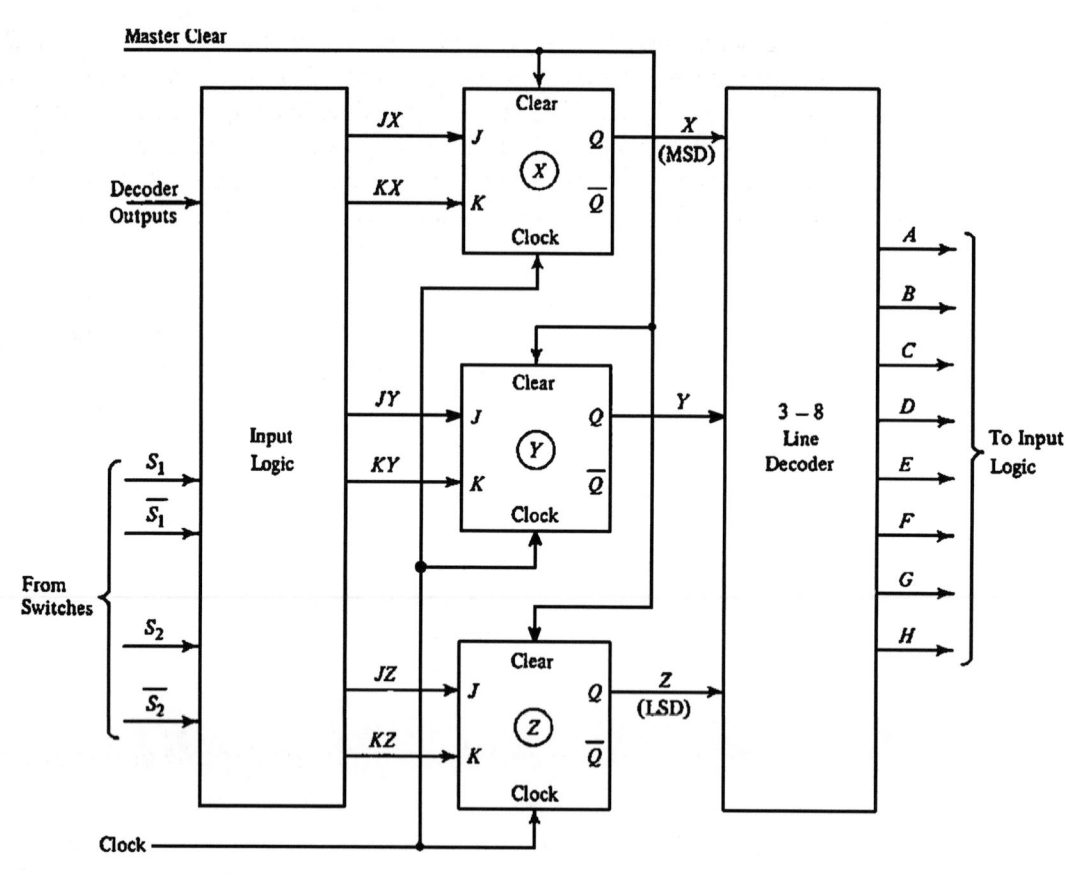

**Exhibit 8.4** Logic sequencer block diagram

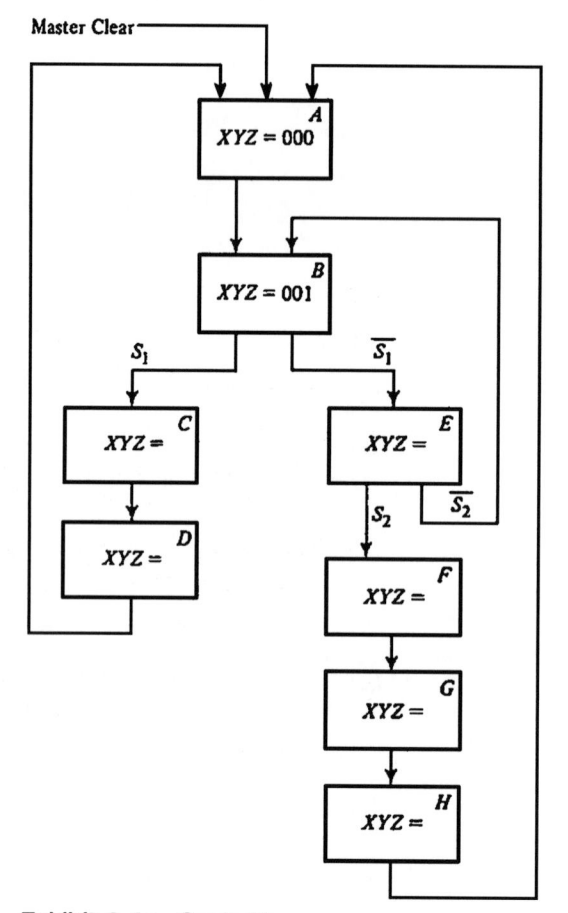

**Exhibit 8.4a** Control loop

External switches $S_1$ and $S_2$ determine jump conditions (i.e., when $S_1$ is ON the sequencer steps from state $B$ to $C$, and when $S_1$ is OFF the sequencer steps from state $B$ to $E$). Use three J-K flip-flops whose decoded outputs represent the eight states. The flip-flops are clocked simultaneously by a synchronous clock. Assume the two switches $S_1$ and $S_2$ remain unchanged during the active clock pulse time. Start with state $A = \bar{X} \cdot \bar{Y} \cdot \bar{Z}$ as shown in Exhibit 8.4 (the sequencer unconditionally jumps to this state when the system is master cleared).

1. Assign an $X, Y, Z$ control flip-flop combination to each state by filling in the Karnaugh map below. Only one control flip-flop at a time may toggle when the sequencer is clocked from one state to the next; thus, states joined by lines in Exhibit 8.4a are adjacent to each other on the Karnaugh map (Exhibit 8.4b). States $A$ and $B$ are already assigned.

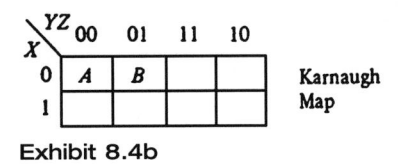

| X \ YZ | 00 | 01 | 11 | 10 |
|---|---|---|---|---|
| 0 | A | B |  |  |
| 1 |  |  |  |  |

Karnaugh Map

**Exhibit 8.4b**

Now transfer these state combinations to Exhibit 8.4a in the same manner as was done for states $A$ and $B$.

2. Write below the five remaining input equations for the three J–K control flip-flops as a function of the eight decodes states and the two switch conditions (the equation for the J-input to flip-flop Z is shown completed).

$$JZ = A$$
$$KZ =$$
$$JY =$$
$$KY =$$
$$JX =$$
$$KX =$$

*Solution*

1. There are two acceptable Karnaugh maps for this problem. One is shown in Exhibit 8.4c.

   The binary code for each state is then entered in the appropriate block of the control loop, as shown. Thus, it is seen that only one flip-flop changes state between adjacent blocks of the control loop. (See Exhibit 8.4c.)

2. The five remaining input equations for the flip-flops are shown below. A $J$ input is set to a logic ONE at the block just prior to the flip-flop switching from a ZERO to a ONE.

The $K$ input is set to a logic ONE at the block just prior to the flip-flop switching from a ONE to a ZERO. These inputs are set up and stable at the time of the next clock pulse.

$$JZ = A$$
$$KZ = C + F$$

$$JY = B \cdot \overline{S_1}$$

$$KY = H + E \cdot \overline{S_2}$$

$$JX = B \cdot S_1 + E \cdot S_2$$

$$KX = D + G$$

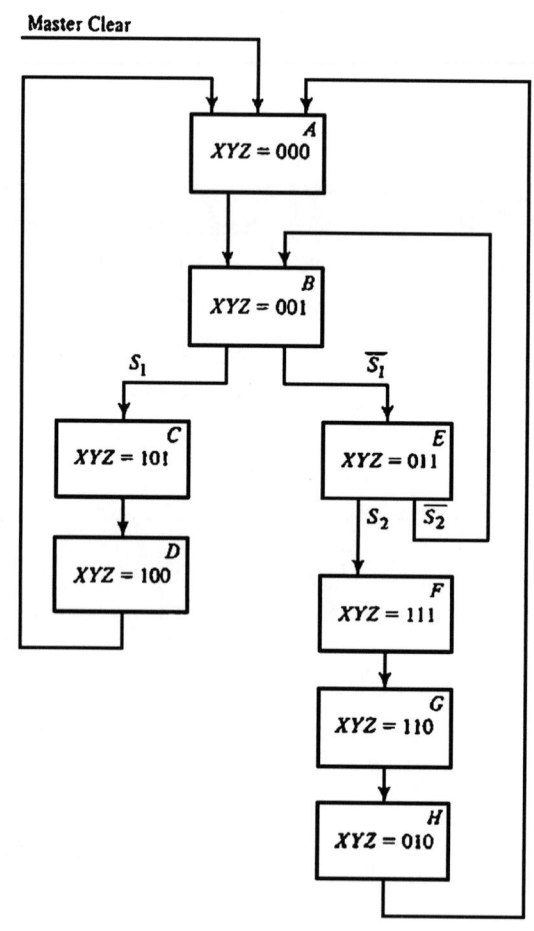

| X\YZ | 00 | 01 | 11 | 10 |
|---|---|---|---|---|
| 0 | A | B | E | H |
| 1 | D | C | F | G |

**Exhibit 8.4c** Control loop

An alternative solution follows:

| X\YZ | 00 | 01 | 11 | 10 |
|---|---|---|---|---|
| 0 | A | B | C | D |
| 1 | H | E | F | G |

**Exhibit 8.4d**

$$KZ = C + F$$

$$JY = B \cdot S_1 + E \cdot S_2$$

$$KY = D + G$$

$$JX = B \cdot \overline{S_1}$$

$$KX = H + E \cdot \overline{S_2}$$

If your answers are correct, go on to Problem 8.5.

If your answers are not correct, review the above solution. For more details, see chapter 7 in reference 1 and chapter 8 in reference 7.

### PROBLEM 8.5  Underlapped Two-Phase Clock Circuit

The *J-K* flip-flop circuit shown in Exhibit 8.5 is commonly used to generate an underlapped two-phase clock. The flip-flop is the master-slave 74107 in which the output switches on the falling edge of the input clock.

Draw a timing diagram showing the relation of the input and output clocks. Also show another method of developing a two-phase clock using a 7474 edge-triggered *D* flip-flop.

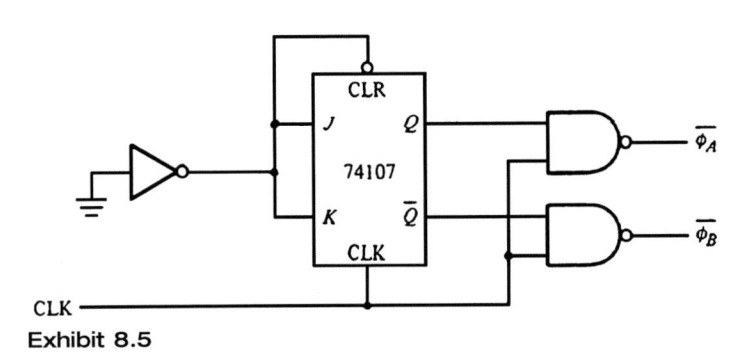

Exhibit 8.5

*Solution*

Exhibit 8.5a shows the timing diagram for the circuit.

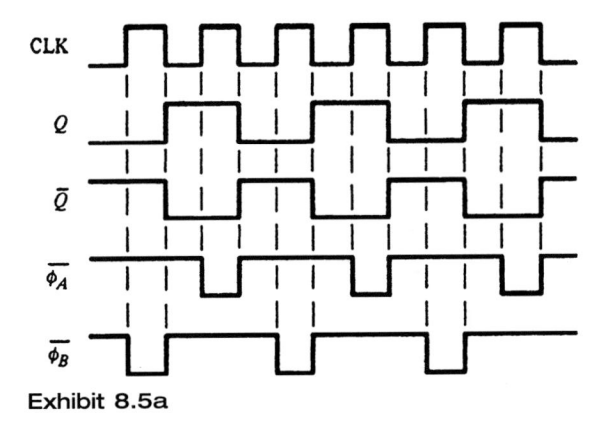

Exhibit 8.5a

Using a *D* flip-flop, Exhibit 8.5b shows an alternative circuit with its timing diagram.

If your solution is correct, go on to Problem 8.6.

If your solution is not correct, review the subject of flip-flops in this chapter. For more details, see chapter 7 in reference 1 and chapter 8 in reference 7.

**244** Chapter 8 Logic

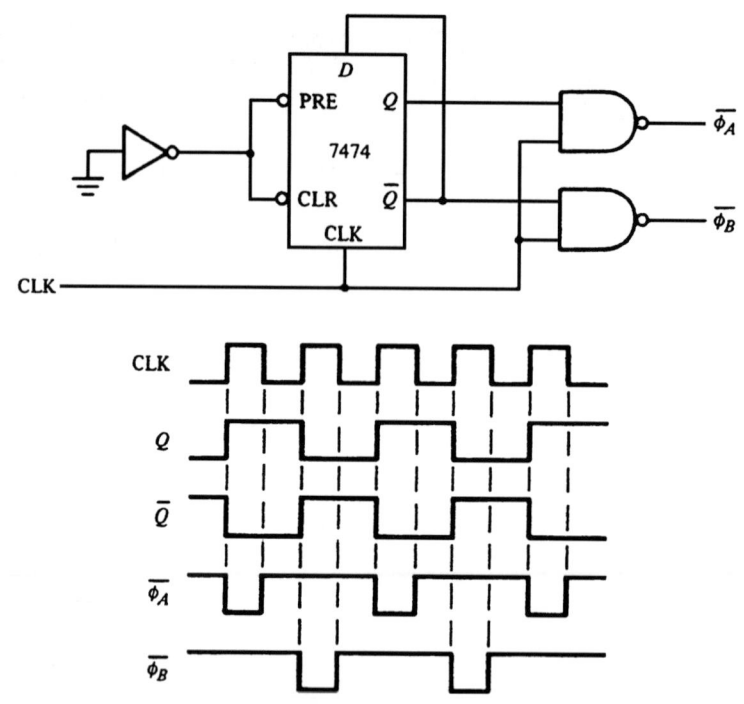

Exhibit 8.5b

## PROBLEM 8.6 Multiphase Clock Circuit, Johnson Counter

The Johnson counter is a popular multiphase clock circuit that is used to generate an even number of non-overlapped clock phases using a shift register with inverted output fed back to the input. Exhibit 8.6 shows a Johnson counter. Draw the timing diagram for the six-phase clock circuit showing the six clock phases, the three shift register outputs, and the input clock.

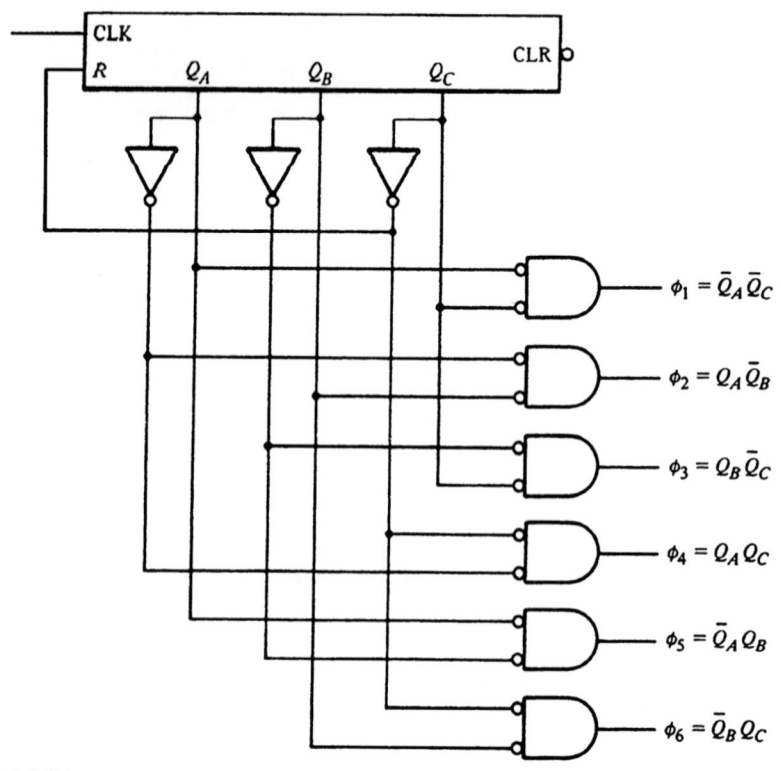

TRUTH TABLE

| $Q_A$ | $Q_B$ | $Q_C$ | $\phi$ |
|---|---|---|---|
| 0 | 0 | 0 | 1 |
| 1 | 0 | 0 | 2 |
| 1 | 1 | 0 | 3 |
| 1 | 1 | 1 | 4 |
| 0 | 1 | 1 | 5 |
| 0 | 0 | 1 | 6 |

$\phi_1 = \bar{Q}_A \bar{Q}_C$

$\phi_2 = Q_A \bar{Q}_B$

$\phi_3 = Q_B \bar{Q}_C$

$\phi_4 = Q_A Q_C$

$\phi_5 = \bar{Q}_A Q_B$

$\phi_6 = \bar{Q}_B Q_C$

Exhibit 8.6

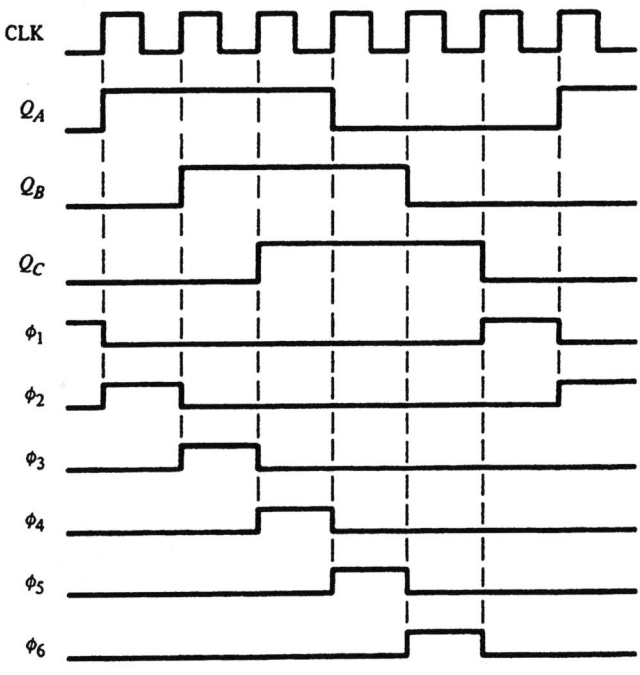

**Exhibit 8.6a**

### Solution

If your solution is correct, go on to the next chapter.

If your solution is not correct, review the preceding problem regarding timing diagrams, the subject of DeMorgan's theorem, and Problem 8.1. For more details, see chapter 7 in reference 1 and chapter 6 in reference 7.

# APPENDIX A

# Recommended References for Further Review

The following texts are cited by number throughout this book as recommended references if you desire additional review on particular topics or problems. Listed editions are the most current at the time of writing of this book. However, earlier editions of the referenced texts should provide equally useful review material.

1. *Electrical Engineering: PE License Review*, 9th Edition, Lincoln D. Jones, Kaplan AEC Education, 2004.

2. *Introductory Circuit Analysis,* 10th Edition, Robert Boylestad, Prentice Hall, 2003.

3. *Electrical Machines, Drives, and Power Systems*, 5th Edition, Theodore Wildi, Prentice Hall, 2000.

4. *Control Systems Engineering,* 4th Edition, Norman S. Nise, John Wiley and Sons, 2004.

5. *Electronic Devices and Circuit Theory,* 8th edition, Robert Boylestad and Louis Nashelsky, Prentice Hall, 2002.

6. *Communication Systems: Analysis & Design,* Harold Stern and Samy Mahmoud, Prentice Hall, 2004.

7. *Digital Fundamentals,* 8th Edition, Thomas Floyd, Prentice Hall, 2003.

# APPENDIX B

# Problems by Topic

| Problem Number | Topic |
|---|---|
| 2.1 | Series-Parallel Resistance |
| 2.2 | Work, Energy, and Power |
| 2.3 | Coulomb's Law |
| 2.4 | Wye-Delta Transformation |
| 2.5 | Schering Bridge |
| 2.6 | Circuit Network—Loop Current Analysis |
| 2.7 | Capacitance Energy and Charge |
| 2.8 | Two-Capacitor Charge Transfer |
| 2.9 | Power, Energy, and Charge |
| 2.10 | Unknown Device |
| 2.11 | Parallel Branches |
| 2.12 | Simple Lag Circuit |
| 2.13 | Simple Lead Circuit |
| 2.14 | Voltmeter Design |
| 2.15 | Impedance Transformation |
| 2.16 | AC and DC Ammeters |
| 2.17 | Series Resonance |
| 2.18 | Passive Circuit Calculation |
| 2.19 | Series-Parallel Resonant Filter |
| 2.20 | Series Resonant Filter |
| 2.21 | RLC Bandpass Filter |
| 2.22 | Damped RLC Circuit |
| 2.23 | Shunt Peaking |
| 2.24 | Thevenin Circuit—Maximum Power |
| 2.25 | Ideal Transformer—Maximum Power |
| 2.26 | Maximum Power Transfer |
| 2.27 | Maximum Power Transfer Corollary |
| 2.28 | RL Transient |
| 2.29 | Double Energy Transient |
| 2.30 | Double Energy Transient |
| 2.31 | Transient Response |
| 2.32 | Insulation |

| Problem Number | Topic |
|---|---|
| 3.1 | Single-Phase kVA and Power Factor |
| 3.2 | Three-Phase Power Factor and Line Current |
| 3.3 | Motor Input Current |
| 3.4 | Phase Sequence (Lamp Test) |
| 3.5 | Unbalanced Load |
| 3.6 | Power Factor Correction (Capacitor) |
| 3.7 | Power Factor Correction (Synchronous Motor) |
| 3.8 | Power Factor (Adjusting Existing Load) |
| 3.9 | Transmission Line |
| 3.10 | Transmission Line Regulation |
| 3.11 | Wattmeter |
| 3.12 | Wattmeter (Unbalanced Load) |
| | |
| 4.1 | Series-Wound DC Motor |
| 4.2 | Shunt-Wound DC Motor |
| 4.3 | Compound (Series-Shunt) Wound Motor |
| 4.4 | Shunt Motor Torque |
| 4.5 | Separately-Excited DC Generator |
| 4.6 | Induction Motor Speed |
| 4.7 | Induction Motor Efficiency |
| 4.8 | Induction Motor Losses |
| 4.9 | Induction Motor Speed |
| 4.10 | Motor Starting-Line Voltage Drop |
| 4.11 | Induction Motor Connections |
| 4.12 | Transformer Efficiency and Regulation |
| 4.13 | Regulation Improvement |
| 4.14 | Transformer Specifications |
| 4.15 | Autotransformer Currents |
| 4.16 | Autotransformer Rating |
| 4.17 | Reactor |
| 4.18 | Magnetic Device |
| | |
| 5.1 | Second-Order System |
| 5.2 | Partial Fractions |
| 5.3 | Inverse Laplace Transform |
| 5.4 | Routh Stability |
| 5.5 | Error and Stability |
| 5.6 | Transfer Function |
| 5.7 | Cancellation Compensation |
| 5.8 | Lead Compensation |
| 5.9 | Bode Analysis |
| 5.10 | Root Locus |
| | |
| 6.1 | Diode Suppression |
| 6.2 | One-Stage Transistor Amplifier |
| 6.3 | Common Base Amplifier |
| 6.4 | Two-Stage Transistor Amplifier, Common Emitter |
| 6.5 | Two-Stage Voltage Gain, $h$-Parameters |

| Problem Number | Topic |
|---|---|
| 6.6 | Current Gain, $Z_{out}$, $h$-Parameters |
| 6.7 | Transistor Amp—Upper and Lower Cutoff Frequency |
| 6.8 | Darlington/Transient Problem |
| 6.9 | Transistor Curves and Load Line |
| 6.10 | Transistor Stability |
| 6.11 | Transistor Specs, Stability |
| 6.12 | Field Effect Transistor Amplifier |
| 6.13 | Operational Amplifier |
| 6.14 | AC/DC Converter |
| 6.15 | Amplifier Class |
| 6.16 | Zener Regulator |
| 6.17 | Ripple Factor |
| | |
| 7.1 | Communication Line Characteristics |
| 7.2 | LF Transmission Line Power |
| 7.3 | Transmission Line Maximum Power |
| 7.4 | Reflection Coefficient |
| 7.5 | Stub Match |
| 7.6 | Normalized Sending Impedance |
| 7.7 | Normalized $Z_R$, $Z_S$, and $K$ |
| 7.8 | Stub Match, Smith Chart |
| 7.9 | Attenuator, L-Section |
| 7.10 | Attenuator, Π-section |
| 7.11 | Vertical Antenna |
| 7.12 | Microwave Antenna |
| 7.13 | Two Half-Wave Antennas |
| 7.14 | Waveguide Power Rating |
| | |
| 8.1 | DeMorgan's Theorem |
| 8.2 | Logic Simplification, Karnaugh Map |
| 8.3 | Timing Diagram, Truth Table |
| 8.4 | Logic Sequencer Input Logic |
| 8.5 | Underlapped Two-Phase Clock Circuit |
| 8.6 | Multiphase Clock Circuit, Johnson Counter |

# INDEX

## A

AC circuits
    parallel branches, 45–46
    unknown device, 43–45
AC motors and generators, 108–128
    asynchronous machines, 111
    autotransformers, 125–128
    induction generator, 112
    polyphase induction motor, 111
    single-phase induction motors, 112–120
    synchronous generators, 109–110
    synchronous motors, 110–111
    two-winding transformers, 120–125
AC/DC circuit converter, 191–193
Alternating current, defined, 4
Ammeters, 50–51
Ampere, 4
Amplifiers
    class, 193–195
    common base, 170–171
    field effect transistor (FET), 187–189
    one-stage transistor, 168–170
    operational circuit applications, 189–193
    two-stage transistor, 171–174
Amplitude modulation (AM), 199
Antennas
    half-wave, 227–228
    microwave, 226–227
    vertical, 225–226
    waveguide power rating, 228
Antiresonance, 26. *See also* Parallel resonance
Apparent power, 74
Armature, 99
Associative laws, 232
Asynchronous machines, 111
Attenuation, 220–224
    attenuator, L-section, 222–223
    attenuator, Π section, 223–224
Autotransformers, 125–128
    currents, 126–127
    rating, 127–128
Average value, waveform, calculating, 29–31

## B

Back emf, 101
Bandwidth, 25
Bias stability, 180
Biasing
    transistor curves and load line, 180–182
    transistor specs, stability, 184–185
    transistor stability, 182–184
Bipolar transistors, 162
Bistable multivibrator, 235
Black box analysis
    general two-port (four-terminal) network, 70
    hybrid parameters, 71
    open-circuit impedance parameters, 71
    short-circuit admittance parameters, 71
Bode analysis, 150–154
Boolean functions, terms and synonyms, 231
Branch, 17

## C

Cancellation compensation, 148–149
Capacitance, 7, 9
Capacitor(s)
    defined, 7
    energy and charge, 40–41
    for power factor correction, 84
    two-capacitor charge transfer, 41–43
CGS (centimeter-gram-second) system, 21
Charge
    capacitors, 40–41
    electric, defined, 4
Circle, areas of in circular mils, 7
Circuit(s)
    AC, 43–46
    AC/DC converter, 191–193
    ammeters, 50–51
    analyzing methods for, 20
    black box analysis, 70–71
    capacitors, 40–43
    compensating circuits, 46–48
    Coulomb's law, 35–36
    impedance transformation, 49–50
    insulation, 69
    Kirchoff's laws, 17
    Laplace transform, 16–17, 46, 139
    magnetic, 21–22
    maximum power, 60–65
    networks, 36–40
    Norton's equivalent, 19–20
    passive, resonance calculation, 52–53
    power supply, 195–197
    resistance, 34–35
    resonance, 51–60
    RLC, damped, 56–58
    series-parallel combinations, 10–11
    short, 96–98
    simple lag, 46–47
    simple lead, 47–48
    superposition, 20
    Thevenin's theorem, 18
    transfer function, 16
    transients, 13–16, 65–68
    transistor equivalent, 162–167
    two-capacitor charge transfer, 41–43
    voltage division, 20
    voltmeters, 48–49
    waveforms, 69–70
    work, energy power, 35
Circuit breakers, 96
Circuit element(s)
    capacitor, 7
    characteristics of, 6
    equations, 12–13
    inductor, 6–7
    reactor, 6
    values, 7–9
Circuit network–loop current analysis, 38–40
Coaxial line parameters, 205
Coefficient of self-induction, 7
Combinational systems, 233
Common base amplifier, 170–171
Common base equivalent circuits, 166
Common collector (emitter-follower) equivalent circuits, 167
Common emitter equivalent circuits, 165
Communications
    antennas, 224–228
    attenuation, 220–224
    line characteristics, 200–201
    low-frequency transmission, 199–203
    radio frequency (RF) transmission, 203–220
Commutator, 99
Compensation, 145–150
    cancellation, 148–149
    lag, 146
    lag-lead, 146
    lead, 146, 149–150
    simple lag, 46–47, 147
    simple lead, 47–48
    system, 147
Complex algebraic notation, 11–12
Complex notation, 12
Complex plane, 136
Composite waveform, 29
Compound-wound DC motor, 105

Connectives, 230
 binary, truth table of, 231
Conservation of charge principle, 42
Continuous current, 4
Control theory, 131–157
 basic feedback systems, terminology, 132–133
 Bode analysis, 150–154
 compensation, 145–150
 partial fractions, 136–139
 poles and zeros, 136
 root locus, 154–157
 second-order systems, 134–135
 singularity functions, 133–134
 stability, 139–144
 transfer function, 144–145
Coulomb, defined, 4
Coulomb's law, 9–10, 35–36
Counter emf, 101
Current
 alternating, 4
 autotransformers, 126–127
 continuous, 4
 defined, 4
 direct, 4
 input, motors, 79
 leading, 73
 pulsating, 4
Current gain $Z_{OUT}$, $h$- parameters, 176–177

## D

Damping factor, 134
Darlington/transient problem, 179–180
DC motors and generators, 99–108
 components of, 99
 compound (series-shunt)-wound motors, 105
 separately-excited DC generator, 107–108
 series-wound motor, 102–104
 shunt-wound motor, 104, 106–107
Delta ($\Delta$) connection, polyphase power, 76
DeMorgan's theorem, 232, 236–237
Dependent sources, 159
Determinant solutions
 second order, 22
 third order, 23
Diode(s)
 circuit model, 161
 classical equation for, 160
 suppression, 161–162
 symbol and characteristics, 160–161
 zener diode, 161
 zener model, 161
Diode circuit model, 161
Diode-logic, 233
Diode-transistor logic (DTL) gates, 233, 234
Direct-coupled transistor logic (DCTL) gates, 233

Direct current, defined, 4
Dissipation, 203
Dissipationless line. *See* Zero dissipation
Double energy transients, 13–14, 66–68
Duality, 232

## E

Efficiency
 motor, 79
 single-phase induction motors, 113–115
 synchronous generators, 109–110
 two-winding transformers, 121–123
Eighth-wave line, 212
Electrical quantities
 charge, 4
 current, 4
 electromotive force, 5
 energy (work), 2–3
 power, 3–4
 voltage, 4–5
Electric charge, 4
Electromotive force (emf), 5
Electronics
 amplifier class, 193–195
 biasing and stability, 180–185
 dependent sources, 159
 diodes, 159–162
 field effect transistors (FETs), 185–189
 operational amplifiers, 189–193
 power supply circuits, 195–197
 transistors, 162–180
Electron-volt, 3
Electrostatic fields, theory of, 9
emf. *See* Electromotive force (emf)
Emitter-coupled logic (ECL), 233
Energy
 application problems, 35, 43
 capacitors, 40–41
 formula for, 2–3
 units of, 3
Equivalent circuits
 common base, 166
 common collector (emitter-follower), 167
 common emitter, 165
 Norton's, 19–20

## F

Faults, 96. *See also* Short circuits
Feedback system
 characteristic equation of, 132
 terminology, 132–133
Field, 99
Field effect transistor (FET), 185–189
 amplifier, 187–189
Figure of merit, $Q$, formula, 24–25
Filters
 RLC bandpass, 56
 series-parallel resonant, 53–55
 series resonant, 55–56

Flip-flops, 235–236
Foot-pound, 4
Form factor, 120
Fourier analysis, 28
Fourier series, 28
Frequency modulation (FM), 199
Fundamentals of Engineering (FE/EIT) examination, viii

## G

Gates, 233–234
Generator, 101

## H

Half-wave antenna, 227–228
Half-wave line, 213
Henry, 7
Horsepower, 4
Hunting, 110
Huntington's postulates, 231

## I

Ideal transformer, 27
 maximum power, 61–62
Imaginary capacitance, 7
Imaginary inductance, 7
Impedance matching
 normalized sending, 216
 normalized $Z_R$, $Z_S$, and $K$, 216, 218
 stub match, 214–215, 219
Impedance transfer, 220
Impedance transformation, 26–27, 49–50
Incident wave, 207
Inductance, 7, 8–9
Induction generator, 112
Induction motors
 polyphase, 111
 single-phase, 112–120
Inductor, 6–7
Insulation, 69
Isotropic, 224

## J

Jacobi's law, 103
Johnson counter, 244–245
Joule, 3, 4

## K

Karnaugh map, 237–238
Kilowatt-hour, 3
Kirchhoff's laws, 17

## L

Lag compensation, 146
Lag-lead compensation, 146
Laplace transform, 16–17, 46, 139

Lead compensation, 146, 149–150
Leading current, 73
Line-to-ground short circuit, 96
Line-to-line short circuit, 96
Load
   adjusting existing, 86–87
   unbalanced, polyphase power, 81–83
   unbalanced, wattmeter, 94–95
Logic
   definition of terms, postulates, properties, 236
   problems, 236–245
   sequencer input logic, 239–243
   simplification, 237–238
Low-frequency transmission
   communication line characteristics, 200–201
   line maximum power, 202–203
   line power, 201–202
Lumped-parameter transmission line, 88

## M

Machinery. *See also* Motor(s)
   AC machines, 108–128
   DC machines, 99–108
   electric motors, by type, 100
   magnetic devices, 128–130
Magnetic circuits, 21–22
Magnetic devices, 128–130
Magnetomotive force, 113
Marginal stability, 139
Maximum power
   ideal transformer, 61–62
   low-frequency transmission line, 202–203
   Thevenin circuit, 60–61
   transfer theorem, 18, 63
   transfer theorem corollary, 19, 64–65
Maxterm, 233
Microwave antenna, 226–227
MKS (meter-kilogram-second) system, 21–22
Modulation, 199
Motor(s)
   AC, 108–128
   action, 101
   DC, 99–108
   efficiency, 79
   electric, by type, 100
   input current, 79
   polyphase induction, 111
   single-phase induction, 112–120
   speed, 112–113, 116–117
   starting-line voltage drop, 117–119
   synchronous, 84–86, 110–111
   torque, 101
Multiphase clock circuit, 244–245

## N

National Council of Examiners for Engineering and Surveying (NCEES), viii. *See also* Principles and Practice of Engineering (PE) examination
Networks
   network–loop current analysis, 38–40
   Schering bridge, 37–38
   wye-delta (Y-Δ) transformation, 36–37
Nodes, 17, 209
Non-repeated poles, 137–138
Norton's equivalent circuit, 19–20

## O

Ohm, 5
Ohm's law, 5–6, 43
One-stage transistor amplifier, 168–170
Open loop transfer function, 145
Open-wire line parameters, 204
Operating point, 180

## P

Parallel branches, AC circuits, 45–46
Parallel resonance, 26
Partial fractions, 46, 136–139
   inverse Laplace transform, 139
   non-repeated poles, 137
   repeated poles, 137–138
Passive circuits, resonance calculation, 52–53
PE exam. *See* Principles and Practice of Engineering (PE) examination
Percentage reactance, 96
Percent resistance, 96
Periodic waveform, 28
Phase angle, 73
Phase modulation (PM), 199
Phase order, 77
Phase sequence, 77, 79–81
Phasor diagram, 73
Poles, 99, 136
   non-repeated, 137
   repeated, 137–138
Pole-zero map, 136
Polyphase induction motor, 111
Polyphase power, 76–83
   briefly described, 76
   delta (Δ) connection, 76
   motor input current, 79
   phase sequence, 77, 79–81
   three-phase power factor and line current, 77–78
   unbalanced load, 81–83
   wye (Y) connection, 76
Potential difference, 4–5, 6
Power
   apparent, 74
   application problems, 35, 43
   defined, 3–4
   polyphase, 76–83
   power factor correction, 83–87
   radio frequency (RF) transmission, 212
   reactive, 74
   resistive, 74
   short circuit calculations, 96–98
   single-phase, 73–75
   transmission line calculations, 87–92
   units of, 4
   wattmeter measurements, 92–95
Power factor correction
   adjusting existing load, 86–87
   capacitor, 84
   synchronous motor, 84–86
Power supply circuits
   ripple factor, 196–197
   zener regulator, 195–196
Prime implicant, 233
Principles and Practice of Engineering (PE) examination
   calculators, use of, xiii
   dates, xi
   development of, x
   exam day preparations, xii–xiii
   exam-taking suggestions, xii
   materials allowed in exam room, xiii–xiv
   procedure, xi–xii
   as a requirement for professional engineer registration, viii
   reviewing for, suggested approach, vii–viii
   structure of, xi
Professional engineer
   state boards of registration, listing of, ix–x
   steps to becoming, viii
Pulsating current, 4

## Q

Quarter-wave line, 213

## R

Radio frequency (RF) transmission
   coaxial line parameters, 205
   eighth-wave line, 212
   half-wave line, 213
   impedance matching, 213–215
   impedance transfer, 220
   normalized sending impedance, 216
   open-wire line parameters, 204
   power, 212
   quarter-wave line, 213
   reflection coefficient, 207, 208–209
   reflection losses at any point, 211
   Smith chart, 215–216, 217, 219
   standing waves, 209–210
   stub match, 214–215, 219

zero dissipation line constants, 205–206
zero-dissipation-line input impedance, 210–211
zero dissipation line voltages and currents, 206–209
Reactive capacitance, 7
Reactive inductance, 7
Reactive magnetic devices, 128–129
Reactive power, 74
Rectified full wave, 31
Rectified half wave, 30
Reflected wave, 207
Reflection coefficient, 207, 208–209
Reflection losses, 211
Relative stability, 140
Repeated poles, 137–138
Resistance
   of a conductor, calculating, 7–8
   defined, 5
   Ohm's law, 5–6
   series-parallel, 34–35
Resistive power, 74
Resistor, defined, 6
Resistor-transistor logic (RTL) gates, 233, 234
Resonance, 24–27
   antiresonance, 26
   damped RLC circuit, 56–58
   defined, 24
   figure of merit, $Q$, 24–25
   impedance transformation, 26–27
   parallel, 26
   passive circuit calculation, 52–53
   RCC circuit, 14
   RC circuit, 14
   RLC bandpass filter, 56
   series, 25, 51–52
   series-parallel resonant filter, 53–55
   series resonant filter, 55–56
   shunt peaking, 58–60
RF transmission. See Radio frequency (RF) transmission
Ripple factor, 196–197
RLC bandpass filter, 56–58
RLC circuit, damped, 56–58
RL circuit, 13–14, 65–66
RMS value, waveform, calculating, 28–29
Root locus, 154–157
Routh stability criterion, 141–142

## S

Scaling modifiers, table of, 3
Schering bridge, 37–38
Second-order systems, 134–135
Self-induced voltage, 6
Self-induction, coefficient of, 7
Separately-excited DC generator, 107–108
Sequential system, 233
Series-parallel combinations, 10–11

Series-parallel resistance, 34–35
Series-parallel resonant filter, 53–55
Series resonance, 25, 51–52
Series resonant filter, 55–56
Series-shunt-wound DC motor, 105
Series-wound DC motor, 102–104
Short circuits
   admittance parameters, black box analysis of, 71
   calculations, 96–98
   line-to-ground, 96
   line-to-line, 96
   three-phase, 96
Shunt peaking, 58–60
Shunt resistance, 26
Shunt-wound DC motor, 104
   torque, 106–107
SI (Système International), 22
Simple lag circuit, 46–47
Simple lag compensator, 147
Simple lead circuit, 47–48
Sine wave, 29
Single energy transients, 13–14
Single-phase induction motors
   connections, 119–120
   efficiency, 113–115
   losses, 115–116
   motor speed, 112–113, 116–117
   motor starting-line voltage drop, 117–119
Single-phase power, 73–75
Singularity functions
   unit impulse, $\delta(t)$, 133
   unit ramp, 134
   unit step, $u(t)$, 133
Slip, 111
Slip rings, 109
Small dissipation, 203
Smith chart, 215–216, 217, 219
Smooth line, 210
Solenoid with nonmagnetic core, formula for, 8
S-plane, 136
Stability
   defined, 139
   error and, 142–144
   marginal, 139
   methods for determining, 140
   relative, 140
   Routh criterion, 141–142
   stable system, example of, 140–141
   transistors, 180–185
Standing wave ratio, 210
Standing waves, 207, 209–210
Stub match, 214–215, 219
Superposition, 20
Symmetrical sine wave, 30
Synchronous generators, 109–110
Synchronous motors
   power factor correction, using, 84–86

and synchronous generators compared, 110
   uses of, 110–111
System compensation, 147
System transfer function, 145

## T

Thevenin circuit, maximum power, 60–61
Thevenin's theorem, 18
Timing diagram, 238–239
Toroid of circular section with nonmagnetic core, formula for, 8
Toroid of rectangular section with nonmagnetic core, formula for, 8–9
Torque
   development of, 101
   shunt motor, 106–107
Transfer function, 16
   open loop, 145
   range $K$ for stable system, 145
   system, 145
Transformers
   autotransformers, 125–128
   ideal, 27, 61–62
   two-winding, 27, 120–125
Transient(s)
   circuit examples, 14–16
   double energy, 13–14, 66–68
   RCC circuit, 15–16
   RC circuit, 14
   response, 68
   RL circuit, 13–14, 65–66
   single energy, 13–14
   VC circuit, 14
   VC/RC circuit, 15
   VL circuit, 14
   VRC circuit, 15
   VRL circuit, 15
Transistor equivalent circuits
   common base, 166
   common collector (emitter-follower), 167
   common emitter, 165
Transistors, 162–180
   amp-upper and lower cutoff frequency, 177–178
   biasing and stability, 180–185
   bipolar, 162
   common base amplifier, 170–171
   current gain $Z_{OUT}$, $h$-parameters, 176–177
   darlington/transient problem, 179–180
   equivalent circuits, 162–167
   field effect (FETs), 185–189
   $h$-parameters, 164
   one-stage amplifier, 168–170
   small signal characteristics, 163
   two-stage transistor amplifier, 171–174
   two-stage voltage gain, $h$-parameters, 174–176

Transistor-transistor logic (TTL) gates, 233, 234
Transmission
  low-frequency, 199–203
  radio frequency (RF), 203–220
Transmission line(s), 87–92
  defined, 87
  dissipation, 203
  lumped-parameter, 88
  regulation of, 91–92
  for single-phase transmission, 87
  for three-phase transmission, 88
Truth table, 238–239
  of binary connectives, 231
Two-capacitor charge transfer, 41–43
Two-stage transistor amplifier, 171–174
Two stage voltage gain, $h$-parameters, 174–176
Two-winding transformers
  efficiency and regulation, 121–123
  ideal, 27
  regulation improvement, 123–124
  specifications, 124–125

## U

Unbalanced load
  polyphase power, 81–83
  wattmeter, 94–95
Underlapped two-phase clock circuit, 243–244

Unity power factor, 24, 26
Unknown device, AC circuits, 43–45
Unknowns, solving for, 22
Unsymmetrical square wave, 30

## V

VC circuit, 14
VC/RC circuit, 15
Vector diagram, 73
Vector sum, 207
Venn diagrams, 232
Vertical antenna, 225–226
VL circuit, 14
Voltage
  defined, 4–5
  division, 20
  drop, motor starting-line, 117–119
  phase sequence, 77
  regulation, 101
  self-induced, 6
Voltmeters, design of, 48–49
VRC circuit, 15
VRL circuit, 15

## W

Watt, 4
Wattmeter
  defined, 92

  single-phase, 93–94
  unbalanced load, 94–95
Watt second, 3
Waveforms, 69–70
  average value, 29–31
  composite, 29
  periodic, 28
  rectified full, 31
  rectified half, 30
  RMS value, 28–29
  sine, 29
  symmetrical sine, 30
  unsymmetrical square wave, 30
Waveguide power rating, 228
Wheatstone bridge, 36
Work, 2–3, 35
Wye (Y) connection, polyphase power, 76
Wye-delta (Y-Δ) transformation, 11, 36–37

## Z

Zener diode, 161
Zener model, 161
Zener regulator, 195–196
Zero dissipation
  defined, 203
  line constants, 205–206
  line voltages and currents, 206–209
  -line input impedance, 210–211
Zeros, 136